Lecture Notes in Mathematics

Edited by A. Dold and B. Eckmann

925

The Riemann Problem, Complete Integrability and Arithmetic Applications

Proceedings of a Seminar
Held at the Institut des Hautes Etudes Scientifiques,
Bures-sur Yvette, France,
and at Columbia University, New York, U.S.A., 1979-1980.

Edited by D. Chudnovsky and G. Chudnovsky

Springer-Verlag
Berlin Heidelberg New York 1982

Editors
David V. Chudnovsky
Gregory V. Chudnovsky
Department of Mathematics
Columbia University
N.Y. 10027, U.S.A.

AMS Subject Classifications (1980): 34B, 34B25, 35P, 35Q, 41A21, 45E

ISBN 3-540-11483-1 Springer-Verlag Berlin Heidelberg New York
ISBN 0-387-11483-1 Springer-Verlag New York Heidelberg Berlin

Printing and binding: Beltz Offsetdruck, Hemsbach/Bergstr.
2141/3140-543210

PREFACE

This volume, "Seminar on the Riemann Problem, Complete Integrability and Arithmetic Applications", contains a series of lectures presented at a seminar of the same title given by D. and G. Chudnovsky and held in 1979-1980 at the Institute des Hautes Etudes Scientifiques (IHES) in Bures-sur-Yvette, France (1979), and at Columbia University in the City of New York, U.S.A. The Seminar speakers examine different aspects of analytic and arithmetic problems arising in various ways from contemporary studies of the Riemann boundary value and monodromy problems. Particular subdivisions of the volume are the following: studies in spectral theory and completely integrable systems (inverse scattering method); the Riemann monodromy problem and statistical mechanics; Padé approximations associated with the Riemann boundary value problem and arithmetical applications to transcendental numbers.

We want to express our profound gratitude to the authors who contributed to this volume for their wonderful presentations at the Seminar and contributions to the diverse and fascinating subject, and for the preparation of manuscripts.

We want to thank the participants of the Seminar at IHES and Columbia University. Professor N. Kuiper (Director of IHES) made it possible for the Seminar to meet at IHES. Our special thanks go to colleagues M. J. Ablowitz, L. Bers, D. Bessis, H. Cornille, J. Frohlich, F. Gürsey, H. Jacquet, R. Jost, A. Neveu, and A. Voros for their invaluable discussions on the subject of the Seminar.

The editors acknowledge with gratitude partial support extended to the editors by CNRS and CEN-Saclay in France and ONR and NSF in the United States.

We warmly thank F. Brown for her constant help during the preparation of this volume and K. March for typing the manuscript.

We open the volume with an introduction in which we try to summarize the seemingly disconnected and various aspects of applications of the Riemann boundary value problem. The purpose of this is to allow immediate access, for students of the subjects as well as for teachers presenting special courses on the Riemann problem, to the contemporary research literature in this rapidly changing field. D. and G. Chudnovsky.

TABLE OF CONTENTS

INTRODUCTION

When one speaks about the Riemann problem one does not necessarily mean the problem of the distribution of zeroes of the Riemann ζ-function $\zeta(s)$. In this volume The Riemann Problem is the Riemann boundary value problem first rigorously considered and solved by Riemann [1] and then by Hilbert [2][1)].

Since the time of Riemann and Hilbert the Riemann boundary value problem has been the most universal, flexible and convenient instrument for the construction of analytic objects with prescribed global properties. For completeness, we present a classical formulation of the Riemann boundary value problem in the $\mathbb{CP}^1$ case. Let us consider in the complex λ-plane, a union Δ of simple smooth contours, which have no common points and bound some finite connected region Δ^+. The complement of $\Delta \cup \Delta^+$ is denoted by Δ^-. We consider an $n \times n$ matrix G_λ, regular everywhere on $\Delta(\lambda \in \Delta)$, and satisfying the Hölder condition. Then the Riemann boundary value problem with a given connection matrix G_λ, $\lambda \in \Delta$ consists of finding the $n \times n$ matrix functions $\varphi^+(\lambda)$, holonomic in Δ^+, and φ^-, holonomic in Δ^-, such that their limit values on Δ satisfy the connection formulas

$$\varphi^+(\lambda_0) = \varphi^-(\lambda_0)G_{\lambda_0} : \lambda_0 \in \Delta.$$

In this formulation the problem is still undetermined since the behavior of φ^- at infinity is not prescribed. If e.g. one puts

$$\varphi^-(\lambda) = \gamma(\lambda) + O(1/\lambda),$$

where $\gamma(\lambda)$ is a (polynomial) principal part of $\varphi^-(\lambda)$ at infinity, then the integral equation defining $\varphi^-(\lambda)$ has the following canonical form

---

1) The names of these two great mathematicians are put together in another reformulation of this problem. However there is no semantic uniformity as J. Plemelj [3] e.g. calls the same problem the Riemann-Klein problem. Nevertheless the mathematical formulation of the Riemann problem seems to be unambiguous.

$$(1) \qquad \varphi^-(\lambda_0) - \frac{1}{2\pi i}\int_\Delta \frac{G_{\lambda_0}^{-1}G_\lambda^{-1}}{\lambda - \lambda_0}\varphi^-(\lambda)d\lambda = \gamma(\lambda),$$

which is an ordinary matrix system of Fredholm equations of the second kind.

It is preferable to formulate the Riemann boundary problem in this way, since the corresponding generalizations to Riemann surfaces can basically be reduced to the given one [3]. For the overall exposition and global implications of the problem see Bers' review [4]. Traditional encyclopedic treatises are Muskhelishvili [5], Vekua [6] and Gakhov [7].

Integral equation (1) describes an extensive class of analytic objects including many important objects from algebraic geometry (vector bundles over Riemann surfaces and period structures of algebraic varieties) and mathamatical physics. The choice of the connection matrix G_λ $(\lambda \in \Delta)$ determines the particular properties of matrix $\varphi(\lambda)$. For example, a special choice of the matrix G_λ which is piecewise constant in λ, leads to the Riemann monodromy problem [1]. The graphical illustration in Figure 1

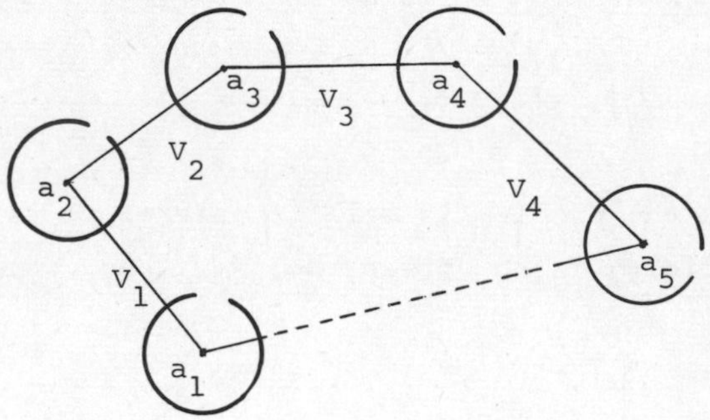

Figure 1

reduces the formulation of the Riemann boundary value problem to the following monodromy problem. Find n multivalued functions $\vec{y}(\lambda) = (y_1(\lambda),\ldots,y_n(\lambda))$ regular everywhere but in $\lambda = a_1,\ldots,a_m,\infty$, and admitting linear substitutions V_j (called monodromy matrices), after analytic continuation along the closed path γ_j which contains inside only one of the singularities a_j from $\{a_1,\ldots,a_m,\infty\}$:

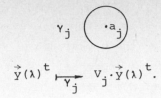

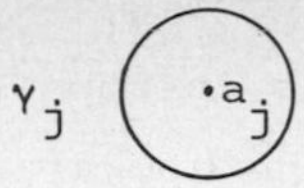

$$\vec{y}(\lambda)^t \underset{\gamma_j}{\longmapsto} V_j \cdot \vec{y}(\lambda)^t .$$

The subgroup of $GL(n,\mathbb{C})$ generated by V_j : $j = 1,2,\ldots,m,\infty$ is called the monodromy group of the system of functions $\vec{y}(\lambda)$. We owe to Riemann [1] the proof that $\vec{y}(\lambda)$ satisfies a system of Fuchsian linear differential equations with rational function coefficients, and we owe to L. Fuchs the definition of the monodromy group and local multiplicities for a Fuchsian system of linear differential equations with rational function coefficients [8], [9]. The Riemann monodromy problem which requires the construction of a system of functions admitting any given monodromy group, was essentially solved by Plemelj [3], Birkhoff [10] and Lappo-Danilevsky [11], though some degenerate cases (of multiply degenerate mod $\mathbb{Z}$ eigenvalues of V_j) are still awaiting a detailed analysis [12]. One realizes that the finite monodromy group case leads to the description of a Riemann surface with a given Galois group. Similarly the description of the moduli of vector bundles of rank r over Riemann surface Γ of genus g leads to a special form of the Riemann boundary value problem for $rg \times rg$ matrix functions [27]. One can easily identify in this Riemann boundary value formululation, the so called Tjurin parameters describing semistable vector bundles. These objects have recently attracted great attention in the solution of multidimensional completely integrable systems [14].

Consideration of the Riemann monodromy problem per se in the beginning of this century, had already led R. Fuchs [15] and L. Schlesinger [9] to the derivation of systems of nonlinear ordinary differential equations, called isomonodromy deformation equations, which describe conditions on a system of Fuchsian linear differential equations satisfied by $\vec{y}(\lambda)$ as having a fixed monodromy group, while the singularities a_j : $j = 1,\ldots,m,\infty$ are varied. These isomonodromy deformation (or Schlesinger) equations are known to contain Painlevé equations with immovable singularities and, according to R. Garnier, also contain classical one dimensional completely integrable systems.

We refer readers, for references and history of the subject, to [14].

Recently, isomonodromy deformation equations have attracted particular attention because of the appearance of Painlevé transcendents and other isomonodromy deformation equations in statistical mechanics, as expressions for correlation functions in the scaling limit (or in general) for the two dimensional Ising model. These pioneering achievements in this field belong to Wu, McCoy and Tracy et al [13] who were the first to connect singular integral equations of the Riemann boundary value type, and Painlevé transcendents directly with correlation functions of integrable models of statistical mechanics. In this volume four papers pursue this direction of investigation. The paper of B. McCoy and K. Perk 1.1 deals with analysis of fine properties of correlation functions for the Ising model, while the paper of T. Miwa and M. Jimbo 1.2 surveys the general algebraic approach suggested and developed in detail in a series of papers by M. Sato, T. Miwa, M. Jimbo and others. D. Abraham, in his paper 1.3, shows the link between correlation functions as solutions of Painlevé equations, and integral equations arising from the inverse scattering method. In this direction we particularly recommend a series of papers by M. Ablowitz, and H. Segur [16] in which the relationship between self-similar solutions of completely integrable systems and equations of Painleve type was established.

The inverse scattering method, which, for the Schrödinger operator $(-\partial^2/\partial x^2 + u(x) - k^2)\varphi = 0$, allows us to reconstruct the potential $u(x)$ from scattering coefficient $r(k)$ via an integral equation of Gelfand-Levitan type, is now a well known method of investigation of the so called "completely integrable" equations of mathematical physics. Among these equations, the most famous is, of course, the Korteweg deVries equation of hydrodynamics

$$\frac{\partial}{\partial t} u = 6u \frac{\partial u}{\partial x} + \frac{\partial^3}{\partial x^3} u.$$

For all these equations there exist infinitely many polynomial first integrals. For the Korteweg deVries (KdV) equation, these first integrals arise in an expansion of the scattering coefficient $r(k)$. The relationship with the Riemann boundary value problem and singular

singular integral equation (1) can be traced to the cornerstone paper on inverse scattering by R. Jost and R. Newton [17]. Recently it was observed that the Riemann boundary value problem and the Riemann monodromy problem for regular, and especially irregular singular cases, is the natural generalization of the inverse scattering method to a larger class of operators We refer here to the paper of Flashka and A. Newell [18] which contains many examples worked out in detail for second order operators. For operators of arbitrary order, and singular spectral problems such as

(2) $$\frac{d\varphi(\lambda)}{dx} = \Sigma_{j=1}^{m}\Sigma_{r=1}^{s} \frac{U_j^{(r)}}{(\lambda-a_j)^2}\varphi(\lambda),$$

and for more traditional spectral problems such as

(3) $$\frac{d\varphi(\lambda)}{dx} = (G\lambda + U)\varphi(\lambda),$$

the inverse scattering method is naturally substituted by the Riemann boundary value problem, with connection matrices G_λ being Stokes multipliers for functions $\varphi(\lambda)$ in the neighborhood of the singularities $\lambda = a_j,\infty$. This provides broad possibilities of generalizing the inverse scattering method to examine wider classes of completely integrable systems. A considerable part of the volume is connected with the study of completely integrable systems. Papers 2.1, 2.2, 2.3 deal with two-dimensional completely integrable systems and their decomposition into a sequence of one-dimensional completely integrable Hamiltonians. This decomposition established in 2.3, presents a canonical sequence of one-dimensional Hamiltonians arising directly from the resolvent expansion of the corresponding spectral problem (3).[1)] Papers 2.7, 2.8, 2.9 of the volume also touch upon the problem of quantization of completely integrable systems. In 2.7 an analysis of spectral problem (2) is supplemented by the study of a symplectic structure of Hamil-

---

[1)] From the historical point of view, it should be noted that the study of the KdV equation was always connected with studies of resolvent expansion for the Schrödinger operator. Thus in the pioneering papers of J. Drach [19] of 1919, all the stationary KdV equations were solved.

tonian systems associated with it. Here for the first time, the λ-plane is substituted by an arbitrary Riemann surface (the case of genus $g = 1$, i.e. an elliptic curve, is considered in detail). Examples of completely integrable lattice systems arising this way are studied in 2.8, where the relationship with finite dimensional Lie groups is indicated. In 2.9 the authors (A. Neveu and editors) deal with the quantized matrix nonlinear Schrödinger equations and study the complete integrability properties of it.

Most of the papers in the volume are connected with the Riemann boundary value problem on $\mathbb{CP}^1$. The paper of J. Bourguignon 2.8 examines a completely integrable system: the self-dual Yang-Mills equations arising from a different projective space. Vector bundles over $\mathbb{P}^3$ and the corresponding Riemann boundary value problem [20] lead to the description of the famous instanton solutions of gauge field theories. Complete integrability of the four-dimensional system under consideration(the self-dual Yang-Mills equations) is still not entirely established, as is the case for many other similar systems of considerable importance in modern physics. For this reason the volume contains a contribution by R. Churchill 2.5 providing rigorous criteria for establishing noncomplete integrability of Hamiltonians of classical mechanics. The results of his paper imply that the complete system of Yang-Mills equations is not completely integrable, as it contains a noncompletely integrable Hamiltonian system with two degrees of freedom with potential $q_1^2 \cdot q_2^2$. The paper of M. Berger 2.6 deals with the definition of complete integrability for general systems of partial differential equations.

At this point we should perhaps stop and explain why the complex variable in the Riemann boundary value problem described above is denoted by λ. In fact, it is a spectral variable which explains the "linearization" of nonlinear problems of mathematical physics into linear integral equations of type (1) by means of inverse spectral transformation. From this point of view, most of the "applied" studies of this volume are related to spectral theory. This relationship is very fruitful, and several papers here deal with the spectral theory (in the λ-plane) of differential operators (the Schrödinger operator, to be precise), important in physics. Trace formulas and ζ-functions

$\zeta_{\mathfrak{D}}(s) = \Sigma \lambda_i^{-s}$ of a differential operator $\mathfrak{D}$ (with λ_i being eigenvalues of $\mathfrak{D}$) are dealt with in 3.1-3.4.[1)] We especially recomment for careful study the exposition of A. Voros 3.2 on the spectral properties of the Schrödinger operator with potential x^M (e.g. the famous case of the quartic potential). This success, based on careful analysis of singularities in the λ-plane, indicates great prospects for the study of more complicated quantum systems. Another example of fine analytic treatment of trace identities in the λ-plane is the letter of L. Bers 3.3, giving a simple proof of trace identities for the matrix Schrödinger operator. Difficulties of spectral analysis, for a Schrödinger operator with slightly nontrivial, say quasiperiodic, potential, can become insurmountable, since topological and diophantine obstacles arise. The paper of S. Aubry 3.5, having its roots in solid state physics and the theory of magnetism, reviews this important area for special potentials, when the spectrum itself becomes a nontrivial topological continuum.

Returning to the Riemann boundary value problem, one sees an important sequence of integers--degrees of polynomials in $\gamma(\lambda)$ (called partial indices of the problem [5], [6]) associated with it. In other words, one has the freedom in the solution of the Riemann boundary value problem, like in the Riemann monodromy problem, of adding integers to local multiplicities at singular points. This transformation can be represented following Riemann [1], as a linear transformation with rational function coefficients:

$$\varphi(\lambda) \mapsto R(\lambda)\varphi(\lambda),$$

where $\varphi(\lambda)$ is an $n \times n$ matrix with entries polynomial (rational) in λ. In the vector case, such linear transformations are known as (generalized) Padé approximations to a system of functions $\vec{y}(\lambda) = (y_1(\lambda),\ldots,y_n(\lambda))$. It might be useful for completeness, to give the corresponding definition in the introduction, though all the

---

1) For the definitions and general properties of ζ-functions of differential operators, see the fundamental papers of Gelfand-Dikij [21] and Seeley [22].

necessary definitions and background material are included in the corresponding sections of the volume (see papers 4.1 and 4.3).

By a one-point Padé approximation to a system of functions $\vec{y}(\lambda)$ with weights $(m_1,\ldots,m_n)$ at a point $\lambda = \lambda_0$, one understands a system of polynomials $P_1(\lambda),\ldots,P_n(\lambda)$ of degrees $m_1,\ldots,m_n$ such that the linear combination

$$R(\lambda) = P_1(\lambda)y_1(\lambda) +\ldots+ P_n(\lambda)y_n(\lambda)$$

(which is called the remainder function) has at $\lambda = \lambda_0$ a zero of order $\{\Sigma_{i=1}^{n} (m_i+1)\} - 1$.

In classical literature, Padé approximations are usually examined for n = 2, when they are closely associated with continued fraction expansions, the classical moment problem and orthogonal polynomials. In the multidimensional case when the geometric interpretation becomes more complex, (there is no unique way to construct a continued fraction expansion any more), only methods of the Riemann boundary value problem and vector bundles over $\mathbb{P}^1$ provide some analytic and algebraic methods to study Padé approximations. Padé approximations are of interest in their turn, not only because of their geometric interpretation, but also because they provide the most useful instrument for approximating (simultaneously approximating) the values of functions $y_i(\lambda)$. This is important for application in physical problems, where $y_i(\lambda)$ may arise from a spectral problem, or in number theory, when construction of rational approximations to functions $y_i(\lambda)$ constitutes the best way to construct diophantine approximations to values of these functions at rational (algebraic) points λ. The papers of J. Nuttall and J. Gammel and J. Nuttall 4.1 and 4.2, give the description of asymptotic properties of generalized Padé approximants and the remainder function, using the fine analytic methods of the Riemann boundary value problem on a Riemann surface. In paper 4.3 Padé approximations are studied in relation with the Riemann monodromy problem and multidimensional interpretation, using methods of algebraic geometry. The particular role of polylogarithmic functions in connection with Feynman integrals and rational approximations to polylogarithmic functions is presented.

Arithmetic analysis of Hermité's [24] original system of Padé[1)] approximations to exponential and logarithmic functions is presented in paper 4.4, where immediate applications to diophantine approximations of classical transcendental numbers are presented. E.g. the paper contains the proof of the following measure of irrationality of π: $|\pi - \frac{p}{q}| > |q|^{-19.88}$ ... for rational integers p,q : $|q| \geq q_0$. Diophantine and rational approximations are dealt with in a more detailed way in paper 4.5, where entirely new criteria of algebraic independence of two numbers are proved. Applications of these results in conjunction with the methods of Padé approximations are extremely promising (e.g. for values of elliptic functions [25], [26]). The paper of K. Prendergast 4.6 indicates, using elliptic functions as a model example, how new rapidly convergent rational approximations can be constructed for solutions of nonlinear ordinary differential equations.

The miscellaneous topics, united around the Riemann problem by their formulation or method of solution, are not exhausted by those presented in this volume. Furthermore, it should be noted that the problems touched upon here are only at an initial stage of development.

---

1) Or, historically speaking, Hermité-Padé approximations.

References

[1] B. Riemann, Oeuvres Mathématiques, Albert Blanchard, Paris, 1968, pp. 353-363.

[2] D. Hilbert, Grundzüge der Integralgleichungen, Leipzig-Berlin, 2-te Aufl., 1924.

[3] J. Plemelj, Problems in the sense of Riemann and Klein. John Wiley, 1964.

[4] L. Bers, Mathematical developments arising from Hilbert problems, Proc. Symp. Pure Math., v. 28, AMS, Providence, 1976, pp. 559-610. N. Katz, ibid., pp. 537-557.

[5] N.I. Muskhelishvili, Singular integral equations, Ed. by J.R.M. Radok, P. Noordhoff, Leiden, 2-nd edition, 1977. N.I. Muskhelishvili, Some basic problems of the mathematical theory of elasticity. Ed. by J.R.M. Radok, P. Noordhoff, 1953.

[6] I.N. Vekua, Systems of singular integrals equations Ed. by J.N. Ferziger, Gordon and Breach, 1967.

[7] F.D. Gakhov, Boundary value problems, Trans. and ed. by J.N. Sneddon, Pergamon Press, 1966. New Russian edition appeared in 1977.

[8] L. Fuchs, Gesammelte Mathematische Werke, Ed. 1-3, Berlin, 1900-1906.

[9] L. Schlesinger, Einführung in die theorie der gewöhnlichen differentialgleichungen auf functionentheoretischer grundlage, 3-aufl., Berling-Leipzig, 1922.

[10] G.D. Birkhoff, Transaction of the American Mathematical Society (1909), pp. 436-460 and (1910), pp. 199-202.

[11] I.A. Lappo-Danilevsky, Mémories sur la théorie des systèmes des equations differéntielles linéaires, Chelsea Publishing Company, 1953.

[12] J.L. Verdier, A. Douady, Seminar on the Riemann problem, ENS, Asterique (to appear).

[13] T. T. Wu, B.M. McCoy, C.A. Tracy and E. Barouch, Phys. Rev. B13 (1976), 316, B.M. McCoy, C.A. Tracy and T.T. Wu, J. Math. Phys. 18 (1977), 1058.

[14] D.V. Chudnovsky, Riemann monodromy problem, isomonodromy deformation equations and completely integrable systems, Cargèse Lectures, June 1969, in Bifurcation phenomena in Mathematical Physics and Related Topics, D. Reidel Publ. Company, 1980, pp. 385-447.

[15] R. Fuchs, C.R. Acad Sci. Paris, v. 141 (1905), pp. 555-558.

[16] M. Ablowitz, H. Segur, Phys. Rev. Lett. 38 (1977), 1103.

[17] R. Jost, R.G. Newton, Nuovo Cimento 1 (1955), pp. 590-622.

[18] H. Flashka and A.C. Newell, Comm. Math. Physic. 76 (1980), 65.

[19] J. Drach, C.R. Acad. Sci. Paris (1919), pp. 47-50 and pp. 337-340.

[20] R.C. Ward, Doctoral Thesis, Oxford, 1977.

[21] I.M. Gelfand, Uspekhi Math. Nauk 11(1956), 191. L.A. Dikij, Isv. Akad. Naul USSR, Scr. Math. 19 (1955), 187; Uspekhi Math. Nauk. 13 (1958), 111.

[22] R.T. Seeley, Proc. Symp. Pure Math., American Math. Society, Providence, 10 (1967), 288.

[23] G.A. Baker Jr., Essentials of Padé approximants, Academic Press, 1975.

[24] C. Hermité, Oeuvres, v.1-3, Gauthier-Villars, 1901-1917.

[25] G.V. Chudnovsky, Algebraic independence of the values of elliptic function at algebraic points, Invent. Math. 61 (1980), pp. 267-290.

[26] G.V. Chudnovsky, Padé approximation and the Riemann monodromy problem, Cargèse Lectures, June 1979, in Bifurcation Phenomena in Mathematical Physics and Related Topics, D. Reidel Publ. Company, 1980, pp. 499-510.

[27] C.S. Seshardi, Generalized multiplicative meromorphic functions on a complex analytic manifold, J. Indian Math. Soc. 21 (1957), pp. 149-178.

CONTINUOUS EXPONENTS OF SPIN CORRELATION FUNCTIONS OF INHOMOGENEOUS LAYERED ISING MODELS

Barry M. McCoy
and
Jacques H. H. Perk

ABSTRACT

In this note we consider the inhomogeneous layered Ising model and we present a general formalism for expressing as a Toeplitz determinant the correlation of two spins in a row possessing reflection symmetry. For the special case where there is one altered row of horizontal bonds E_1' in an otherwise homogeneous lattice we discuss the correlations in detail and show that the critical exponents β and η depend on E_1'.

I. INTRODUCTION

The study of layered inhomogeneous Ising lattices was initiated some time ago by McCoy and Wu[1,2] and by Fisher and Ferdinand.[3] In ref. 1 and 3 properties of a half plane of Ising spins near the boundary were studied and it was found[1] that the surface magnetization vanishes as $|T_c-T|^{\frac{1}{2}}$ [as compared to $|T_c-T|^{1/8}$ for the bulk] and that at T_c the correlation $\langle\sigma_{00}\sigma_{0N}\rangle$ of two spins on the boundary behaves as N^{-1} for large N [as opposed to $N^{-\frac{1}{4}}$ for the bulk].

For the case of random layers in an Ising model it was subsequently shown[4] that the correlation of two spins in a row parallel to the random layering has many peculiar properties which are not even describable in terms of the usual critical exponent language. It is thus clear that the study of inhomogeneous Ising lattices will lead to many new properties not seen in the homogeneous lattice of Onsager[5]. In particular there must be many extensions[6] of the relations between Painlevé functions and Ising model correlations.[7]

In this note we first present in Sec. 2 the formalism needed to study the correlation $<\sigma_{00}\sigma_{0N}>$ in a layered Ising model specified by

$$\mathcal{E} = - \sum_{m=-\mathcal{M}}^{\mathcal{M}} \sum_{n=-\mathcal{N}}^{\mathcal{N}} \{E_1(m)\sigma_{m,n}\sigma_{m,n+1} + E_2(m)\sigma_{m,n}\sigma_{m+1,n}\} \tag{1.1}$$

where periodic boundary conditions are assumed and the horizontal [vertical] interactions $E_1(m)$ [$E_2(m)$] are allowed to vary from row to row. In sec. 3 we will then specialize to the case

$$E_2(m) = E_2 > 0,\ E_1(0) = E_1',\ E_1(m) = E_1 > 0,\ \text{for } m \neq 0 \tag{1.2}$$

and calculate the critical exponents β, η, and the asymptotic behavior for $N>>1$ for both $T>T_c$ and $T<T_c$.[8]

II. CALCULATION OF THE TOEPLITZ DETERMINANT FOR $<\sigma_{00}\sigma_{0N}>$

We study $<\sigma_{00}\sigma_{0N}>$ by means of the transfer matrix and Clifford algebra techniques of Onsager[1] and Kaufman.[9,10] It can be seen that these techniques are closely related to the methods of Pfaffians.[11]

For the interaction energy (1.1) we define a transfer matrix T in terms of

$$T_{2m} = \exp \sum_{n=-\mathcal{N}}^{\mathcal{N}} H_m \sigma_n^x \sigma_{n+1}^x \quad , \tag{2.1}$$

$$T_{2m+1} = \exp \sum_{n=-\mathcal{N}}^{\mathcal{N}} V_m^* \sigma_n^z \quad , \tag{2.2}$$

where

$$H_m = E_1(m)/kT, \qquad V_m = E_2(m)/kT , \tag{2.3}$$

and V_m^* satisfies

$$1 = \sinh 2V_m \sinh 2V_m^* = \cosh 2V_m \tanh 2V_m^*$$
$$= \tanh 2V_m \cosh 2V_m^* , \tag{2.4}$$

as

$$T = T_0^{\frac{1}{2}}(T_1T_2 \cdots T_{2m}T_{2m+1})(T_{-2m}T_{-2m+1} \cdots T_{-2}T_{-1})T_0^{\frac{1}{2}}. \tag{2.5}$$

We may then write the correlation of two spins in row zero as

$$\langle\sigma_{00}\sigma_{0N}\rangle = \frac{\mathrm{Tr}\,\sigma_0{}^x\sigma_N{}^x T}{\mathrm{Tr}\;T} . \tag{2.6}$$

We introduce the Clifford algebra operators γ_k as

$$\sigma_n{}^x = 2^{\frac{1}{2}}\,\gamma_{2n}\prod_{k=-n}^{n-1}(2i\gamma_{2k}\gamma_{2k+1}) , \tag{2.7a}$$

$$\sigma_n{}^y = -2^{\frac{1}{2}}\gamma_{2n+1}\prod_{k=-n}^{n-1}(2i\gamma_{2k}\gamma_{2k+1}) , \tag{2.7b}$$

and

$$\sigma_n{}^z = 2i\gamma_{2n}\gamma_{2n+1} , \tag{2.7c}$$

with

$$\gamma_k\gamma_\ell + \gamma_\ell\,\gamma_k = \delta_{k\ell} . \tag{2.8}$$

In terms of these operators

$$\sigma_0{}^x\,\sigma_N{}^x = \prod_{k=1}^{N}(2i\;\gamma_{2k-1}\;\gamma_{2k}), \tag{2.9}$$

$$T_{2m} = \exp\,[\,\sum_n 2i\;H_m\;\gamma_{2n+1}\gamma_{2n+2}] \equiv \exp[i\;H_m\;\underline{\gamma}\cdot\underline{\underline{A}}\cdot\underline{\gamma}] \tag{2.10}$$

and

$$T_{2m+1} = \exp[\sum_n 2i\;V_m^*\gamma_{2n}\;\gamma_{2n+1}] \equiv \exp[iV_m^*\;\underline{\gamma}\cdot\underline{\underline{B}}\cdot\gamma] \tag{2.11}$$

where the antisymmetric matrices $\underline{\underline{A}}$ and $\underline{\underline{B}}$ have the nonvanishing elements

$$A_{2n-1,2n} = -A_{2n,2n-1} = 1 , \tag{2.12a}$$

$$B_{2n,2n+1} = -B_{2n+1,2n} = 1 . \tag{2.12b}$$

We are only concerned with expressions with an even number of operators γ_k. Then we may modify the boundary condition between $-\mathcal{N}$ and $+\mathcal{N}$ so that $\underline{\underline{A}}$ and $\underline{\underline{B}}$ are 2×2 block-cyclic matrices. For such matrices we may introduce a Fourier transform in the horizontal direction as

$$\hat{X} \equiv \hat{X}(\phi) \equiv (2\mathcal{N}+1)^{-1} \sum_{k=-\mathcal{N}}^{\mathcal{N}} \sum_{\ell=-\mathcal{N}}^{\mathcal{N}} e^{i\phi(\ell-k)} \begin{pmatrix} X_{2k-1,2\ell-1} & X_{2k-1,2\ell} \\ X_{2k,2\ell-1} & X_{2k,2\ell} \end{pmatrix} . \qquad (2.13)$$

In particular

$$\hat{A}(\phi) = \begin{pmatrix} 0 & 1 \\ -1 & 0 \end{pmatrix} \text{ and } \hat{B}(\phi) = \begin{pmatrix} 0 & -e^{-i\phi} \\ e^{i\phi} & 0 \end{pmatrix} . \qquad (2.14)$$

We may now calculate the correlation function using the Wick theorem to write

$$\langle\sigma_{00}\sigma_{0N}\rangle = \underset{1\leq k<\ell\leq 2N}{\mathrm{Pf}} \; G_{k\ell} \qquad (2.15)$$

where

$$G_{k\ell} = 2i \frac{\mathrm{Tr}\; \gamma_k \gamma_\ell T}{\mathrm{Tr}\; T} . \qquad (2.16)$$

The major task is to find an efficient evaluation of $G_{k\ell}$. We first use the anticommutation relation (2.8) to write

$$2i\delta_{k\ell} = G_{k\ell} + G_{\ell k} \qquad (2.17)$$

and thus using the cyclicity of the trace

$$2i\delta_{k\ell} = G_{k\ell} + 2i \frac{\mathrm{Tr}(T^{-1}\gamma_k T)\gamma_\ell T}{\mathrm{Tr}\; T} . \qquad (2.18)$$

In the last term we use the formula

$$e^{-\underline{\gamma}\cdot\underline{\underline{C}}\cdot\underline{\gamma}} \underline{\gamma} \; e^{\underline{\gamma}\cdot\underline{\underline{C}}\cdot\underline{\gamma}} = e^{2\underline{\underline{C}}} \underline{\gamma} \qquad (2.19)$$

where $\underline{\underline{C}}$ is any antisymmetric matrix and we then treat the expression as a 2×2 block cyclic matrix and Fourier transform using (2.13) to obtain the 2×2 matrix equation

$$2i1 = (1 + \hat{T}) \quad \hat{G} = \hat{G}(1 + \hat{T}) \tag{2.20}$$

where

$$\hat{T} = \hat{T}_0^{\frac{1}{2}} \hat{T}_1 \hat{T}_2 \cdots \hat{T}_{2m} \hat{T}_{2m+1} \hat{T}_{-2m} \cdots \hat{T}_{-2} \hat{T}_{-1} \hat{T}_0^{\frac{1}{2}} \tag{2.21}$$

and

$$\hat{T}_{2m} = \exp(2iH_m \hat{A}) = \hat{T}^{\dagger}_{2m} , \tag{2.22}$$

$$\hat{T}_{2m+1} = \exp(2iV_m^* \hat{B}) = \hat{T}^{\dagger}_{2m+1} . \tag{2.23}$$

We note from (2.20) that

$$\hat{G} = 2i(1+\hat{T})^{-1}. \tag{2.24}$$

and also that a product of matrices $\hat{T}_m$ is always of the form

$$\hat{D} = \begin{pmatrix} x & iy \\ -iy^* & x^* \end{pmatrix} \tag{2.25}$$

with

$$\det \hat{D} = 1 . \tag{2.26}$$

We now restrict our attention to transfer matrices which are reflection symmetric

$$\hat{T}_m = \hat{T}_{-m}. \tag{2.27}$$

We can then build up any transfer matrix by starting with $\hat{T} = 1$ (or $\hat{G} = i1$) and inserting factors in pairs of $\hat{T}_m$ and $\hat{T}_{-m}$. This will give us a recursion relation. If we call the transfer matrix with two new factors inserted $\hat{T}'$ we have by construction

$$\hat{T}' = \hat{D} \hat{T} \hat{D}^{\dagger} . \tag{2.28}$$

Then using (2.20)

$$\hat{G}' - 2i1 = -\hat{G}'\hat{T}' = -\hat{G}' \hat{D} \hat{T} \hat{D}^{\dagger} \tag{2.29}$$

and thus

$$(\hat{G}'-2i1)\hat{D}^{\dagger -1}\hat{G} = -\hat{G}'\hat{D}\hat{T}\hat{G}$$
$$= \hat{G}'\hat{D}(\hat{G} - 2i1) \qquad (2.30)$$

where we have used (2.20) again to obtain the last line. This constitutes a recursion relation between $\hat{G}$ and $\hat{G}'$. To solve the relation we first note that if

$$\hat{G} = \begin{pmatrix} i & K \\ -K^* & i \end{pmatrix} \qquad (2.31a)$$

and

$$\hat{G}' = \begin{pmatrix} i & K' \\ -K'^* & i \end{pmatrix} \qquad (2.31b)$$

then (2.30) is satisfied if

$$K' = \frac{y + x K}{x^* + y^* K} . \qquad (2.32)$$

Thus the form of $\hat{G}$ is invariant under recursion and the form also holds at the outset for $\hat{T} = 1$, $\hat{G} = i1$, $K = 0$. Using this form of $\hat{G}$ in (2.15) we obtain

$$\langle\sigma_{00}\sigma_{0N}\rangle = \det |a_{i-j}| , \quad i,j = 1,\cdots,N , \qquad (2.33a)$$

and

$$a_n = (2\pi)^{-1}\int_0^{2\pi} d\theta e^{-in\theta} K(e^{i\theta}) . \qquad (2.33b)$$

We now proceed to use the recursion relation (2.32) to study the specific case (1.2). Consider first the homogeneous lattice $H_m = H, V_m = V$. In the thermodynamic limit $\mathcal{M} \to \infty$ we must have invariance for adding more rows. Thus we have the fixed point equation for the kernel C

$$C = C' = \frac{y+xC}{x^*+y^*C} \qquad (2.34)$$

where the matrix $\hat{D}$ is

$$\hat{D} = \exp(iH\hat{A})\exp(2iV^*\hat{B})\exp(iH\hat{A}) \tag{2.35}$$

which, upon explicit multiplication, gives

$$x = x^* = \cosh 2H \cosh 2V^* - \cos\phi \sinh 2H \sinh 2V^* ,$$

$$y = \sinh 2H \cosh 2V^* - \cos\phi \cosh 2H \sinh 2V^* + i \sin\phi \sinh 2V^* . \tag{2.36}$$

This is readily solved to give

$$C^2 = y/y^* \quad \text{or} \quad C = (y/y^*)^{\frac{1}{2}} . \tag{2.37}$$

The sign is determined by $C(\pi) = +1$ since for $\phi = \pi$ $\hat{A} = \hat{B}$ and $\hat{G}$ can be directly calculated from (2.24). From (2.37) it follows immediately that

$$C(e^{i\theta}) = \left[\frac{(1-\alpha_1 e^{i\theta})(1-\alpha_2 e^{-i\theta})}{(1-\alpha_1 e^{-i\theta})(1-\alpha_2 e^{i\theta})} \right]^{\frac{1}{2}} \tag{2.38}$$

where

$$\alpha_1 = z_1(1-z_2)/(1+z_2), \quad \alpha_2 = z_1^{-1}(1-z_2)/(1+z_2) \tag{2.39}$$

and

$$z_i = \tanh E_i/kT. \tag{2.40}$$

We may now calculate K for (1.2) where

$$H_0 = H', \quad H_m = H \text{ for } m \neq 0, \text{ and } V_m = V . \tag{2.41}$$

The transfer matrix for this system is obtained from that of the pure system using (2.28) with

$$\hat{D} = \exp i(H'-H)\hat{A} . \tag{2.42}$$

Therefore

$$x = \cosh(H'-H) \tag{2.43a}$$

and

$$y = \sinh(H'-H) \tag{2.43b}$$

and from (2.32) we instantly find the desired result

$$K = \frac{\tanh(H'-H) + C}{1+C\tanh(H'-H)} . \tag{2.44}$$

Finally, as another application of the formalism consider changing the row of vertical bonds between rows -1 and 0 and between 0 and +1 to E_2'. Then we have

$$V_0 = V_{-1} = V', \quad V_m = V \text{ for } m \neq 0, -1 \tag{2.45a}$$

and

$$H_m = H \cdot \tag{2.45b}$$

We relate the transfer matrix of this system to that of the pure case using (2.28) with

$$\hat{D} = \exp(iH\hat{A}) \exp[2i(V'^* - V^*)\hat{B}]\exp[-iH\hat{A}]. \tag{2.46}$$

Multiplying out we have

$$\begin{aligned} x &= \cosh 2(V'^* - V^*) - i\sin\phi \cosh 2H \sinh 2(V'^* - V^*), \\ y &= [-\cos\phi + i \sin\phi \cosh 2H] \sinh 2(V'^* - V^*) , \end{aligned} \tag{2.47}$$

where, again from (2.32),

$$K = \frac{x+yC}{y^*+x^*C} . \tag{2.48}$$

III. PROPERTIES OF $\langle\sigma_{00}\sigma_{0N}\rangle$

In the previous section we found that for the lattice (1.2) the correlations of two spins in the row of the impurities (zero) is

$$\langle\sigma_{00}\sigma_{0N}\rangle = \det|a_{i-j}| \quad , \ i,j = 1, \cdots, N, \tag{3.1}$$

where

$$a_n = (2\pi)^{-1} \int_0^{2\pi} d\theta \, e^{-in\theta} K(e^{i\theta}), \tag{3.2}$$

$$K(e^{i\theta}) = [C(e^{i\theta})+\kappa]/[1+\kappa C(e^{i\theta})] \tag{3.3}$$

where

$$\kappa = \tanh(E_1' - E_1)/kT \tag{3.4}$$

and $C(e^{i\theta})$ is given by (2.38).

To highlight the relation between the cases $\kappa = 0$ and $\kappa \neq 0$ we contrast the analytic properties of K and C. Consider first the index of K defined by

$$\text{Ind } K(e^{i\theta}) = (2\pi i)^{-1}[\ln K(e^{2\pi i}) - \ln K(e^{i0})] \tag{3.5}$$

When $0<\alpha_2<1$, $(T<T_c)$, the index of K is zero and when $1<\alpha_2<\infty$, $(T>T_c)$, the index is -1. This is identical with the index of C. At $\alpha_2 = 1$, $(T=T_c)$, $K(\xi)$ is not single valued on $|\xi| = 1$ and we have

$$K(e^0) = K^*(e^{2\pi i}) = (i+\kappa)/(1+i\kappa). \tag{3.6}$$

Next consider the analyticity properties for $\alpha_2 \neq 1$. The kernel $K(\xi)$ has square root branch points at α_1, α_2, α_2^{-1}, and α_1^{-1}, [as does $C(\xi)$]. However, in contrast to $C(\xi)$, $K(\xi)$ is in general finite and non-zero at these branch points. In addition $K(\xi)$ will have zeroes when

$$C(\xi) + \kappa = 0 \tag{3.7a}$$

and poles when

$$1 + \kappa\, C(\xi) = 0. \tag{3.7b}$$

Consider first $T<T_c$ $(\alpha_2<1)$. Then on $|\xi| = 1$, $C(\xi)$ is never real and negative so if $0<\kappa<1$ there are neither zeroes nor poles on the first sheet. However, if $\kappa<0$ then (3.7) does have solutions and we find that $K(\xi)$ has zeroes at β_1^{-1} and β_2 where

$$\left.\begin{matrix}\beta_1^{-1}\\ \\ \beta_2\end{matrix}\right\} = \tfrac{1}{2}(\alpha_1-\kappa^2\alpha_2)^{-1}\ \{(1-\kappa^2)(1+\alpha_1\alpha_2) \pm [(1-\kappa^2)^2(1-\alpha_1\alpha_2)^2 + 4\kappa^2(\alpha_1-\alpha_2)^2]^{\frac{1}{2}}\} \tag{3.8}$$

and poles at β_1 and β_2^{-1}. The β_i have the properties

i) if $\kappa = 0$ then $\beta_i = \alpha_i$,

ii) if $-1 < \kappa < 0$ then

$$-1 < \beta_1 < \alpha_1 < \alpha_2 < \beta_2 < 1. \tag{3.9}$$

When $T>T_c$ $(\alpha_2>1)$ there are poles and zeroes for both signs of κ. If $0 < \kappa < 1$ then $K(\xi)$ has a zero at β_2 and a pole at β_2^{-1} where

$$1 < \beta_2 < \alpha_2 < \alpha_1^{-1} \tag{3.10}$$

and β_2 is still given by (3.8). Similarly, if $-1 < \kappa < 0$ then $K(\xi)$ has a zero at β_1^{-1} and a pole at β_1 where

$$-1 < \beta_1 < \alpha_1 < \alpha_2^{-1} < 1. \qquad (3.11)$$

We now proceed to study $\langle\sigma_{00}\sigma_{0N}\rangle$ by use of the techniques developed to study the case $\kappa = 0$.[12,13,14]

1) Spontaneous Magnetization

The spontaneous magnetization in the zeroth row is defined as

$$\mathcal{M}_0^2 = \lim_{N\to\infty} \langle\sigma_{00}\sigma_{0N}\rangle. \qquad (3.12)$$

This may be calculated by using Szegö's theorem[12] in the form

$$\ln \mathcal{M}_0^2 = \frac{1}{8}\int_0^{2\pi} \frac{d\theta_1}{2\pi}\int_0^{2\pi} \frac{d\theta_2}{2\pi}\left[\frac{\ln K(e^{i\theta_1}) - \ln K(e^{i\theta_2})}{\sin \frac{1}{2}(\theta_1-\theta_2)}\right]^2. \qquad (3.13)$$

When $T = T_c$, $(\alpha_2 = 1)$, this double integral diverges logarithmically at $\theta_1 \sim 0$, $\theta_2 \sim 2\pi$ [and $\theta_1 \sim 2\pi$, $\theta_2 \sim 0$] due to the discontinuity in $K(\xi)$, (3.6). The amplitude of the divergence is proportional to the square of the discontinuity in $\ln K(\xi)$. Thus we find as $T \to T_c^-$, $(\alpha_2 \to 1^-)$,

$$\mathcal{M}_0 \to f(\kappa_c)(1-\alpha_2^{\,2})^{\beta(\kappa_c)} \qquad (3.14)$$

where

$$\beta(\kappa_c) = \tfrac{1}{2}\{\pi^{-1} \arccos[2\kappa_c/(1+\kappa_c^2)]\}^2$$

$$= \tfrac{1}{2}\{\pi^{-1} \arccos \tanh 2(E_1'-E_1)/kT_c\}^2 \qquad (3.15)$$

and

$$\ln f(\kappa) = \frac{1}{8}\int_0^{2\pi} \frac{d\theta_1}{2\pi}\int_0^{2\pi} \frac{d\theta_2}{2\pi} \{\sin\tfrac{1}{2}(\theta_1-\theta_2)\}^{-2}$$

$$\times \{[\ln K_0(e^{i\theta_1})-\ln K_0(e^{i\theta_2})]^2 + 2\beta(\theta_1-\theta_2)^2\} \qquad (3.16)$$

where $K_o(e^{i\theta})$ is given by (3.3) with $C(e^{i\theta})$ replaced by

$$ie^{-i\theta/2}[(1-\alpha_1 e^{i\theta})/(1-\alpha_1 e^{-i\theta})]^{\frac{1}{2}}. \tag{3.17}$$

The exponent $\beta(\kappa_c)$ depends continuously on κ_c and agrees with the related calculation of Bariev.[15] Note that $\beta(+1) = 0$, $\beta(0) = 1/8$, $\beta(-1) = 1/2$ and that $\ln f(\kappa)$ vanishes when $\kappa = +1$ and diverges when $\kappa \to -1$. These latter two limits reflect the fact that when $E_1' \to +\infty(-\infty)$ the spins in row zero are all parallel (antiparallel).

2) Correlation at T_c

When $T = T_c$ the leading term as $N \to \infty$ may be studied by calculating the ratio of $\langle\sigma_{00}\sigma_{0N}\rangle$ to the $N\times N$ Cauchy determinant generated by

$$\tilde{K}(e^{i\theta}) = \exp\{-i(\theta-\pi)[2\beta]^{\frac{1}{2}}\}. \tag{3.18}$$

This comparison kernel is determined by requiring that its discontinuity at $\theta = 0, 2\pi$ be the same as (3.6). The $N \to \infty$ behavior of the $N\times N$ Cauchy determinant generated by (3.18) is[11]

$$A(\kappa_c)N^{-2\beta(\kappa_c)} \tag{3.19}$$

where

$$A = (1-2\beta)\exp\{-[2\beta\gamma_E + \sum_{n=1}^{\infty}(2\beta)^n[\zeta(2n-1)-1]n^{-1}\} \tag{3.20}$$

with γ_E denoting Euler's constant and $\zeta(n)$ is the zeta function. The ratio of $\langle\sigma_{00}\sigma_{0N}\rangle$ to the Cauchy determinant is now easily calculated following the method of Wu[13] and we find as $N \to \infty$

$$\langle\sigma_{00}\sigma_{0N}\rangle \sim f(\kappa_c)A(\kappa_c)N^{-2\beta(\kappa_c)}, \tag{3.21}$$

where $\ln f(\kappa)$ is given by (3.16).

3) Large N Behavior for $T<T_c$

Any $N\times N$ Toeplitz determinant whose generating function $K(\xi)$ has index zero and is continuous and nonzero for $|\xi| = 1$ may be studied as $N \to \infty$ as was the pure Ising case[14] $\kappa = 0$. This is carried out by first calculating the Wiener-Hopf splitting

$$K(\xi) = P_<^{-1}(\xi)Q_<^{-1}(\xi^{-1}) \tag{3.22}$$

where $P_<(\xi)$ and $Q_<(\xi)$ are both analytic for $|\xi|<1$ and continuous and nonzero for $|\xi|\leq 1$. The function $P_<(\xi)$ may be written as

$$P_<(\xi) = e^{G_+(\xi)} \tag{3.23}$$

where for $|\xi|<1$ we have the explicit formula

$$G_+(\xi) = -\frac{1}{2\pi i}\oint_{|\xi'|=1} d\xi' \frac{\ln K(\xi')}{\xi'-\xi} . \tag{3.24}$$

Furthermore, our kernel has the property that

$$K(\xi^{-1}) = K^{-1}(\xi) , \tag{3.25}$$

so

$$P_<(\xi)Q_<(\xi) = 1. \tag{3.26}$$

The methods of ref. 14 then give

$$\langle\sigma_{00}\sigma_{0N}\rangle = \mathcal{M}_0^2[\exp - \sum_{n=1}^{\infty} F_<^{(2n)}] \tag{3.27}$$

where

$$F_<^{(2n)} = n^{-1}(2\pi)^{-2n} \prod_{j=1}^{n} \oint d\xi_{2j-1}\xi_{2j-1}^{N} P_<(\xi_{2j-1})P_<(\xi_{2j-1}^{-1})$$
$$\times (\xi_{2j-1}-\xi_{2j})^{-1} \oint d\xi_{2j}\xi_{2j}^{-N} P_<^{-1}(\xi_{2j})P_<^{-1}(\xi_{2j}^{-1})(\xi_{2j}-\xi_{2j+1})^{-1}, \tag{3.28}$$

where $\xi_{2n+1} \equiv \xi_1$, $|\xi_{2j+1}|< 1$, and $|\xi_{2j}|>1$.

The integral (3.24) can calculated explicitly in terms of integrals over Jacobi's theta functions. However, the principal qualitative features follow immediately from the behavior of the n=1 term of (3.27) and these in turn follow from the analyticity properties $K(\xi)$ discussed previously. There are 3 regions:

a) $0 < \kappa < 1$

Here the behavior of $F^{(2)}$ is dominated by the square root of $P_<(\xi_1^{-1})$ at $\xi_1 = \alpha_2$ and the square root of $P_<(\xi_2)$ at $\xi_2 = \alpha_2^{-1}$. Near $\xi_1 = \alpha_2$ for example we expand $P_<(\xi_1^{-1}) \sim P_<(\alpha_2^{-1}) + (\xi_1-\alpha_2)^{\frac{1}{2}}C$ and find from (3.28) that

$$\langle\sigma_{00}\sigma_{0N}\rangle \sim \mathcal{M}_0^2\{1 + a_1(\kappa,T)N^{-3}\alpha_2^{\ 2N}\} , \tag{3.29}$$

where a_1 diverges as $\kappa \to 0$. Here the correlation length is independent of κ.

For $-1<\kappa<0$ there are two subregions. The dominant contribution to the ξ_2 integral comes from the pole of $P_<^{-1}(\xi_2)$, [pole of $K(\xi)$], at β_2^{-1}. However, in the ξ_1 integrand the leading singularity can be either the square root at α_2 or the pole of $P_<(\xi_1^{-1})$, [zero of $K(\xi^{-1})$], at β_1. The dividing line between these 2 cases occurs at $\kappa = -\lambda$ where

$$\lambda = \left[\frac{(1+\alpha_2^2)(\alpha_1+\alpha_2)}{2\alpha_2(1+\alpha_1\alpha_2)}\right]^{\frac{1}{2}} . \tag{3.30}$$

b) $-\lambda < \kappa < 0$

In this region we see from the expression for β_i, (3.8), that $-\beta_1<\alpha_2$ and hence the leading behavior is

$$\langle\sigma_{00}\sigma_{0N}\rangle \sim \mathcal{M}_0^2\{1+a_2(\kappa,T)N^{-3/2}(\alpha_2\beta_2)^N\} . \tag{3.31}$$

Here the approach to the $N \to \infty$ limit is monotonic since both α_2 and β_2 are positive, but since β_2 depends on κ, the correlation length depends on κ. In addition the factor $a_2(\kappa,T)$ vanishes as $\kappa \to 0$.

c) $-1 < \kappa < -\lambda$

In this final case $-\beta_1 > \alpha_2$ and the leading behavior is

$$\langle\sigma_{00}\sigma_{0N}\rangle \sim \mathcal{M}_0^2\{1 + (-1)^N a_3(\kappa)(\beta_2|\beta_1|)^N\} . \tag{3.32}$$

Here we have made explicit the fact that the approach to the limit is oscillatory. This oscillation is expected because a large negative value of E'_1 tends to make the spins in the zeroth row antiallign. Note that if $T \to T_c (\alpha_2 \to 1)$, then $\lambda \to 1$ and hence the local antiferromagnetic behavior is eventually swamped by the ferromagnetic critical behavior of the bulk.

Finally we note that in all cases, even though the correlation length may depend on κ the critical exponent ν is always one. Therefore from the previous discussion of the exponents β and η we see that from the low-temperature side the hyperscaling relation

$$(d-2+\eta)\nu = 2\beta \tag{3.33}$$

is satisfied with d = 2.

4) Large N behavior for $T > T_c$.

When $T > T_c$ we may also study $\langle\sigma_{00}\sigma_{0N}\rangle$ by generalizing the studies[13,14] of the $\kappa = 0$ case. The kernel $K(\xi)$ has index minus one so we define a shifted kernel

$$K_>(e^{i\theta}) = -e^{i\theta}K(e^{i\theta}) \tag{3.34}$$

and make the factorization

$$K_>(\xi) = P_>^{-1}(\xi)Q_>^{-1}(\xi^{-1}), \tag{3.35}$$

where again $P_>(\xi)$ and $Q_>(\xi)$ are analytic for $|\xi|<1$ and continuous and nonzero for $|\xi|\leq 1$, and $P_>(\xi)Q_>(\xi) = 1$. We find[14]

$$\langle\sigma_{00}\sigma_{0N}\rangle = \mathcal{M}_{0>}^2(\kappa) \sum_{n=0}^{\infty} g^{(2n+1)} \exp\left(-\sum_{n=1}^{\infty} F_>^{(2n)}\right) \tag{3.36}$$

where $\mathcal{M}_{0>}^2$ is obtained by applying Szegö's theorem (3.13) to $K_>$, $F_>^{(2n)}$ is obtained from (3.28) with $P_<$ replaced by $P_>$ and N by N+1, and

$$g^{(2n+1)} = (2\pi i)^{-1} \oint d\xi_0 \xi_0^{N-1} P_>(\xi_0)P_>(\xi_0^{-1})$$
$$\times \prod_{\ell=1}^{2n} (2\pi i)^{-1} \oint d\xi_\ell \xi_{\ell-1}\xi_\ell^N (1-\xi_{\ell-1}\xi_\ell)^{-1}$$
$$\times \prod_{\ell=1}^{n} P_>^{-1}(\xi_{2\ell-1})P_>^{-1}(\xi_{2\ell-1}^{-1})P_>(\xi_{2\ell})P_>(\xi_{2\ell}^{-1}). \tag{3.37}$$

We first comment on the qualitative behavior as $N \to \infty$. This is obtained by analyzing the behavior of $g^{(1)}$ and using the analyticity properties of $K_>(\xi)$. There are again three regions:

a) $0 < \kappa < 1$

In this case the leading singularity in the integrand of $g^{(1)}$ is the pole in $P_>(\xi^{-1})$ at $\xi = \beta_2^{-1}$. Thus

$$\langle\sigma_{00}\sigma_{0N}\rangle \sim \mathcal{M}_0^2(\kappa)\, a_{1>}(\kappa,T)\beta_2^{-N}, \tag{3.38}$$

where $a_{1>}(\kappa,T) \to 0$ as $\kappa \to 0$.

b) $-\lambda^{-1} < \kappa < 0$

Here the leading singularity is the square root in $P_>(\xi^{-1})$ at $\xi = \alpha_2^{-1}$ and we find

$$<\sigma_{00}\sigma_{0N}> \sim \mathcal{M}_{0>}^2 a_{2>}(\kappa,T) N^{-3/2} \alpha_2^{-N} , \tag{3.39}$$

where $a_{2>}(\kappa,T)$ diverges as $\kappa \to 0$.

c) $-1 < \kappa < -\lambda^{-1}$

Here the pole in $P_>(\xi^{-1})$ at $\xi = \beta_1$ dominates the integral and we have

$$<\sigma_{00}\sigma_{0N}> \sim \mathcal{M}_{0>}^2 a_{3>}(\kappa)(-1)^N |\beta_1|^N , \tag{3.40}$$

which is an oscillatory antiferromagnetic behavior analogous to region c for $T<T_c$. This region also disappears as $\alpha_2 \to 1$.

Finally we must comment on the behavior of (3.36) in the scaling region $T \to T_c^+$. The quantity $\mathcal{M}_{0>}^2$ computed via Szegö's theorem vanishes as

$$\mathcal{M}_{0>}^2 \sim |T-T_c|^{2\beta(-\kappa)} , \tag{3.41}$$

where $\beta(\kappa)$ is given by (3.15). Moreover the splitting function $P_>(\xi)$ contains a temperature dependent multiplicative factor which behaves as

$$|T-T_c|^{\beta(\kappa)-\beta(-\kappa)} , \tag{3.42}$$

as $T \to T_c$. Therefore as $T \to T_c$ the full expression $\mathcal{M}_{0>}^2 \sum_{n=0}^{\infty} g^{(2n+1)}$ contains the factor $|T-T_c|^{2\beta(\kappa)}$. Furthermore for $-1< \kappa < 1$ the correlation length for $T>T_c$ diverges as $|T-T_c|^{-1}$. Therefore from the high temperature side the hyperscaling relation, as given by (3.33), also holds, although in a more subtle fashion than for $\kappa = 0$.

Acknowledgements

This work has been supported in part by National Science Foundation Grants PHY-79-06376 and DMR 79-08556.

REFERENCES

1. B. M. McCoy and T. T. Wu, Phys. Rev. 162, 436 (1967).
2. B. M. McCoy and T. T. Wu, Phys. Rev. Letts. 21, 549 (1968) and Phys. Rev. 176, 631 (1968).
3. M. E. Fisher and A. E. Ferdinand, Phys. Rev. Letts. 19, 169 (1967); A. E. Ferdinand and M. E. Fisher, Phys. Rev. 185, 832 (1969).
4. B. M. McCoy, Phys. Rev. Letts. 23, 383 (1969); Phys. Rev. 188, 1014 (1969).
5. L. Onsager, Phys. Rev. 65, 117 (1944).
6. Some recent work on inhomogeneous Ising lattices has been made by J.H.H.Perk, Phys. Letts. (to be published).
7. E. Barouch, B.M.McCoy and T.T.Wu, Phys. Rev. Letts. 31, 1409 (1973).
8. Our results are presented in B.M.McCoy and J.H.H.Perk, Phys. Rev. Letts. 44, 840 (1980).
9. B. Kaufman, Phys. Rev. 76, 1232 (1949); B. Kaufman and L. Onsager, Phys. Rev. 76, 1244 (1949).
10. J.H.H.Perk and H.W.Capel, Physica 89A, 265 (1977).
11. For an explanation of these techniques see "The Two-Dimensional Ising Model" by B.M.McCoy and T. T. Wu, Harvard University Press, Cambridge, Mass. 1973.
12. G. Szegö, Communications du Séminaire Mathématique de l'Université de Lund, tome supplémentaire dédié à Marcel Riesz, 1952, p. 228.
13. T. T. Wu, Phys. Rev. 149, 380 (1966).
14. T. T. Wu, B.M.McCoy, C. A. Tracy and E. Barouch, Phys. Rev. B13, 316 (1976).
15. R. Z. Bariev, Zh. Eksp. Teor. Fiz. 77, 1217 (1979), Sov. Phys. JETP 50, 613 (1979).

Institute for Theoretical Physics
State University of New York at Stony Brook
Stony Brook, New York 11794

Introduction to Holonomic Quantum Fields

by

Tetsuji Miwa and Michio Jimbo

The purpose of this paper is to present an introduction to the theory of what we call the holonomic quantum fields; to be specific we give a brief survey to each of our series of short notes entitled "Studies on Holonomic Quantum Fields, I-XVI" [1] as well as of the earlier preprint [2].

We would like to express our heartiest gratitude to Professor M. Sato, who has introduced this subject to us who has been and is working with us with his constant enthusiasm toward science.

The materials treated in [1], [2] are grouped as follows:

(1) Algebraic preliminaries (V and part of I III, Iv, X).

(2) Ising model ([2],I) and the deformation theory of Euclidean Dirac-equation (II, III, IV; VII, VIII, IX).

(3) Riemann-Hilbert problem (VI).

(4) Higher dimensions (XI, XII, XIII, XIV).

(5) Impenetrable bosons (XVI) and double scaling limit of XY model (XV).

A full account for (1), (2), (3) and (5) are found in [3], [4], [5] and [6], respectively.

Field theory of the 2-dimensional Ising model in the scaling limit [2].

1. Matrix elements of the Ising spin above the critical temperature are calculated in the infinite lattice. The reference states are those in the Fock space of free fermions originally introduced by Onsager.

2. Taking the scaling limit of the Ising spin, we obtain a local field operator $\varphi(x)$ in 2-dimensional space-time. $\varphi(x)$ is expressed as the following normal product of free fermion $\psi(p)^+ = \theta(p^o)\psi(\vec{p})^+ + \theta(-p^o)\psi(-\vec{p})$ ($p = (p^o,\vec{p})$ is on the mass shell $p^2 = (p^o)^2 - (\vec{p})^2 = m^2$).

$$(1) \quad \varphi(x) = :\psi(x)\cdot\exp\left(\frac{1}{2}\iint \underline{dp_1 dp_2} 2i\frac{p^o - p^{o\prime}}{\vec{p}+\vec{p}'+i0}\psi(p_2)^+\psi(p_1)^+ e^{i(p_1+p_2)x}\right):,$$

$$\psi(x) = \int \underline{dp}\ \psi(p)\ e^{ipx},$$

$$\underline{dp} = \frac{1}{2\pi} \frac{d\vec{p}}{2\sqrt{\vec{p}^2+m^2}} .$$

3. The asymptotic fields φ^{in}, φ^{out} for $\varphi(x)$ are calculated. There is no particle production, and the S matrix in the n-particle sector is found to be

(2) $$S = (-1)^{\frac{n(n-1)}{2}} .$$

4. We have checked the generalized unitarity relation for our field operator $\varphi(x)$.

Studies on Holonomic Quantum Fields I.

1. Without recourse to the lattice theory, direct construction of the spin operators $\varphi_F(x)$ $(T \leq T_c)$ and $\varphi^F(x) = \varphi(x)$ in (1) $(T \geq T_c)$ in the continnum is presented. Guiding principle is the theory of Clifford group. $\varphi_F(x)$ is introduced as the Clifford group element which induces the following "rotation" $T_{\varphi_F(x)}$ in the orthogonal space $W = \{w(x) = (w_+(x), w_-(x)) \mid \frac{\partial w_\pm}{\partial x^\pm} = \pm m w_\mp\}$ of solutions to free neutral Dirac equation.

(3) $$T_{\varphi_F(x)}(w^+ + w^-) = w^+ - w^- \quad \text{if} \quad w^\pm \text{ belongs to}$$

$$W_x^\pm = \{w \in W \mid w(x') = 0 \quad \text{for} \quad (x - x')^2 < 0,\ x'^1 - x^1 \lessgtr 0\}.$$

2. The symplectic (or Bosonic) version of the above construction is also given.

3. The Landau singularities of the n point Green's functions for the above operators are confined to those corresponding to graphs with no internal vertices. The order of the leading singularity is determined.

Studies on Holonomic Quantum Fields II.

1. The following monodromy problem is discussed for solutions

$w = w(z)$ of the Euclidean Dirac equation:

(4-1) $$\left(m - \begin{bmatrix} & \partial_z \\ \partial_{\bar z} & \end{bmatrix}\right) w = 0,$$

$$(z \in \mathbb{R}^2 \cong \mathbb{C}).$$

(4-2) w is multi-valued real analytic in $\mathbb{R}^2 - \{a_1, \ldots a_n\}$,

(4-3) w changes its sign when prolonged once around the branch points a_j $(j = 1 \ldots, n)$,

(4-4) $$|w(z)| = O(|z - a_j|^{-1/2}) \quad \text{at} \quad z = a_j,$$
$$= O(e^{-2m|z|}) \quad \text{at} \quad z = \infty.$$

The space $W^{\text{strict}}_{a_1,\ldots,a_n}$ of such wave functions w is shown to be n-dimensional.

2. The iso-monodromy properties (4) give rise to the following linear system of differential equations for a column vector $w = \begin{bmatrix} w_1 \\ \vdots \\ w_n \end{bmatrix}$ formed from an arbitrary basis $w_1, \ldots, w_n$ of $W^{\text{strict}}_{a_1,\ldots,a_n}$:

(5-1) $$\left(m - \begin{bmatrix} & \partial_z \\ \partial_{\bar z} & \end{bmatrix}\right) w = 0$$

(5-2) $$\left(z\partial_z - \bar z \partial_{\bar z} + \frac{1}{2}\begin{bmatrix} 1 & \\ & -1 \end{bmatrix}\right) w = (Bm^{-1}\partial_z - \bar B m^{-1} \partial_{\bar z} + E) w$$

(5-3) $$dw = (\Phi m^{-1}\partial_z + \bar\Phi m^{-1}\partial_{\bar z} + \Psi) w.$$

Here B, $\bar B$, E (resp. Φ, $\bar\Phi$, Ψ) are $n \times n$ matrices of functions (resp. 1 forms) in $a_1, \ldots, a_n$.

3. The integrability conditions of the linear system (5) (suitably normalized) give rise to the following non linear completely integrable system of differential equations.

(6.1) $$dF = -[F, \theta] + m^2[dA, {}^t G \bar A G] + m^2 [A, {}^t G d\bar A G]$$

(6-2) $$dG = -G\theta + \bar{\theta}G$$

where

(6-3) $$A = \begin{bmatrix} a_1 & & \\ & \ddots & \\ & & a_n \end{bmatrix}, \qquad B = \sqrt{G}mA\sqrt{G}^{-1},$$

$$F = \sqrt{G}^{-1}E\sqrt{G} \quad \text{and } [A,\theta] + [dA,F] = 0, \qquad \theta_{\mu\mu} = 0.$$

If n = 2 (6) reduces to a Painleve equation of the third kind.

Studies on Holonomic Quantum Fields III, IV.

1. The local expansions for products $\psi(x)\varphi_F(a)$ and $\psi(x)\varphi^F(a)$ are given. Here $\psi(x) = \begin{bmatrix} \psi_+(x) \\ \psi_-(x) \end{bmatrix}$ is the free neutral Dirac field.

2. By construction our field operators satisfy the following peculiar commutation relations.

(7-1) $$\varphi_F(x)\psi(x') = \pm\psi(x')\varphi_F(x)$$

(7-2) $$\varphi^F(x)\psi(x') = \mp\psi(x')\varphi^F(x)$$

(7-3) $$\varphi_F(x)\varphi^F(x') = \pm\varphi^F(x')\varphi_F(x) \quad \text{for} \quad x'^+ \gtrless x^+, \quad x'^- \lessgtr x^-.$$

3. If we define

(8) $$\hat{w}^j_{F,n}(x) = \langle\psi(x)\varphi_F(a_1)\ldots\varphi^F(a_j)\ldots\varphi_F(a_n)\rangle/\tau_n$$

(9) $$\tau_n = \langle\varphi_F(a_1)\ldots\varphi_F(a_n)\rangle$$

then $(\hat{w}^1_{F,n}(x),\ldots,\hat{w}^n_{F,n}(x))$ forms a basis of $W^{strict}_{a_1,\ldots,a_n}$ of (4).

4. $\omega = d\log\tau_n$ is expressed in terms of solutions to (6) as follows.

(10) $$\omega = \frac{1}{2}\,\mathrm{tr}\Big(\frac{1}{2}T\theta - \frac{1}{2}F\theta + m^2(-{}^t\bar{G}AG + \bar{A})dA\Big)$$

$$+ \text{ complex conjugate},$$

$$T = (1-G)(1+G)^{-1}.$$

The exact formula for 2-point function in terms of a Painlevé transcendent [ ] is generalized in this way.

Studies on Holonomic Quantum Fields V.

1. We start with a finite dimensional orthogonal space $(W,\langle\ ,\ \rangle)$. Given a bilinear form $(w,w')\mapsto \langle ww'\rangle$ such that $\langle ww'\rangle + \langle w'w\rangle = \langle w,w'\rangle$, we construct a linear isomorphism

(11) $$\mathrm{Nr}: A(W) \overset{\sim}{\rightarrow} \Lambda(W)$$

between the Clifford algebra $A(W)$ and the Grassmann algebra $\Lambda(W)$ over W.

2. The closure in $A(W)$ of the Clifford group

(12) $$G(W) = \{g \in A(W) \mid \exists g^{-1},\ gWg^{-1} = W\}$$

is characterized by

(13) $$\overline{G(W)} = \{g \in A(W) \mid \mathrm{Nr}(g) = cw_1\dots w_k \exp(\rho/2),$$
$$c \in \mathbb{C},\quad w_1,\dots,w_k \in W,\quad \rho \in \Lambda^2(W)\}.$$

Formula for computing $w_1,\dots,w_k$ and ρ for $g \in G(W)$ from the orthogonal transformation

(14) $$T_g: w \longmapsto gwg^{-1}$$

is given.

3. The product formula for $\mathrm{Nr}(g_1\dots g_n)$ for $g_1,\dots,g_n \in \overline{G(W)}$ is given.

Studies on Holonomic Quantum Fields, VI.

1. We construct explicitly a field operator $\varphi(a;L)$ satisfying the following commutation relations with m component one dimensional free field $\psi^{(i)}(x)$ $(i = 1,\dots,m;\ x \in \mathbb{R})$.

(15) $$(\varphi(a;L)\psi^{(1)}(x),\dots,\varphi(a;L)\psi^{(n)}(x))$$
$$= \begin{cases} (\psi^{(1)}(x)\varphi(a;L),\dots,\psi^{(n)}(x)\varphi(a;L)) & (x > a) \\ (\psi^{(1)}(x)\varphi(a;L),\dots,\psi^{(n)}(x)\varphi(a;L))e^{2\pi iL} & (x < a) \end{cases}$$

where L is an $m \times m$ skew symmetric matrix.

2. We construct the solution $Y(x; {}^{a_1 \ldots a_n}_{L_1 \ldots L_n})$ to Riemann's monodromy problem with branch points $a_1, \ldots, a_n$ and local exponents $L_1, \ldots, L_n$.

$$\text{(16)} \quad Y(x; {}^{a_1 \ldots a_n}_{L_1 \ldots L_n})_{ij} = -2\pi i(x_0 - x) \frac{\langle \psi^{(i)}_{(x_0)} \psi^{(j)}_{(x)} \varphi(a_1; L_1) \ldots \varphi(a_n; L_n) \rangle}{\langle \varphi(a_1; L_1) \ldots \varphi(a_n; L_n) \rangle},$$

where $Y(x_0; {}^{a_1 \ldots a_n}_{L_1 \ldots L_n}) = 1$.

<u>Studies on Holonomic Quantum Fields, VII, VIII.</u>

The results of II, III and IV are generalized to the case of general local exponents other than $-\frac{1}{2}$.

<u>Studies on Holonomic Quantum Fields, IX.</u>

The symplectic (or Bosonic) version of I, II, III and IV is given. Especially a field operator $\varphi^B(x) = \begin{bmatrix} \varphi^B_+(x) \\ \varphi^B_-(x) \end{bmatrix}$ which obeys the Fermi statistics is constructed. The n point functions satisfy the following simple relation

$$\text{(17)} \qquad \langle \varphi_B(a_1) \ldots \varphi_B(a_n) \rangle \langle \varphi_F(a_1) \ldots \varphi_F(a_n) \rangle = \sqrt{\det \cosh H}$$

where $G = e^{-2H}$ (6).

<u>Studies on Holonomic Quantum Fields, X, XI.</u>

1. We give a variational formula for $\langle g \otimes g^{-1} \rangle$ where g is the Clifford group element which induces the classical scattering $T[A]: w_{in} \to w_{out}$ for the 2-dimensional Weyl equation

$$\text{(18)} \qquad \left(\frac{\partial}{\partial x^+} - A(x)\right) w(x) = 0.$$

2. As an application we derive a variational formula for the block Toeplitz determinant.

Remark. Recently we realized that the above mentioned result on Toeplitz determinant was known by Widom [8].

Studies on Holonomic Quantum Fields, XII.

The results of X is generalized to the case of the Dirac equation with arbitrary space-time dimension.

Studies on Holonomic Quantum Fields, XIII.

1. We formulate the analogue of Riemann-Hilbert boundary value problem in the case of s dimensional Euclidean Dirac equation:

(19-1) $$(-\not\partial_x + m)w(x,x') = \delta^s(x - x'). \quad (x,x' \notin \Gamma),$$

(19-2) $$|w(x,x')| = O(e^{-m|x|}) \quad (x \to \infty),$$

(19-3) $$w(\xi^+ w') = M(\xi)w(\xi^-,x').$$

Here ξ is the point on the boundary Γ of a bounded domain $D^+ \subset \mathbb{R}^s$, and $w(\xi^{\pm},x')$ denote the boundary values from $\genfrac{}{}{0pt}{}{\text{in}}{\text{out}}$-side of Γ. $w(x,x')$ is an N component vector and $M(\xi)$ is an $N \times N$ matrix.

2. The variational formula for $w(x,x')$ is given

(20) $$\delta w(x,x') = \int_\Gamma d\sigma(\xi)\Sigma_{\mu=1}^{s} \delta\rho^\mu(\xi) \times w(x,\xi^+)(n_\mu \not\partial - \not n \partial_\mu)M(\xi)w(\xi^-,x').$$

Here the vector field $\Sigma\delta\rho^\mu(\xi)\partial_\mu$ represents the variation of Γ, and n is the outer unit normal to Γ.

Studies on Holonomic Quantum Fields, XIV.

As the massless case of (19), the continuum monodromy version of Schlesinger's theory is given, which reproduces the original Schlesinger's equation in the limit of $M(\xi)$ in (19) being a step function. The results of II are also reproduced along the same line.

Studies on Holonomic Quantum Fields, XV.

The double scaling limit of 1 dimensional XY model is studied.

The corresponding isomonodromy deformation theory considered in momentum space contains an irregular singularity of rank 1 at $p = \infty$. The linear system is of the form,

$$(21) \qquad dY = \Omega Y$$

$$\Omega = \Sigma_s A_s d\log(p - c_s) + d(pA_\infty) + \theta$$

$$[\theta, A_\infty] = [\Sigma_s A_s, dA_\infty]$$

$$A_\infty : \text{diagonal.}$$

The corresponding non-linear system takes the form,

$$(22) \qquad dA_s = -\Sigma_{s'(\neq s)} [A_s, A_{s'}] d\log(c_s - c_{s'}) - [A_s, d(c_s A_\infty) + \theta].$$

In terms of solution matrices $\{A_s\}$ of (22), the logarithmic derivative of n point function is given by

$$(23) \qquad \omega = \frac{1}{2} \operatorname{trace}(\Sigma_s A_s d(c_s A_\infty) + \frac{1}{2}\theta \, \Sigma A_s + \frac{1}{2} \Sigma_{s\neq s'} A_s A_{s'} d\log(c_s - c_{s'}))$$

$$+ \text{const.} \, \Sigma_{s\neq s'} \, d\log(c_s - c_{s'}).$$

Studies on Holonomic Quantum Fields, XVI.

It is shown that the one particle reduced density matrix $\rho(x)$ for impenetrable bosons, or equivalently the 2 point function for the quantum non-linear Schrodinger equation with infinite coupling and with finite particle density, satisfies a third order non-linear differential equation

$$(24) \qquad \left(x\frac{d^2\sigma}{dx^2}\right)^2 = -4\left(x\frac{d\sigma}{dx} - 1 - \sigma\right)\left(x\frac{d\sigma}{dx} + \left(\frac{d\sigma}{dx}\right)^2 - \sigma\right)$$

where $\sigma(x) = x\frac{d}{dx}\log \rho(x)$.

Acknowledgements.

We wish to thank D. and G. Chudnovsky for giving us an opportunity to talk on these subjects in their Seminar at I.H.E.S.

References

[1] M. Sato, T. Miwa and M. Jimbo, Proc. Japan Acad. 53A(1977), 6-10(I), 147-152(II), 153-158(III), 183-185(IV), 219-224(V),

54A(1978), 1-5(VI), 36-41(VII), 221-225(VIII), 263-268(IX)(*), 309-313(X).

55A(1979), 6-9(XI), 73-77(XII), 115-120(XIII)(*), 157-162(XIV)(*), 267-272(XV)(*).

XVI(*): to appear.

[2] M. Sato, T. Miwa, M. Jimbo, Field theory of the two dimensional Ising model in the scaling limit, preprint RIMS 207 (1976).

[3] M. Sato, T. Miwa, M. Jimbo, Publ. RIMS 14, 223-267 (1978).

[4] M. Sato, T. Miwa, M. Jimbo, preprint, RIMS 260, 263 & 267 (1978).

[5] M. Sato, T. Miwa, M. Jimbo, Publ. RIMS 15, 201-278 (1978).

[6] M. Jimbo, T. Miwa, Y. Mori and M. Sato Density matrix of impenetrable bosons and the fifth Painleve transcendent (1979), submitted to Physica D.

[7] E. Barouch, B.M. McCoy and T.T. Wu, Phys. Rev. Lett. 31, 1409-1411 (1973).
T.T. Wu, B.M. McCoy, C.A. Tracy and E. Barouch, Phys. Rev. B13, 316-374 (1976).

[8] H. Widom, Advances in Math. 13, 284-322 (1974), Proc. Amer. Math. Soc. 50, 168-173 (1975). Advances in Math. 21, 1-29 (1976).

(*) With author(s) M. Jimbo (IX), M. Jimbo and T. Miwa (XIII,XIV), M. Jimbo, T. Miwa and M. Sato (XV) M. Jimbo, T. Miwa, Y. Mori and M. Sato (XVI).

Research Institute for Mathematical Sciences, Kyoto University, Kyoto, Japan.

PLANAR ISING FERROMAGNET: CORRELATION FUNCTIONS AND THE INVERSE SCATTERING METHOD

by

D.B. Abraham*

INTRODUCTION

Over the last few years there has been considerable progress in understanding the thermodynamic and microscopic properties of the planar Ising ferromagnet. This models phase transitions in such diverse systems as binary alloys, mixtures, one component fluids [1]. Of recent interest has been the observation that in the critical region the planar Ising ferromagnet is in fact an exactly solvable Euclidean field theory model with one space dimension [2].

The approach to this field adopted by the author resides in an important review article by Schultz, Mattis and Lieb [3], where the relationship between the planar Ising model and two free Fermi fields is established. The essence of recent progress is contained in a review [4] and original sources [5]; it will not be repeated here. Rather, the main aim, and one consistent with the title of the seminar, will be to show a rather direct analytical link with the theory of non linear systems and the inverse scattering problem. We shall show that, whether the scaling limit be taken or not, the two-point function can be expressed in terms of the solution of an integral equation of Fredholm type. In the scaling limit, to be defined later, this equation may be recast in terms of a Zakharov-Shabat [6] scheme. This work may thus be viewed as a complementation of the original and remarkable association of the Ising scaling limit with a Painlevé transcendent, due to Wu and co-workers [7]. An advantage of the present approach is that the same thread runs through the determination of n - point functions, the theory

---

*On leave from Oxford University

of phase separation or symmetry breakdown and that of the surface properties of the planar Ising ferromagnet.

2. DEFINITIONS

Consider a subset Λ of $\mathbb{Z}^2$ of the form $\Lambda = \{(n,m);\ 1 \le n \le N,\ 1 \le m \le M\}$. Assign to each point $\underline{r}$ of $\mathbb{Z}^2$ a classical spin variable $\sigma(\underline{r}) = \pm 1$. Besides its magnetic connotation, we may think of $\mathbb{R}^2$ as divided into a periodic array of contiguous square cells. Then the variable $\sigma(\underline{r})$ distinguishes between two molecular species for the binary alloy or mixture, or between single occupation (multiple being forbidden) and vacuum in the lattice gas model. A configuration of the lattice Λ is a specification of $\sigma(\underline{r})$ for each $\underline{r} \in \Lambda$, denoted $\{\sigma\}$; it has an associated energy given by

$$
\begin{aligned}
E_\Lambda\{\sigma\} = & - \sum_{|\underline{i}-\underline{j}|=1} J(\underline{i}-\underline{j})\ \sigma(\underline{i})\ \sigma(\underline{j}) \\
& - H \sum_{\underline{i}} \sigma(\underline{i}) \\
& - \sum_{i \in \partial\Lambda} H(i)\ \sigma(i)\ . \qquad (2.1)
\end{aligned}
$$

The $J(x)$ are non-negative couplings and the H and $H(\underline{i})$ are magnetic fields (or fugacities) the latter acting solely on spins in the surface, denoted $\partial\Lambda$. The statistical mechanics of the model is given by the canonical probability measure

$$P_{\Lambda,B(\Lambda)}\ (\{\sigma\}) = Z_\Lambda^{-1}\ \exp - \beta\, E_\Lambda\{\sigma\} \qquad (2.2)$$

where Z_Λ, called the partition function, normalises the measure. The system is supposed to be in equilibrium with a large heat bath at a temperature T, with $\beta = 1/kT$, k being Boltzmann's constant in suitable energy units. The label $B(\Lambda)$ specifies the boundary conditions $H(i)$ on Λ. The reader who wants a general account of the connection between Z_Λ and thermodynamics and a review of the properties of (2.2) which relate to phase transitions should consult lecture notes by Martin-Löf [8] and a review by Gallavotti [9].

If we think of the states of columns of spins as an evolution of the state in say the first column of Λ, then (2.1) and (2.2) show that we have a Markov chain. This non-trivial simplification results from the nearest-neighbour character of the interactions in the (1,0) direction. The n-point correlation function for (2.1) and (2.2) is defined by

$$\rho\{(\underline{r})_n \mid \Lambda, B(\Lambda)\} = \langle \prod_1^n \sigma(\underline{r}_j) \rangle_{\Lambda, B(\Lambda)} \tag{2.3}$$

where $\langle \quad \rangle_{\Lambda,B(\Lambda)}$ denotes expectation with respect to (2.2), and the coordinate notation $(\underline{r})_n$ means

$$(\underline{r})_n = (\underline{r}_1, \ldots, \underline{r}_n) \tag{2.4}$$

and $\underline{r}_j = (x_j, y_j)$ in Cartesian form. Provided there are no interactions between the first and last columns of spins, (2.3) may be written as

$$\rho\{(\underline{r})_n \mid \Lambda, B(\Lambda)\}$$

$$= \Big(a,\ U^{x_1}\ \sigma(y_1)\ U^{x_2-x_1}\ \sigma(y_2)\ \quad U^{(x_n-x_{n-1})}\ \sigma(y_n)$$

$$U^{N-x_n}\ b\Big) \,/\, (a,\ U^N b) \tag{2.5}$$

where $x_j > x_{j-1}$; and a and b are vectors in the space of configurations of a column and are given in terms of the H(i) for left and right extreme columns respectively. The column label in each spin variable has been suppressed. The key feature in (2.5) is that U is a $2^M \times 2^M$ matrix and matrix multiplication is implied. A further crucial simplification incorporated in (2.5) is that the H(i) acting on the top and bottom spin in each column are column-independent so that the same matrix can be used throughout.

The next step is to define a Hilbert space H_M and operators the representatives of which reproduce (2.5). The matrix U is thus obtained from the transfer operator V which may be chosen self-adjoint. Thus (2.5) can be decomposed spectrally to give

$$\rho\ \{(\underline{r})_n \mid \Lambda, B(\Lambda)\}$$

$$= \sum_{j_1 \dots j_{n+1}} (a,\ \phi(j_1)) \prod_{k=1}^{n} (\phi(j_k),\ \sigma(y_n)\ \phi(j_{k+1}))$$

$$(\phi(j_{n+1}),\ b)\ \exp - \{x_1\ \gamma(j_1) + \sum_2^n (x_k - x_{k-1})\ \gamma(j_k)$$

$$+ (N - x_n)\ \gamma(j_{n+1})\}$$

$$/ \sum_{j_1} (a,\ \phi(j_1))(\phi(j_1),\ b)\ \exp\,(-N\ \gamma(j_1)) \qquad (2.6)$$

where $\gamma(j) > 0$, the operator V being normalised to have unit maximum eigenvalue. In the following we shall consider lattices Λ with no surface fields on internal columns and, moreover, with cyclic boundary conditions. Then the $\phi(j)$ may be taken as eigenvectors simultaneously for V and for translation within the column. As Schultz, Mattis and Lieb showed, there are then two invariant subspaces for V which are coupled by odd products of spin operators; in each space, V has eigenvectors which are of free-Fermi type and its eigenvalues are easily given. The remaining problem is to determine the translational character of the $\phi(j)$ and to unravel the structure of the matrix elements which appear in (2.6). This has now been accomplished for the limiting case $M \to \infty$ [5]. Each intermediate state label j is replaced by a multiple $(\omega_{1j}, \dots, \omega_{nj}) = (\omega_j)_n$ with $\omega_{jk} \in [0, 2\pi]$. Such a variable describes an intermediate Fermi particle, which we may associate with a vertex of a family of graphs, F, providing a natural setting for (2.6). Let G be in F; its vertex set V may be partitioned into subsets V_k (some of which may be empty) corresponding to each intermediate state label j_k, $k = 1, \dots, n+1$. V_k itself contains vertices v_{kj}, $j = 1, \dots, m_k$, corresponding to the intermediate Fermi particles labelled ω_{kj}, with weight

$$w(\omega_{kj}) = \exp\{-(x_{k+1} - x_k)\ \gamma(\omega_{kj}) + i\ \mathrm{sgn}\ k\ (y_{k+1} - y_k)\ \omega_{kj}\}\ . \qquad (2.7)$$

The Onsager [10] function $\gamma(\omega)$ is given by

$$\cosh \gamma(\omega) = \cosh 2K_1^*\ \cosh 2K_2 - \sinh 2K_1^*\ \sinh 2K_2\ \cos \omega \qquad (2.8)$$

with $\gamma(\omega) > 0$ for $\omega \in [0, 2\pi]$, $K_1 = \beta J(1, 0)$, $K_2 = \beta J(0, 1)$ and K_j^* given by the involution $\exp - 2K_j^* = \tanh K_j$, with K_j^* real if $K_j \geq 0$. The Fermi particles may be considered to scatter off the spins $\sigma^x(y)$, but in fact, without conservation of particle number, as described by the matrix elements in (2.6). We shall conclude this section by writing down the limiting form of (2.6) as $M \to \infty$; it is hoped that the graph-theoretical ideas above will make it easier to visualise. We have

$$\rho \left\{(r)_n \mid N, a, b\right\} = Z(N, a, b)^{-1} W\left((\underline{r})_n \mid N, a, b\right) \tag{2.9}$$

where

$$Z(N, a, b) = \sum_{m=0}^{\infty} \frac{1}{m!\,(2\pi)^m} \int_0^{2\pi} \int d(\omega)_m\, F_a\big((\omega)_m\big)\, F_b\big((\omega)_m^*\big) \exp - N \sum_1^m \gamma(\omega_j) \tag{2.10}$$

and

$$W\big((r)_n \mid N, a, b\big) = \sum_{m_1 \dots m_{n+1} = 0}^{\infty} \prod_1^{n+1} \frac{1}{m_j!\,(2\pi)^{m_j}} \int_0^{2\pi} \int d(\omega_1)_{m_1} \dots d(\omega_{n+1})_{m_{n+1}}$$

$$\left(\prod_{v \in V} w(v)\right) F_a\left[(\omega_1)_{m_1}\right] F_b\left[(\omega_{n+1})_{m_{n+1}}\right]^* \prod_{k=1}^{n} F^x\left[(e^{i\omega})_{m_k} \,\middle|\, (e^{i\omega})_{m_{k+1}}\right]. \tag{2.11}$$

Equations (2.9), (2.10) and (2.11) give the n-point functions for an infinite strip of width N with arbitrary boundary conditions on the edges of the strip described by the labels a and b. The special case of a torus with one dimension finite is given by taking $F_u\big((\omega)_m\big) = 1$, $u = a,b$ in (2.10) for all $m \geq 0$; in (2.11), replace $(\omega_{n+1})_{m_{n+1}}$ by $(\omega_1)_{m_1}$, remove $(\omega_{n+1})_{m_{n+1}}$ and associated prefactors.

3. MATRIX ELEMENTS

In this section the results of recent computations of the factors F will be given: First consider the elements $F_x\{(e^{i\omega})_m \mid (e^{i\omega})_{m,n}\}$ with $0 \le m \le n$. Their behaviour depends crucially on the spin of $T - T_c$, where T_c is the critical temperature, given by $\gamma(0) = 0$ from (2.8). At this point (2.7) and (2.11) show that the correlation length diverges, a well-known criterion of criticality. If lengths are rescaled by $|r'| = |r|/\gamma(0)$, then as $\gamma(0) \to 0$ the vertex weights become

$$w_s(p_{kj}) = \exp\left[-(x'_{k+1} - x'_k)(1 + p_{kj}^2)^{\frac{1}{2}} + i \operatorname{sgn} k\, p_{kj}(y'_{k+1} - y'_k)\right] . \quad (3.1)$$

These are appropriate for the Euclidean field theory, but quite different field theories are obtained as $T \to T_c\pm$.

Let us define parameters A and B by

$$A = \coth K_1^* \coth K_2 , \quad B = \coth K_1^* \tanh K_2 . \quad (3.2)$$

We use the notation

$$\Delta_J(\omega)_I = (\omega)_{I/J} \quad J \subset I \quad (3.3)$$

$$= \phi \quad \text{otherwise}$$

for index sets I and J. Then we have the results in Table 1.

Knowledge of the Pfaffian structure of the matrix elements enables one to complete the graphical picture of the Schultz-Mattis-Lieb approach. First we focus attention on the two-point function with toroidal boundary conditions in the limit $N \to \infty$. From (1.5), (1.6) we have $m_1 = m_3 = 0$, $m_2 \ge 0$. The vertex set V for a $G \in \mathcal{F}$ thus has one element in its partition.

1. For $T > T_c$, $|V| = 2n$, n integer. Each vertex has degree precisely 2, and E(G) is a union of disjoint cycles, each having even length, and edge weight $f_>(\omega_1,\omega_2)$, with $\varepsilon = -1$ in Table 1. This structure is entirely appropriate for application of the linked cluster theorem [11]: we get

TABLE 1. STRUCTURE OF MATRIX ELEMENTS

$T > T_c$:

$$F^x\left((e^{i\omega})_m \mid (e^{i\omega})_{2n+1}\right)$$

$$= \sum_{1}^{2n+1} (-1)^j f(\omega_j) F\left(\Delta_j (e^{i\omega})_m \mid (e^{i\omega})_{2n+1}\right) \tag{1}$$

$$F\left((e^{i\omega})_m \mid (e^{i\omega})_{2n}\right)$$

$$= \sum_{2}^{2n} (-1)^j f_{\pm}(\omega_1,\omega_j) F\left(\Delta_{1j} (e^{i\omega})_m \mid (e^{i\omega})_{2n}\right) . \tag{2}$$

Take $f_+(\omega_1,\omega_j)$ (resp. $f_-(\omega_1,\omega_j)$) for $2 \le j \le m$ (resp. $m+1 \le j \le 2n$).

$$f_{\pm}(\omega_1,\omega_2) = (g_>(\omega_1)\, g_>(-\omega_1) \pm g_>(\omega_2)\, g_>(-\omega_2)) \, [g_>(-\omega_1)\, g_>(-\omega_2)$$
$$(1 + \exp - i(\omega_1+\omega_2))]^{-1} \tag{3}$$

with $g_>(\omega) = [(e^{i\omega} - B^{-1})(e^{i\omega} - A)]^{-\frac{1}{2}}$ (4a)

$$f(\omega) = -(A/B)^{\frac{1}{2}}\, g_>(-\omega) \tag{4b}$$

$$F(\phi) = [1 - (\sinh 2K_1 \sinh 2K_2)^2]^{\frac{1}{8}} / \cosh K_1^* . \tag{5}$$

$T < T_c$:

$$F^x\left((e^{i\omega})_m \mid (e^{i\omega})_{2n}\right) = F\left((e^{i\omega})_m \mid (e^{i\omega})_{2n}\right) \tag{6}$$

given by (2) and (3) above but with $g_>$ replaced by

$$g_<(\omega) = [(e^{i\omega} - A)/(e^{i\omega} - B)]^{\frac{1}{2}} \tag{7}$$

in (3), and (5) replaced by

$$F(\phi) = [1 - (\sinh 2K_1 \sinh 2K_2)^{-2}]^{\frac{1}{8}} . \tag{8}$$

The matrix element in (1) with $2n+1$ replaced by $2n$ vanishes identically, as does that in (6) with $2n$ replaced by $2n+1$.

$$\rho((\underline{r})_2) = (m^*)^2 \exp \tfrac{1}{2} \operatorname{Tr} \log(1 + \mathbb{K}_<^2) \tag{3.4}$$

where the kernel $\mathbb{K}_<$ is a mapping from $\mathbb{L}^2([0,2\pi], d\mu)$ to itself with

$$d\mu(\omega) = d\omega/2\pi \sinh \gamma(\omega) \tag{3.5}$$

and

$$(\mathbb{K}_< f)(\omega_1) = i \int_{-\pi}^{\pi} d\mu(\omega_2) f(\omega_2)\, e_<(\omega_1,\omega_2) \exp\Big[-x(\gamma(\omega_1) + \gamma(\omega_2)) + iy(\omega_1 + \omega_2)\Big] / 2 \tag{3.6}$$

with

$$e_<(\omega_1,\omega_2) = \Big(p(\omega_1)\, q(\omega_1) - p(\omega_2)\, q(\omega_1)\Big) / 2 \sin((\omega_1 + \omega_2)/2)$$

with (3.7)

$$p(\omega) = (2\cos\omega - A - 1/A)^{\frac{1}{2}}\ , \qquad q(\omega) = (2\cos\omega - B - 1/B)^{\frac{1}{2}}\ .$$

Van Hove has warned against unwarranted application of the linked cluster theorem [12]. The problem is that, whereas (2.11), is trivially convergent, (3.4) might not be. Nevertheless, we can easily check that (3.4) converges for all y, provided $x \neq 0$ by use of the Cauchy-Schwartz inequality on each term in the expansion of the log, using $\|e_<\| \leq 1$. The case $x = y = 0$ can be handled using the elliptic mapping of Onsager [10]; $\mathbb{K}$ becomes a <u>difference</u> kernel, but this property regretably does not persist for $\underline{r} \neq 0$.

For $T > T_c$, $|V|$ is odd. In this case, $E(G)$ consist of a single chain with an odd number of vertices in disjoint union with a union of disjoint cycles, as before. The linked cluster theorem is again rigorously applicable and yields

$$\rho((\underline{r})_2) = (h,\ (1 + \mathbb{K}_>^2)^{-1} h) \exp \tfrac{1}{2} \operatorname{Tr} \log (1 + \mathbb{K}_>^2) \tag{3.8}$$

where $\mathbb{K}_>$ is defined by (3.6) with $e_<$ replaced by

$$e_>(\omega_1,\omega_2) = (\sinh\gamma(\omega_2) - \sinh\gamma(\omega_1))/2 \sin((\omega_1 + \omega_2)/2) \tag{3.9}$$

and

$$h(\omega) = \exp(-x\,\gamma(\omega) + i\omega y)/2 \tag{3.10}$$

with the scalar product (,) defined by

$$(f,g) = \int_{-\pi}^{\pi} f(\omega)^* \, g(\omega) \, d\mu(\omega) \quad . \tag{3.11}$$

Equation (3.8) is highly suggestive: solve the Fredholm problem

$$(1 + \mathbb{K}_>^{\,2})g = h \tag{3.12}$$

then the first factor on the right of (3.8) is simply (h,g). By differentiation with respect to x or y the trace term can also be related to a Fredholm problem: let

$$\left[(1 + \mathbb{K}^2)\mathbb{L}\right] (\cdot,\cdot) = \mathbb{K}(\cdot,\cdot) \tag{3.13}$$

where the second argument is a parameter in (3.13). Then, for instance, solution of (3.13) gives

$$\tfrac{1}{2} \operatorname{Tr} \log(1 + \mathbb{K}^2) = - \int_y^{\infty} du \operatorname{Tr} \frac{\partial \mathbb{K}}{\partial u} \mathbb{L} \quad . \tag{3.14}$$

Thus numerical solution of (3.12) and (3.13) enables one to calculate $\rho((\underline{r})_2)$ precisely using (3.4) and (3.8). This result quite possibly will lead to some general results on Töplitz determinants [13].

The structure developed above simplifies considerably in the scaling limit. But we shall return to this after describing results for surface properties.

4. RESULTS FOR SURFACES

A magnetic field J_0 applied to the left-hand surface column may be modelled by imagining a zeroth column A "ghost" spins, with applied field $+\infty$ acting on each, coupled by horizontal interactions of magnitude J_0 to the first column. This rather artificial prescription is necessary for technical reasons. That such a system can provide tests of surface scaling ideas has recently been pointed out by de Gennes and Fisher [14]. Subsequently Au Yang and Fisher [15] investigated the surface free energy and magnetisation in the first column, and their dependence on a field applied at the other end of the lattice, provided both fields have the same spin. These results were also obtained by the author, but with arbitrary field on each end, leading to interesting domain wall behaviour [16]. But this is of peripheral interest to the main theme which is to develop the magnetisation profile and pair correlation function near the surface [17, 18].

Let the magnetisation profile be $m(p)$, where p is the distance from the surface in units of the lattice spacing. Then (2.6) gives

$$m(p) = \sum_0^\infty \frac{1}{n!(2\pi)^n} \int_{-\pi}^{\pi} \cdots \int d(\omega)_n \, M((\omega)_n) F_x((\omega)_n) \exp\left(-p \sum_1^n \gamma(\omega_j)\right) \tag{4.1}$$

where $M((\omega)_n)$ is the "surface" matrix element. The structure of (4.1) has n even (resp. odd), for $T < T_c$ (resp. $T > T_c$) $F_x((\omega)_n)$ is given by Table 1. $M((\omega)_n)$ has an analogous Pfaffian structure.

$\underline{T < T_c}$:

$$M((\omega)_{2n}) = \sum_2^n (-1)^j \, f_0(\omega_1, \omega_j) \, M(\Delta_{ij}(\omega)_{2n}) \tag{4.2}$$

with initial condition

$$M(\phi) = 1 \tag{4.3}$$

and

$$f_0(\omega_1, \omega_j) = 2\pi i \, \delta(\omega_1 + \omega_j) \, g(\omega_1) \tag{4.4a}$$

where

$$g(\omega) = [1 - q(\omega) \cot(\delta^*(\omega)/2)]/[1 + q(\omega) \tan(\delta^*(\omega)/2)] \tag{4.4b}$$

with $q(\omega) = \sin\omega \, e^{-2K_2} \sinh 2K_0^*/(\cosh 2K_0^* - \sinh 2K_0 \cos\omega)$

$$e^{i\delta^*(\omega)} = \left(\frac{B}{A}\right)^{\frac{1}{2}} \left[\frac{(e^{i\omega} - A)(e^{i\omega} - B^{-1})}{(e^{i\omega} - A^{-1})(e^{i\omega} - B)}\right]^{\frac{1}{2}} . \tag{4.5}$$

$\underline{T > T_c}$:

$$M((\omega)_{2n+1}) = \sum_{j=1}^{2n+1} (-1)^j f_0(\omega_j) M(\Delta_j(\omega)_{2n+1}) \tag{4.6}$$

with

$$f_0(\omega) = \pi\delta(\omega) \exp 2(K_2 - K_0^*) \left[e^{-2K_2} \left(e^{4K_0^*} - 1 \right) + \sinh 2K_1^* \right.$$
$$\left. / \sinh \gamma(0) \right] . \tag{4.7}$$

Qualitatively speaking, $m(p)$ decays exponentially on a length scale of the bulk correlation length $1/\gamma(0)$, but for $T < T_c$,

$$m(p) \sim m^* + \alpha\, e^{-2p\gamma(0)}/p^{3/2} \tag{4.8}$$

whereas for $T > T_c$,

$$m(p) \sim \alpha'\, e^{-p\gamma(0)} \tag{4.9}$$

where α and α' depend on the parameters of the problem, but are independent of p. The change over to a quasi-one-dimensional decay in (4.9) is most striking.

Analogous results can be obtained for the correlation function between spins at (1,1) and at (p+1, q+1), denoted $\rho_e((p,q))$. We have

$$\rho_e(p,q) = m(1)\, m(p)$$
$$+ \sum_{n=2}^{\infty} \sum_{j,k=1}^{n} \theta(j,k) \frac{1}{n!(2\pi)^n} \int_{-\pi}^{\pi} \int d(\omega)_n\, F_x((\omega)_n)$$
$$\exp\left(-p \sum_1^n \gamma(\omega_j)\right) \exp - iq(\omega_j + \omega_k) \frac{e^{-i\omega_j}}{A(\omega_j)A(\omega_k)} M(\Delta_{jk}(\omega)_n) \tag{4.10}$$

where n is even (resp. odd) to $T < T_c$ (resp. $T > T_c$),

$$\Theta(j,k) = [1 - \delta(j,k)] \operatorname{sgn} (j-k) \tag{4.11}$$

and the function $A(\omega)$ is given by

$$A(\omega) = (\cosh 2K_0^* - \sinh 2K_0^* \cos \omega)\, e^{K_2} \cos(\delta^*(\omega)/2)$$

$$+ \sin\omega \sinh 2K_0^*\, e^{-K_2} \sin(\delta^*(\omega)/2) \,. \tag{4.12}$$

The functions $M((\omega)_n)$ and $F((\omega)_n)$ are as above.

The conclusion which will be drawn in (4.22) et seq. below is that all these functions can be expressed in terms of linear Fredholm problems.

Another interesting boundary condition on the first column which turns out to be solvable is with field $-J_0$ for column positions 1 to $s+1$, but $+J_0$ elsewhere, with $J_0 < J_1$. Thus we have a domain wall D pinned in the surface at $(\frac{1}{2},\frac{1}{2})$ and $(\frac{1}{2}, s+\frac{1}{2})$. The fact that $J_0 < J_1$ means that the domain wall is preferentially bound to the surface, but with concomitant loss of entropy. (The reader who is familiar with lattice statistical mechanics may care to equate <u>domain wall</u> with <u>long contour</u> on the dual lattice Λ^*.) It is remarkable that D undergoes a phase transition from a low-temperature state, in which D is bound to the surface with modified incremental free energy, to an intermediate state in which D develops large fluctuations; the associated incremental free energy is then independent of J_0, and is given by Onsager's value $\gamma(0)$.

With the notation A for the boundary condition with all fields J_0, and B(s) for the modified boundary condition, the magnetisation profile at position $(x, s/2)$ is

$$m(x, s/2) = m(x) + (Z_B(s)/Z_A) \sum_{n=1}^{\infty} \sum_{j,k=1}^{2n} \theta(j, k)$$

$$\frac{1}{(2n)!\,(2\pi)^{2n}} \int_{-\pi}^{\pi} \int d(\omega)_{2n}\, e^{i(\omega_j-\omega_k)s/2}\, e^{-i(\delta^*(\omega_j)+\delta^*(\omega_k))/2}$$

$$\times (A(\omega_j)A(\omega_k))^{-1}$$

$$M(\Delta_{jk}(\omega)_{2n})\, F_x((\omega)_{2n}) \exp\left(-x \sum_{1}^{2n} \gamma(\omega_j)\right) \tag{4.13}$$

where $Z_B(s)$ and Z_A are partition functions. The incremental free energy is

$$\tau = \lim_{s\to\infty} \frac{1}{s} \lim_{\Lambda\to\infty} \log (Z_B(s)/Z_A) \tag{4.14}$$

with

$$\lim_{\Lambda\to\infty} Z_B(s)/Z_A = \delta(s) + \frac{1}{\pi} \int_0^{2\pi} d\omega\, e^{is\omega}\, C(\omega)/A(\omega) \tag{4.15}$$

where

$$iC(\omega) = (\cosh 2K_0^* + \sinh 2K_0^* \cos \omega)\, e^{-K_2} \sin (\delta^*(\omega)/2)$$

$$+ \sinh 2K_0^*\, e^{K_2} \sin \omega \cos (\delta^*(\omega)/2) \; . \tag{4.16}$$

The limit $s \to \infty$ in (4.13) is taken by looking at the singularity structure in the complex ω plane. There are branch points at $e^{\pm i\omega} = A,B$. In addition $A(\omega)$ has simple poles at

$$\omega = iv_0 + 2n\pi \,, \quad n = 0, \pm 1, \ldots \tag{4.17}$$

with

$$2 \cosh v_0 = (B + 1/B) + 2 - (w + 1/w) \tag{4.18}$$

where

$$w = \exp 2K_2(\cosh 2K_1 - \cosh 2K_0)/\sinh 2K_1 \; . \tag{4.19}$$

Further, the poles must satisfy

$$2 \sinh \gamma(\omega) = (w - 1/w) \; . \tag{4.20}$$

Careful consideration of the branch cut structure shows that there are no such poles if $w < 1$. From (4.18) it is clear that, for $w > 1$, the simple pole dominates in the limit $s \to \infty$, so that x dependence enter on a scale $1/\gamma(iv_0)$, giving the binding effect alluded to above. Defining $J_0 = aJ_1$, the equation $w = 1$ defines a locus of points $T_R(a)$ for $0 < a \leq 1$. The region $w > 1$ for given a corresponds to $0 \leq T < T_R(a)$. When $T > T_R(a)$, we have the result

$$\lim_{s\to\infty} m(x, s/2) = - m(x) \tag{4.21}$$

so that at $x \to \infty$, the state with spontaneous magnetisation $-m^*$ is attained, on the scale of the usual bulk correlation length. The transition to $+m^*$ only occurs if z is scaled as αs^δ, with $\delta \geq \frac{1}{2}$. This result is reminiscent of Gallavotti's theorem [19]: we expect to find the domain wall which is pinned at points spaced s units apart at a distance $\sim s^{\frac{1}{2}}$ from any finite bounded set of points in the plane. This, again, is reminiscent of the central limit theorem, as developed in [20].

The representations of $m(x)$ and $m(x, s/2)$ in (4.1) and (4.13) can be simplified enormously by use of the linked cluster theorem: for $T < T_c$, we obtain

$$m(x) = m^* \exp \operatorname{Tr} \log (1 + \mathbb{J}) \tag{4.22}$$

where

$$\mathbb{J}(\omega_1\omega_2) = -(g(\omega_1)g(\omega_2))^{\frac{1}{2}} f_-(\omega_1,\omega_2) \exp -x(\gamma(\omega_1) + \gamma(\omega_2)) \tag{4.23}$$

and

$$m(x, s/2) = \{(a_s, (1 + \mathbb{J})^{-1} a_{-s}) + 1\} m(x) \tag{4.24}$$

where

$$a_s(\omega) = \frac{1}{2\pi}\int_{-\pi}^{\pi} d\omega_1 \, e^{-i\delta^*(\omega_1)/2} A(\omega_1)^{-1} e^{i\omega_1 s/2} f_-(\omega_1,\omega)$$

$$g(\omega)^{\frac{1}{2}} \exp - x\gamma(\omega_1) \exp(-x\gamma(\omega)/2)\cdot(Z_B(s)/Z_A)^{\frac{1}{2}} .$$

This is obviously a remarkable simplification. Notice in particular that the s dependence resides entirely in $a_s(\omega)$. Also, for $T < T_R(a)$ a new correlation length $(1/\gamma(iv_0))$ emerges; this diverges at $T_R(\omega)$. The results here are outlined in [17], and will be developed at length in another publication.

5. SCALING LIMIT

Consider first the magnetisation profile $m(x)$ near a surface subjected to a field J_0. We shall take the scaling limit $x \to \infty$, $\gamma(0) \to 0$ such that $s = \gamma(0)x$ is <u>fixed</u>. The field variable will also be scaled:

$$\hat{J}_0 = (J_0/(\gamma(0))^{\frac{1}{2}})\, e^{2K_2(c)} \quad . \tag{5.1}$$

Since $\gamma(0) \sim |T - T_c|$, this is precisely the type of scaling proposed by Au Yang, de Gennes and Fisher [14,15].

We develop the scaling limit of (4.1). Define a momentum variable $p \in \mathbb{R}$ by the limiting behaviour of $\omega/\gamma(0)$, $\omega \in [0,2\pi]$. Then introduce a 'rapidity' Θ by $p = \sinh \Theta$. Equation (4.1) then becomes

$$m(s) = m^* \exp v(s) \tag{5.2}$$

with

$$v(s) = \sum_1^\infty \frac{(-1)^{n+1}}{n} \frac{1}{(2\pi)^n} \int_{-\infty}^{\infty} \cdots \int d(\Theta)_n \prod_1^n \exp(-2s \cosh \Theta_j)$$

$$\frac{1}{\cosh \Theta_j + \cosh \Theta_{j+1}} G(\Theta_j) \tag{5.3}$$

with

$$G(\Theta) = \frac{\hat{J}_0^2 - (1 + \cosh \Theta)}{1 + \hat{J}_0^2 (1 + \cosh \Theta)/\sinh^2 \Theta} \quad . \tag{5.4}$$

Evidently

$$\frac{\partial v}{\partial s} = \sum_1^\infty (-1)^n \frac{1}{(2\pi)^n} \int_{-\infty}^{\infty} \cdots \int d(\Theta)_n \prod_1^n G(\Theta_j) \exp(-2s \cosh \Theta_j)$$

$$\prod_1^{n-1} \frac{1}{\cosh \Theta_j + \cosh \Theta_{j+1}} \ , \tag{5.5}$$

This can be related to a Fredholm problem directly, for computational purposes; but further insights come from using the identity

$$\int_0^\infty e^{-su}\, ds = 1/u \tag{5.6}$$

valid for $\operatorname{Re} u > 0$, we get

$$\frac{\partial v}{\partial s} = - \sum_{1}^{\infty} (-1)^{n+1} \int_0^{\infty} \cdots \int d(s)_n \prod_{2}^{n-1} \mathbb{L}_s(s_j, s_{j+1}) \, \mathbb{L}_s(0, s_1) \, \mathbb{L}_s(s_n, 0)$$

$$+ \frac{1}{\pi} \int_0^{\infty} G(\Theta) \exp(-2s \cosh \Theta) \, d\Theta \tag{5.7}$$

where

$$\mathbb{L}_s(s_1, s_2) = \frac{1}{2\pi} \int_{-\infty}^{\infty} d\Theta \exp[(s_1 + s_2 + 2s) \cosh \Theta] \; G(\Theta) \; . \tag{5.8}$$

Equation (5.7) may now be put into Fredholm form

$$\frac{\partial v}{\partial s} = - \left((1 + \mathbb{L}_s)^{-1} u_s \right) (0) \tag{5.9}$$

with

$$u_s(y) = \mathbb{L}_s(y, 0) \; . \tag{5.10}$$

Thus we have to determine w_s from

$$w_s + \mathbb{L}_s w_s = u_s \tag{5.11}$$

which, written out explicitly, is

$$w_s(s_1) + \int_0^{\infty} ds_2 \, \frac{1}{2\pi} \int_{-\infty}^{\infty} d\Theta \; G(\Theta) \exp(-(2s + s_1 + s_2) \cosh \Theta) \; w_s(s_2)$$

$$= \frac{1}{2\pi} \int_{-\infty}^{\infty} d\Theta \; G(\Theta) \exp - (2s + s_1) \cosh \Theta \; . \tag{5.12}$$

If we make the identifications

$$M(u) = \frac{1}{2\pi} \int_{-\infty}^{\infty} d\Theta \; G(\Theta) \exp - u \cosh \Theta \tag{5.13}$$

$$K(x, u + x) = v_{x/2}(u) \tag{5.14}$$

and set $x' - x = u \geq 0$, $x'' - x = v \geq 0$, then (5.12) becomes

$$K(x, x') + M(x + x') + \int_x^{\infty} dx'' \; K(x, x'') \, M(x + x'') = 0 \tag{5.15}$$

with $x' > x$. This is a Marchenko equation; thus we have a direct connection to the <u>inverse spectral problem</u> [21].

The corresponding results for $T > T_c$ is

$$m(s) = e^{-s} \frac{t}{(1+1/\hat{J}_0^2)} \int_0^\infty e^{-x}\, dx\, v_s(x) \,. \tag{5.16}$$

The scaling results for the 2-point correlation function in the bulk are remarkably simple: define

$$F_{\gtrless}(s) = \lim_{\substack{t\to 0\pm \\ |r|\to\infty}} t^{-\frac{1}{4}} \rho_{\gtrless}(|\underset{\sim}{r}|/\gamma(0) = s) \,. \tag{5.17}$$

Then the limit has been shown to exist, pointwise, and

$$F_<(s) = \exp \tfrac{1}{2} \operatorname{Tr} \log (1 + \mathbb{B}_s^2) \tag{5.18}$$

$$F_>(s) = (e, (1 + \mathbb{B}_s^2)^{-1} e)\, F_<(s) \tag{5.19}$$

where

$$\mathbb{B}_s(x,y) = i \tanh \frac{(x-y)}{2} \exp\left[-s(\cosh x + \cosh y)/2\right] \tag{5.20}$$

maps $\mathbb{L}^2(\mathbb{R})$ to itself and

$$e_s(x) = \exp(-s \cosh x)/2 \tag{5.21}$$

with the usual scalar product for Lebesgue measure.

Now consider the Fredholm problem

$$(1 + i\mathbb{B}_s)\, g_s = e_s \,. \tag{5.22}$$

Then a little manipulation shows that

$$F_<(s) = \exp - \tfrac{1}{2} \int_s^\infty du \int_{\mathbb{R}} \sinh x\, e_u(x)\, g_u(x)\, dx \tag{5.23}$$

and

$$F_>(s) = F_<(s) \int_{\mathbb{R}} e_s(x)\, g_s(x)\, dx \,. \tag{5.24}$$

This system provides an alternative computational device to that employed by Wu et al, which is based on the Painlevé system [7].

The analogous Gelfand-Levitan-Marchenko scheme is a little more complicated. Consider the system of equations

$$\left.\begin{aligned} f_{1,s} + J_s\, f_{2,s} &= h_s \\ f_{2,s} + L_s\, f_{1,s} &= 0 \end{aligned}\right\} \tag{5.25}$$

with kernels J_s and L_s specified by

$$\left.\begin{aligned} J_s(x,y) &= K_0(s+x+y) \\ L_s(x,y) &= (K_2 - K_0)(s+x+y) \end{aligned}\right\} \tag{5.26}$$

and

$$h_s(x) = K_0(s+x) \tag{5.27}$$

where K_j are Bessel functions. Then

$$\left.\begin{aligned} F_<(s) &= \exp - \tfrac{1}{2} \int_s^\infty du\, f_{1,u} \\ F_>(s) &= f_{2,s}(0)\, F_<(s) \ . \end{aligned}\right\} \tag{5.28}$$

Equations (5.25), (5.26) and (5.27) can be rewritten in the Zakharov-Shabat scheme [22] by transformations analogous to those above for the half-plane case.

Notes:

McCoy and Wu [23] have derived a system of non-linear partial difference equations which extend their original Painlevé work [7] outside the scaling region.

The ideas formulated here originated in part at a Nato Summer School on Non-linear Phenomena in Mathematics and Physics organised by Bardos and Bessis in 1979.

Acknowledgements

The author thanks D. and G. Chudnovsky very much for helpful discussions and for the opportunity to present this work in their seminar. He is also most grateful to P.D.F. Ion for his interest.

REFERENCES

1. M.E. Fisher, Rept. Progr. Physics 30, 615 (1967).

2. See, for instance, J.B. Kogut, Rev. Mod. Phys. 51, 659 (1979).

3. T.D. Schultz, D.C. Mattis and E.H. Lieb, Rev. Mod. Phys. 36, 856 (1964).

4. D.B. Abraham, in Springer Lecture Notes in Physics, ed. D. Iagolnitzer, 1980.

5. D.B. Abraham, Commun. Math. Phys. 59, 17 (1978); 60, 181 (1978); 60, 205 (1978).

6. M.J. Ablowitz, D.J. Kaup, A.C. Newell and H. Segur, Studies in Appl. Math. 53, 249 (1974), and ref. 22 for originating article by Zakharov and Shabat.

7. E. Barouch, B.M. McCoy and T.T. Wu, Phys. Rev. Letts 31, 1409 (1973); T.T. Wu, B.M. McCoy, C.A. Tracy and E. Barouch, Phys. Rev. B13, 316 (1976); T.T. Wu, C.A. Tracy and B.M. McCoy, J. Math. Phys. 18, 1058 (1977).

8. A. Martin-Löf, Springer Lecture Notes in Physics 101, (1979).

9. G. Gallavotti, Riv. Nuovo Cimento 2, 133 (1972).

10. L. Onsager, Phys. Rev. 65, 117 (1944).

11. See, for instance, G.E. Uhlenbeck and G.W. Ford, Lecture in Statistical Mechanics, American Mathematical Society, 1963.

12. L. van Hove, Lectures on the Many Body Problem (ed. E.R. Caianiello) New York: Academic Press (1967).

13. For a review, see M.E. Fisher and R.E. Hartwig, Adv. in Chem. Phys. 15, 333 (1969).

14. P.G. de Gennes and M.E. Fisher, C.R. Acad. Sci. Paris 287, 207 (1978).

15. H. Au Yang and M.E. Fisher, Phys. Rev.

16. D.B. Abraham, unpublished results.

17. D.B. Abraham, Phys. Rev. Letts 44, 1165 (1980).

18. R.Z. Bariev, Theor. and Math. Physics 40, 95 (1979) discusses an edge with $J_0 = 0$.

19. G. Gallavotti, Commun. Math. Phys. 27, 103 (1972).

20. D.B. Abraham and P. Reed, Commun. Math. Phys. 49, 35 (1976).

21. A.C. Scott, F.Y.F. Chu and D.W. McLaughlin, Proc. IEEE 61, 1443 (1973) and many references therein.

22. V.E. Zakharov and A.B. Shabat, Sov. Phys. J.E.T.P. 34, 62 (1972).

23. B.M. McCoy and T.T. Wu, Phys. Rev. Letts 45, 675 (1980).

Department of Mathematics
University of Melbourne
Parkville, Victoria 3052
Australia

Infinite Component σ-models and Instanton Solutions

by

D.V. Chudnovsky

Abstract: We consider SO(N) and SU(N) σ-models in the case of finite N and $N \to \infty$. In the finite-component case, we propose certain set of instanton and elliptic-instanton solutions.

The results given here are part of Colloquium lecture given on October 16, 1978 at the Department of Mathematics and Physics at Yale University.

The σ-model is one of the most interesting two-dimensional systems which is completely integrable. Several types of σ-models are connected with different compact Lie groups. Of course, the most attractive feature of σ-models is their similarity with Yang-Mills theories in 4 dimensions. Probably we can investigate exactly the instanton interactions and the existence of conservation laws only for σ-models although we still don't know them in Yang-Mills theory. We'll describe different ways of introducing σ-models, as well as their instanton and soliton solutions. From some points of view, σ-models corresponding to SO(N) or SU(N) are especially interesting for $N \to \infty$.

1. The first way to introduce σ-models is to connect them with a commutativity condition, where the equations arise as the condition for two linear operators to commute (or, in other words, the condition that then exist large systems of common eigenfunctions for these two operators.

In the two-dimensional case of space variable x and time t, and we also put $2\xi = t - x$, $2\eta = t + x$. There are two possibilities to introduce commutativity condition: one belongs to A. Polyakov and another to V. Zakharov[7]. For gauge theories in (x,t) we consider a commutativity condition such as

(1) $$i\Psi_\xi = U\Psi, \qquad i\Psi_\eta = V\Psi,$$

where complex $N \times N$ matrix functions, U, V, Ψ depend on ξ, η. (1) is equivalent to

$$U_\eta - V_\xi - i[U,V] = 0. \tag{2}$$

Of course, (1)-(2) are invariant under the transformation

$$U \to \tilde{U} = fUf^{-1} + if_\xi f^{-1}, \qquad V \to \tilde{V} = fVf^{-1} + if_\eta f^{-1}, \quad \Psi \to \tilde{\Psi} = f\Psi$$

for an arbitrary nonsingular matrix function f. In order to define proper two-dimensional systems it is necessary to introduce the spectral parameter λ. We put, in the simplest case

$$U = U_0 + \frac{U_1}{\lambda + 1}, \qquad V = V_0 + \frac{V_1}{\lambda - 1} \tag{3}$$

where U_0, U_1, V_0, V_1 are independent of λ. If we now chose the guage f in which

$$\tilde{U}_0 = 0, \qquad \tilde{V}_0 = 0, \qquad \tilde{U}_1 = A, \qquad \tilde{V}_1 = -B,$$

we obtain the main system of equation

$$A_\eta = \frac{i}{2}[A,B], \qquad B_\xi = -\frac{i}{2}[A,B]. \tag{4}$$

Equation (4) is equivalent to the condition of consistency for the system

$$i\Psi_\xi = \frac{A}{\lambda + 1}\Psi, \qquad i\Psi_\eta = -\frac{B}{\lambda - 1}\Psi. \tag{5}$$

2. Another formulation of the system (4) as the commutativity condition is based on the Lax representation for (4). This formalism is due to K. Pohlmeyer[1], A. Neveu[9], L. Takhtadzhyan[8]. We consider two operators

$$L = i\begin{pmatrix} A & 0 \\ 0 & 0 \end{pmatrix} \frac{d}{dx} + \begin{pmatrix} Q_1 & Q_2 \\ Q_3 & 0 \end{pmatrix},$$

$$M = i\begin{pmatrix} I & 0 \\ 0 & -I \end{pmatrix}\frac{d}{dx} + \begin{pmatrix} 0 & 2BQ_2 \\ 2Q_3B & 0 \end{pmatrix}$$

for diagonal matrix $A = \mathrm{diag}(a_1,\ldots,a_N)$, $a_i \neq 0: i = 1,\ldots,N$, $B = A^{-1}$ and $N \times N$ matrices Q_1, Q_2, Q_3 where

$$Q_{1jj} = 0: j = 1,\ldots,N, \qquad Q_3BQ_2 = B.$$

Then (4) can be reduced to the classical Lax representation

$$L_t = i[L,M].$$

The system (4) leads to field theories, e.g. to σ-models, connected with Lie groups. Let's suppose that we have at any point (ξ,η) some element $g(\xi,\eta)$ of the Lie group G, where G is considered as subgroup of complex matrices. We discuss the equations of the motion

(6) $$g_{\xi\eta} = \frac{1}{2}(g_\xi g^{-1} g_\eta + g_\eta g^{-1} g_\xi)$$

and the corresponding action

$$S = \int d\xi d\eta\, \frac{1}{2}\,\mathrm{Tr}\left(\frac{\partial}{\partial\xi} g \frac{\partial}{\partial\eta} g^{-1}\right).$$

The equation (6) can be reduced to the form (4). We define

(7) $$A = ig_\xi g^{-1}, \qquad B = ig_\eta g^{-1}.$$

Then in the notations (7), our system (6) is equivalent to the system (4):

(4') $$A_\eta - B_\xi - i[A,B] = 0, \qquad A_\eta + B_\xi = 0.$$

From a physical (and mathematical) point of view it is reasonable to restrict ourselves to the two most important cases

$$G = SO(N) \quad \text{and} \quad G = SU(N).$$

Moreover it is necessary to consider also some reduction of the principle field g in (6). For example, we obtain one of the interesting systems (σ-models) under the restriction

(8) $$g^2 = 1.$$

The system (6) is thereby reduced to

(9) $$g_{\xi\eta} = \frac{1}{2}(g_\xi g g_\eta + g_\eta g g_\xi).$$

The restriction (8) is consistent with the equation (9).

Let's see to what systems of equations the systems (8), (9) can be reduced for $G = SO(N)$, $SU(N)$. We present g in the form

(10) $$g = 1 - 2P$$

where $P^2 = P$, i.e. P is a projection operator. We can put

(10') $$g = 1 - 2P_k,$$

where k is the dimension of the image of P: the number k is unchanged in the evolution. The system (9) with the restriction (8) becomes

(11) $$[(P_k)_{\xi\eta}, P_k] = 0: k = 1,\dots,[\tfrac{N}{2}]$$

with the action

$$S_k = \frac{1}{2}\int d\xi d\eta \; Tr(P_{k\xi}P_{k\eta}).$$

In other words, each of the systems (11) can be considered as the field which is defined on the space of projections for a given dimension K. This space in the real case is the Grassman real variety $\Gamma^{\mathbb{R}}_{N,k}$ and in the complex case is the complex Grassman variety $\Gamma^{\mathbb{C}}_{N,k}$:

$$\Gamma^{\mathbb{R}}_{N,k} = \frac{SO(N)}{O(k)SO(N-k)}, \qquad \Gamma^{\mathbb{C}}_{N,k} = \frac{SU(N)}{U(k)SU(N-k)}.$$

The most important case is $\underline{k = 1}$, when we have $\mathbb{R}P^{N-1}$ or $\mathbb{C}P^{n-1}$. In this case the projector P_1 can be written in the form

(12) $$(P_1)_{ij} = u_i u_j^* \quad \text{and} \quad \Sigma_{i=1}^{N} u_i u_i^* = 1.$$

Thus we have two σ-models

A) Real $\mathbb{R}P^{N-1}$ σ-model

(13) $$\vec{u}_{\xi\eta} + (\vec{u}_\xi, \vec{u}_\eta)\vec{u} = 0; \qquad (\vec{u},\vec{u}) = 1;$$

B) Complex $\mathbb{C}P^{N-1}$ σ-model

(14) $$\vec{u}_{\xi\eta} + \frac{1}{2}((\vec{u}_\xi,\vec{u}_\eta) + (\vec{u}_\eta,\vec{u}_\xi) + (\vec{u}_{\xi\eta},\vec{u}) - (\vec{u},\vec{u}_{\xi\eta}))\vec{u}$$

$$- (\vec{u},\vec{u}_\xi)\vec{u}_\eta - (\vec{u},\vec{u}_\eta)\vec{u}_\xi + 2(\vec{u},\vec{u}_\eta)(\vec{u},\vec{u}_\xi)\vec{u} = 0$$

and $(\vec{u},\vec{u}) = 1$.

Here we put $\vec{u} = (u_1,\ldots,u_N)$ and $(\vec{a},\vec{b}) = \Sigma_{i=1}^{N} a_i b_i^*$ in the complex case.

Of course, we can treat these σ-models as equations on the sphere S^{N-1}. We find in the complex case $\mathbb{C}P^{N-1}$ that for any solution $\vec{u} = (u_i : i = 1,\ldots,N)$ of (14) and gauge transformation

(15) $$\vec{u}' = e^{\sqrt{-1}\Lambda}\vec{u}, \qquad u_i' = e^{\sqrt{-1}\Lambda}u_i : i = 1,\ldots,N$$

the vector $\vec{u}'$ is also the solution of (14) for arbitrary real function $\Lambda = \Lambda(\xi,\eta)$.

According to the tradition the two solutions $\vec{u}$, $\vec{u}'$ of (14) connected by (15) are considered to be the same.

In other words $\mathbb{C}P^{N-1}$ differs from complex N-dimensional sphere by factorization by U(1). However we can consider equations on

$$\frac{SO(N)}{SO(N-k)} \quad \text{or} \quad \frac{SU(N)}{SU(N-k)}.$$

The corresponding equations take the form

(16) $$(\vec{u}^{\alpha},\vec{u}^{\beta}) = \delta_{\alpha\beta}$$

and

(17) $$\vec{u}^{\beta}_{\xi\eta} + \Sigma^{k}_{\alpha=1}((\vec{u}^{\alpha}_{\eta},\vec{u}^{\beta}_{\xi}) + (\vec{u}^{\alpha}_{\xi},\vec{u}^{\beta}_{\eta}))\vec{u}^{\alpha} = 0$$

for $\vec{u}^{\alpha}$: $\alpha = 1,\ldots,k$.

From the point of view of instantons of the $\mathbb{C}P^{N-1}$ σ-model is interesting. Before considering $\mathbb{C}P^{N-1}$ in more detail we want to mention:

The infinite component σ-model corresponding to the cases SO(N) and SU(N) for $N \to \infty$ can be defined in complete duality with the finite component case. This means simply that we treat $\vec{u}$ as elements of an arbitrary Hilbert space H and $(\vec{a},\vec{b})$ is simply a scalar product in H. Probably the most interesting example is $L^2(\Omega;\mu)$. In other words we consider the functions $u(\xi,\eta,\alpha)$: $\alpha \in \Omega$, where

$$(u,w) = \int_{\Omega} u(\alpha)w^*(\alpha)d\mu.$$

Such system will be called $\mathbb{R}P^{\Omega}$ or $\mathbb{C}P^{\Omega}$. It should be noted that $\mathbb{C}P^{\Omega}$ is invariant under a much larger group of transformation than gauge.

If $\varphi: \Omega \to \Omega$ is one-to-one, then for real $\Lambda(\xi,\eta,\alpha)$ and $V(\alpha)$ such that $|V(\alpha)|^2 = \varphi'(\alpha)$ if

$$u'(\xi,\eta,\alpha) = V(\alpha)e^{i\Lambda(\xi,\eta,\alpha)}u(\xi,\eta,\varphi(\alpha))$$

then u' satisfies $\mathbb{C}P^{\Omega}$.

The simplest equation is $\mathbb{R}P^{\Omega}$ and has the form

$$u_{\xi\eta} + \int_{\Omega} u_{\xi}u_{\eta}d\mu u = 0.$$

The equation $\mathbb{R}P^{\Omega}$ is stable under the transformation

$$u'(\xi,\eta,\alpha) = V(\alpha)u(\xi,\eta,\varphi(\alpha))$$

for

$$V(\alpha)^2 = \varphi'(\alpha).$$

The chiral model (13) of $\mathbb{R}P^{N-1}$ for N = 3 is simply sin-Gordon

$$\alpha_{\xi\eta} = \sin\alpha$$

or

$$\alpha_{xx} - \alpha_{tt} = \sin\alpha.$$

Here $\alpha = \arccos(\vec{u}_\xi \cdot \vec{u}_\eta)$. The same sin-Gordon can be also deduced from the general system (6) for SO(2)[8].

One very interesting Pohlmeyer-Lund-Regge[1],[2] system arises from the system (6) for SU(2). It has the form

$$\beta_{\xi\eta} + \frac{1}{\sin\alpha}(\alpha_\eta\beta_\xi + \alpha_\xi\beta_\eta) = 0;$$

$$\alpha_{\xi\eta} + \sin\alpha - \frac{\sin(\alpha/2)}{2\cos^3(\alpha/2)}\beta_\xi\beta_\eta = 0.$$

Again for $\beta = 0$ we get sin - Gordon. The precise form of SO(N)-models (6) were investigated for N = 2,3,4,5 only. Again in SO(4) case we get Pohlmeyer-Lund-Regge system.

The only σ-model for SO(N) were <u>instanton</u> solutions were found is SO(3)-σ-model. D'Adda, Lüscher, di Vecchia[3] have proved that SO(3) system is equivalent to $\mathbb{C}P^1$ (N = 2). Simultaneously the same authors have found instantons in $\mathbb{C}P^{N-1}$ for arbitrary N. Their results are especially interesting because they describe in great detail the instanton behavior in 2-dimensional cases giving good insight for 4-dimensional Yang-Mills theory.

In order to describe instantons and self-dual (anti-self-dual) σ-models it is more convenient to adopt the notations of the field theory.

As before $\mathbb{C}P^{N-1}$ is the space of all equivalence classes $[\bar{u}]$ of complex vectors $(u_1,\ldots,u_N) \neq 0$, where $\bar{u}$ and $\bar{u}'$ are equivalent if

$$\bar{u}' = u\bar{z} \qquad \text{for} \quad \lambda \in \mathbb{C}.$$

We are considering the fields $[\bar{u}](x)$ for $\bar{x} = (x_1, x_2)$ where $2\xi = x_1 + ix_2$, $2\eta = x_1 - ix_2$. Of course we can consider as equivalent only fields of complex unit vectors

$$(u_1(x), \ldots, u_N(x)); \qquad |u_1|^2 + \ldots + |u_N|^2 = 1,$$

where two fields $[\bar{u}']$ and $[\bar{u}]$ are related by a guage transformation

(15) $$u_j(x) = e^{i\Lambda(x)} u_j(x).$$

Under a gauge transformation (15) the composite field

(16) $$-A_\mu = \frac{i}{2} u^* \overleftrightarrow{\partial}_\mu u = \frac{i}{2}\{u^*_\alpha \partial_\mu u_\alpha - (\partial_\mu u^*_\alpha) u_\alpha\}$$

transforms like an Abelian gauge field

(17) $$A'_\mu = A_\mu - \partial_\mu \Lambda:$$

$\mu = 1,2$. We have the action corresponding to $\mathbb{C}P^{N-1}$ σ-model:

$$S = f_1 \int d^2x (D_\mu \bar{u})^* D_\mu \bar{u}$$

for

(18) $$D_\mu = \partial_\mu + iA_\mu.$$

Here f_1 may be chosen e.g. as $f_1 = \frac{N}{2f}$ for some constant $f > 0$.

The system of nonlinear equations for $\bar{u}$ was written below in (14). In the short notations (18) in (x_1, x_2)-space-time the system (14) can be written in the form

(19) $$D_\mu D_\mu \bar{u} + ((D_\mu \bar{u})^* D_\mu \bar{u})\bar{u} = 0; \qquad |\bar{u}|^2 = 1.$$

The main interest in the $\mathbb{C}P^{N-1}$ is their topological nontriviality. Indeed we can define a topological charge (the winding number) Q of the fields $[\bar{u}](x)$ satisfying natural boundary conditions. Indeed, if

(20) $$[\bar{u}](x) \to [\bar{u}^\infty] \quad \text{as} \quad |x| \to \infty,$$

then by definition

(21) $$u_j(x) \to g(\frac{x}{|x|})u_j^\infty \quad \text{for} \quad |g| = 1 \quad \text{and} \quad |x| \to \infty.$$

The direction dependent phase $g(\frac{x}{|x|})$ defines a mapping from the circle at infinity into U(1). Its winding number (topological charge) Q is

(22) $$Q = \frac{1}{2\pi}\int d^2x\epsilon_{\mu\nu}\partial_\mu A_\nu \qquad \epsilon_{12} = +1.$$

Under boundary conditions (20)-(21) Q is an integer (instanton number) Let's describe the traditional physical way of defining self-dual equations. We represent the topological density q(x) in two ways. It is very easy to check that we have for

$$q(x) = \frac{1}{2\pi}\epsilon_{\mu\nu}\partial_\mu A_\nu = \frac{1}{2\pi}[\partial_1 A_2 - \partial_2 A_1]$$

the representation

(23) $$q(x) = \frac{i}{2\pi}\epsilon_{\mu\nu}(D_\mu\bar{u})^*D_\nu\bar{u} = \frac{i}{2\pi}[D_1\bar{u}^*D_2\bar{u} - D_2\bar{u}^*D_1\bar{u}].$$

Now looking at the action S of the field theory we obtain thanks to the Cauchy-Schwartz inequality:

(24) $$S \geq 2f_1\pi|Q|.$$

Here $S = 2f_1\pi|Q|$ i.e. $(D_\mu\bar{u})^*D_\mu\bar{u} = |\epsilon_{\mu\nu}(D_\mu\bar{u})^*D_\nu\bar{u}|$ if and only if self-dual [or anti-self dual] equations are satisfied:

(25) $$D_\mu\bar{u} = {}^{+}_{(-)}i\epsilon_{\mu\nu}D_\nu\bar{u}$$

or

(25') $$D_1\bar{u} = iD_2\bar{u} \qquad (D_1\bar{u} = -iD_2\bar{u}).$$

By definition, the finite action solutions of these self-dual equations (25') [anti-self dual] are called instantonts [anti-instantonts]. Instantonts are absolute minima of the action S [if topological charge is defined, of course] and therefore instantonts [and anti-instantonts] immediately satisfy our second-order field equations (14) or (19):

$$D_\mu D_\mu \bar{u} + (D_\mu \bar{u}^* D_\mu \bar{u})\bar{u} = 0; \qquad |\bar{u}|^2 = 1.$$

Now we come to the most remarkable property of instanton solutions which is typical for all completely integrable gauge systems [like Yang-Mills]. This is the <u>linearization</u> of equations, defining instantons. For Yang-Mills fields which are connected with any classical group, the reduction of instantons to the problem of linear algebra was made by Atiyah, Hitchin, Drinfeld, Manin [Phys. Lett., <u>65A</u>,285 (1978)] and Drinfled, Manin [Funct. Anal. 1978, N. 3].

However the most general form of k-instanton is unknown for $k > 2$: we don't know even the singularity manifold for general k-instanton. From this point of view the situation with σ-model is of great interest. Self-dual equations for $\mathbb{C}P^{N-1}$ are indeed linearized to Cauchy-Riemann equations.

To linearize (25) in a proper way we define an atlas of holomorphic charts (U_j,φ_j): $j = 1,\ldots,N$ is given by

$$U_j = \{[u] \in \mathbb{C}P^{N-1} : u_j \neq 0\}$$

(26) $$\varphi_j : U_j \to \mathbb{C}^{N-1} \qquad \text{by}$$

$$\varphi_j([u]) = \frac{1}{u_j}(u_1,\ldots,u_N) \equiv (w_1^{(j)},\ldots,w_N^{(j)})$$

(i.e. $w_j^{(j)} = 1$). If $[u] \in U_j \cap U_k$, then the coordinates are related by

$$(w_1^{(j)},\ldots,w_N^{(j)}) = \frac{1}{w_k^{(j)}}(w_1^{(k)},\ldots,w_N^{(k)}).$$

Of course for an arbitrary field $\bar{z}(x)$ we have

$$(27) \qquad \bar{u}(x) = e^{i\Lambda(x)} \frac{\bar{w}^{(j)}(x)}{|\bar{w}^{(j)}(x)|}, \qquad |\bar{w}| = |w_1|^2 + \ldots + |w_N|^2,$$

for some gauge $\Lambda(x)$. Now we can write the self duality [anti-self duality] equations (25) in the form:

$$(28) \qquad (\partial_1 \underset{(+)}{-} i\partial_2)u_\alpha = u_\alpha \Sigma_{\beta=1}^N u_\beta^* (\partial_1 \underset{(+)}{-} i\partial_2)u_\beta$$

for $|\bar{u}| = 1$. Then, substituting (27) into (28) we obtain linearization of self-duality equations (28):

$$(29) \qquad \partial_1 \bar{w}^{(j)} = \underset{(-)}{+} i\partial_2 \bar{w}^{(j)}$$

(since $w_j^{(j)} = 1$). Defining a complex variable $s = x_1 - ix_2$ we see that smooth solutions of (28) are the holomorphic [anti-holomorphic] mappings from the complex s-plane into $\mathbb{C}P^{N-1}$. Of course, $w_\alpha^{(j)}(x)$ are not necessarily entire functions of s (or $\bar{s}$), but any singularity should be removable by an appropriate change of charts (and the possible singularities of $w_\alpha^{(j)}$ are isolated poles).

This assertion may be deduced from the general results of Atiyah, Ward and Drinfeld, Manin, but for σ-model in $N = 2$ was found by Eichenherr (Ph.D. Heidelberg, 1978) and Lüscher[3],[4].

Proposition 1: The most general smooth solution of the self-duality equation

$$(25') \qquad D_1\bar{u} = iD_2\bar{u}, \qquad |\bar{u}|^2 = 1$$

is given by

$$(30) \qquad u_\alpha(x) = e^{i\Lambda(x)} \frac{w_\alpha(x)}{|\bar{w}(x)|}$$

where $\Lambda(x)$ is a real function, $\bar{w}(x)$ is a meromorphic vector-function of $s = x_1 - ix_2$ and $w_j = 1$ for some j.

<u>Proposition</u> 2: <u>The finite action (instanton) solutions of</u> (25') <u>are given by</u> (30) <u>where</u> $w_\alpha(x)$ <u>are rational functions of</u> s <u>and</u> Q <u>the topological charge of</u> $\bar{u}$ <u>is equal to the number of poles of</u> $\bar{w}$ <u>(including those at</u> $s = \infty$).

Let's give the description of these instantonts. One instanton solution can be written in the form: For the $\mathbb{C}P^\Omega$ σ-model the one instanton solution has the following form

$$(31) \qquad u(x,\alpha) = \frac{cu(\alpha) + v(\alpha)\{x_1 - ix_2 - a_1 + ia_2\}}{(c^2 + (x_1-a_1)^2 + (x_2-a_2)^2)^{1/2})} .$$

Here $c > 0$ is the scale size of the instanton, a_μ is its position: $\mu = 1,2$ and the two constant vectors $\vec{u},\vec{v} \in L^2(\Omega)$ satisfy

$$(32) \qquad \|\vec{u}\| = \|\vec{v}\| = 1; \qquad (\vec{u},\vec{v}) = 0.$$

In the case of $\mathbb{C}P^{N-1}$ (i.e. ℓ_N^2) the number of parameters for Q-instantons in $2N(Q + 1) - 1$.

In a similar way it is possible to construct periodic (or doubly-periodic) "instantons". E.g. the "one instanton" periodic solution is of the form

$$u(x,\alpha) = \frac{cu(\alpha) + v(\alpha)\{sh(2x_1 - 2a_1) - i\sin(2x_2 - 2a_2)\}}{(c^2 + sh^2(x_1-a_1) + \sin^2(x_2-a_2))\}^{1/2}}$$

for

$$\bar{u},\bar{v} \in L^2(\Omega), \qquad \|\bar{u}\| = 1, \qquad \|\bar{v}\| = \frac{1}{2}, \qquad (\bar{u},\bar{v}) = 0.$$

We can write similar (complicated) expressions for elliptic instantons. Of course in this case the topological number is finite only in the fundamental domain

$$Q' = \frac{1}{2\pi}\int_F d^2x\, \varepsilon_{\mu\nu}\partial_\mu A_\nu .$$

We consider Jacobi elliptic function $\operatorname{sn}(x,k)$ for $0 < k^2 < 1$ having real period $K(k)$ and imaginary period $K' = K(k')$, $k^2 + k'^2 = 1$.

The simplest expression for "one-instanton" where $Q' = 1$ and $u(x,\alpha)$ is $4K$-periodic in x_1 and $4iK'$-periodic in x_2 is e.g. the following

$$u(x,\alpha) = \frac{\lambda u(\alpha)(c_2'^2 + k^2 s_1^2 s_2'^2) + v(\alpha)(s_1 d_2' + i s_2' c_2' c_1 d_1)}{(\lambda^2 (c_2'^2 + k^2 s_1^2 s_2'^2)^2 + s_1^2 d_2'^2 + c_1^2 d_1^2 s_2'^2 c_2'^2)^{1/2}}.$$

for

$$s_1 = \operatorname{sn}(x_1 - a_1;k); \qquad s_2' = \operatorname{sn}(x_2 - a_2;k');$$

$$c_1 = \operatorname{cn}(x_1 - a_1;k); \qquad c_2' = \operatorname{cn}(x_2 - a_2;k');$$

$$d_1 = \operatorname{dn}(x_1 - a_1;k); \qquad d_2' = \operatorname{dn}(x_2 - a_2;k').$$

The most interesting is the case of complex multiplication by $\sqrt{-1}$, i.e. $k^2 = k'^2 = \frac{1}{2}$, when $z(x,\alpha)$ is periodic in both x_1, x_2 with the period

$$\omega = \frac{[\Gamma(1/4)]^2}{\pi^{1/2}}.$$

This description of the instanton is so simple because $D = 2$. For Yang-Mills theories it is more complicated and involves meromorphic functions (rational, elliptic functions) not of complex, but of quaternoin variable. [Gursay, Manin,[5]...].

For $F_{\mu\nu} = \partial_\mu A_\nu - \partial_\nu A_\mu + [A_\mu, A_\nu]$ and self [anti-self] dual equations

$$F_{\mu\nu} = \overset{+}{(-)} F^*_{\mu\nu}.$$

Then, e.g. t'Hoeft instantons have the form

$$A_\mu = \frac{1}{4} e_\mu a - \frac{1}{4} \overline{a}\overline{e}_\mu \qquad (\overline{a} = e_\nu a_\nu)$$

for

$$a = (\Box F)(DF)^{-1}$$

$$D = e_\mu \partial_\mu \qquad \text{and} \qquad \Box = \partial_\mu \partial_\mu$$

and F is holomorphic of the form

$$F = \mathrm{Tr}\, \Gamma(x - \Lambda)^{-1}$$

for Γ, Λ being diagonal.

σ_j are Pauli matrices, $e_j = i\sigma_j$.

References

[1] K. Pohlmeyer, Comm. Math. Phys. 46, 207 (1976).

[2] F. Lund, T. Regge, Phys. Rev. D14, 1524 (1976); F. Lund, Phys, Rev. Lett. 38, 1175 (1977).

[3] A. D'Adda, P. di Vecchia and M. Lüscher, Nucl. Phys. B146, 63 (1978).

[4] H. Eichenherr, Nucl. Phys. B146, 215 (1978).

[5] F. Gursey, "Second Workshop on Current Problems in High Energy Particle Theory", Ed. Domokos and S. Rovese-Domokos, 179, John Hopkins Univ., Baltimore (1978). F. Gursey and H.C. Tze, to be published.

[6] M.F. Atiyah, N.J. Hitchin, V.G. Drinfeld, Yu.I. Manin, Phys. Lett. 65A, 285 (1978).

[7] V.E. Zakharov, A.V. Mikhailov, JETP 74, 1953 (1978).

[8] A.S. Budagov, L.A. Takhtadzhyan, Doklady Akad. Nauk USSR 235, 805 (1977).

[9] A. Neveu, N. Papanicolaou, Comm. Math. Physics 57, 31 (1978).

Department of Mathematics
Columbia Univesity
New York, NY
USA

Infinite Component Two-dimensional Completely Integrable Systems of KdV Type

by

D.V. Chudnovsky

Abstract. We investigate the infinite-dimensional generalization of the non-linear Schrodinger equation. We consider both non-stationary and stationary subsystems. The close one-to-one connection with the inverse scattering method for the Schrodinger operator with different classes of potentials is established.

§1. Introduction.

For some completely integrable systems in (x,t)-dimensions there are natural multicomponent variants. The most famous example is the multicomponent non-linear Schrodinger equation (MNLS)[1]

(1.1) $$i\vec{\varphi}_t = -\vec{\varphi}_{xx} + \vec{\varphi}\cdot \Sigma_{k=1}^{n}|\varphi_k|^2 \text{ for } \vec{\varphi} = (\varphi_1,\ldots,\varphi_n).$$

Such an n-component system has two special features: a) the stationary system (1.1) corresponds to the stationary n-th order higher KdV[2] and b) the n-soliton solution for the non-linear Schrodinger (NLS) is reduced to a 1-soliton solution of an n-th component MNLS[1]. The wide class of the solutions of the MNLS can be easily constructed using Baker[3] eigenfunctions which correspond to hyperelliptic curves [5] (for $n > 1$ some non-hyperelliptic curves may arise as well[6]).

The most important feature of the MNLS is its strong connection with KdV (i.e. with the Schrodinger operator [4],[5]). It is not artificial to consider the components φ_j of $\vec{\varphi}$ to be eigenfunctions of the time-dependent Schrodinger operator with the potential $|\vec{\varphi}|^2$. Moreover for finite-band (and arbitrary C^∞ periodic) potentials we can present each potential in the form $-|\vec{\varphi}|^2$ with $\vec{\varphi}$ being the solution of the stationary subsystem (1.1)[8]-[11]:

$$-\vec{\varphi}_{xx} = k\vec{\varphi} - |\vec{\varphi}|^2\vec{\varphi}.$$

It is possible to continue this analogy between the MNLS and the

$$(2.1)\qquad \begin{aligned} i\varphi_j &= -\varphi_{jxx} + \varphi_j \Sigma_{k=1}^{n} \Psi_k\varphi_k \quad j = 1,\dots,n \\ -i\Psi_j &= -\Psi_{jxx} + \Psi_j \Sigma_{k=1}^{n} \varphi_k\Psi_k : \quad j = 1,\dots,n. \end{aligned}$$

This system arises in a natural way from the Lax representation

$$(2.2)\qquad iL_t = [L,M]$$

for the (n+1)X(n+1) matrix differential operator L of order one

$$(2.3)\qquad L = \begin{pmatrix} 1 & & & 0 \\ & -1 & & \\ & & \ddots & \\ 0 & & & -1 \end{pmatrix} \frac{d}{dx} + \begin{pmatrix} 0 & \varphi_1 \cdots \varphi_n \\ \Psi_1 & \\ \vdots & \\ \Psi_n & 0 \end{pmatrix}$$

and M being the (n+1)X(n+1) matrix differential operator of order two[8]-[10].

By n-component non-linear Schrodinger (NLS_n) we udnerstand the system (2.1) with $\Psi_i = \varphi_i^*$: $i = 1,\dots,n$[1]:

$$(2.4)\qquad i\varphi_{jt} = -\varphi_{jxx} + \varphi_j \Sigma_{k=1}^{n} |\varphi_k|^2 : \quad j = 1,\dots,n.$$

An important approach in the completely integrable systems, connected with the infinite-dimensional Hamiltonian form, is based on the existence of the resolventa expansion. This idea was realized first by J. Drach[5].

J. Drach starts from the equation for the diagonal of the resolventa for $-\varphi'' + (u+\zeta)\varphi = 0$ (Sturm-Liouville problem). The equation for the resolvent $\bar{R}(x,z) \in \mathbb{C}[[z^{-1}]]$ has the form:

$$(2.5)\qquad \begin{aligned} -\bar{R}''' + 4(u+\zeta)\bar{R}' + 2u'\bar{R} &= 0, \quad \zeta = z^2; \\ \bar{R} &= \Sigma_{k=0}^{\infty} \bar{R}_k z^{-k}, \end{aligned}$$

where $\bar{R}_k = \bar{R}_x [u,u',u'',\dots]$ are differential operators in u (polynomials in $u, u_x, u_{xx}, \dots$).

After the integration of the equation (2.5) J. Drach obtains the equation of the second order

Schrodinger operator with potentials rapidly decreasing on infinity[10]. However in this case, we consider the MNLS for $n \to \infty$, expressing the potential in terms of the integral on the spectral parameter from the squares of Jost functions.

This naturally suggests the introduction of the MNLS in the case $n \to \infty$ (more precisely by considering φ as the function of x,t and a new variable k).

One reason for the investigation of this system is its connection with the uniqueness of the inverse scattering problem for Schrodinger with different types of potentials. In this case we have some dynamic system defined on the set of all potentials. Complete integrability of such a dynamic system giving infinitely many conservation laws is a good reason for the existence and uniqueness of the solution of the inverse scattering problem. In other words, the system, like the infinite component non-linear Schrodinger gives complete integrability of a one component equation like Korteweg-de Vries. It is highly probable, that starting from an arbitrary (x,t)-dimensional completely integrable system having a Lax representation, we can construct (x,t,k)-completely integrable system where "quasipotentials" will be the squares of the eigenfunctions of the corresponding linear problem with k as a spectral parameter[13]. The (x,t,k)-system, interesting in itself, becomes the only useful tool for the clever construction of the inverse scattering of differential operators of arbitrary dimension and order. However we have no intention to present this general formal scheme as it is too algebraic and not very analytic. This is done in our separate paper[14].

Our main object here is the infinite component non-linear Schrodinger which is both time dependent and stationary. We present its natural connection with the KdV and with the infinite component modified Korteweg-de Vries equation (MKdV).

§2.

First of all we present the multicomponent non-linear-like Schrodinger equation. This equation in n-components $(\varphi_1,\dots,\varphi_n),(\Psi_1,\dots,\Psi_n)$ has the form

(2.6) $$-2\bar{R}\bar{R}'' + (\bar{R}')^2 + 4(u+\zeta)\bar{R}^2 = c(\zeta)$$

for $c(\zeta) \in \mathbb{C}[[\zeta^{-1}]]$ with constant coefficients. For $c(\zeta) \equiv 1$ we obtain the standard resolvent

(2.7) $$R(x,z) = \Sigma_{k=1}^{\infty} R_k[u]z^{-2k+1}$$

Here

$$R_1 = 1/2, \quad R_2 = \frac{1}{4}u, \quad R_3 = \frac{1}{16}(3u^2-u''),\ldots ,$$

and we have one of the most famous identities "Drach-Burchnall-Chaundy-Lenard"

(2.8) $$-R_k''' + 4uR_k' + 4R_{k+1}' + 2u'R_k = 0.$$

These quantities $R_k[u]$ are the grounds for the proof of the complete integrability of the Korteweg-de Vries equation[2],[4],

$$u_t = 6\, uu_x - u_{xxx}.$$

Lax showed[4] that the KdV equation is equivalent to the following operator identity

$$\frac{dL_2}{dt} = [L_3,L_2]$$

for $L_2 = -d^2/dx^2 + u(x,t)$, $L_3 = -4d^3/dx^3 + 3(u\frac{d}{dx} + \frac{d}{dx}u)$. In general for the functional[2]

$$I_n = R_n[u,u',\ldots]dx$$

the non-linear equation

(2.9) $$u_t = \frac{\partial}{\partial x}\frac{\delta I_n}{\delta u}$$

is equivalent to the Lax representation

$$\frac{dL_2}{dt} = [L_k, L_2]$$

for the operator L_k in $\frac{d}{dx}$ of odd order k. The equaLions (2.9) are called higher order KdV equations.

§3. The Infinite Component Non-linear Schrodinger.

The most natural way to introduce the infinite component system is to consider an index k of φ_k as a variable. We'll see later that this corresponds to the introduction of the spectral parameter k as the new variable.

We consider the Hilbert space $L^2(\Omega, d\mu_k)$ where $\Omega \subseteq R$ and $d\mu_k$ is a measure on Ω. We treat the functions $u(k)$, $\varphi(k)$: $k \in \Omega$ as elements of $L^2(\Omega, d\mu_k)$.

In $L^2(\Omega, d\mu_k)$ there exists a natural definition of a scalar product

$$(u,v) = \int_\Omega u(k)v(k)d\mu_k$$

for $u(k), v(k) \in L^2(\Omega, d\mu_k)$ and $|u|^2 = (u,u^*)$.

The most general system has the form

$$\text{(3.1)} \qquad \begin{aligned} i\varphi_t &= -\varphi_{xx} + \varphi\int_\Omega \varphi\Psi \, d\mu_k \\ -i\Psi_t &= -\Psi_{xx} + \Psi\int_\Omega \varphi\Psi \, d\mu_k, \end{aligned}$$

for $\varphi = \varphi(x,t,k)$, $\Psi = \Psi(x,t,k)$. As a natural generalization of the non-linear Schrodinger we have one equation

$$\text{(3.2)} \qquad i\varphi_t = -\varphi_{xx} + \varphi\int_\Omega |\varphi|^2 d\mu_k.$$

In the stationary cases of the systems (3.1)-(3.2) we write it in the form

$$\text{(3.3)} \qquad \begin{aligned} f_{xx} &= kf - f\int_\Omega fg \, d\mu_k \\ g_{xx} &= kg - g\int_\Omega fg \, d\mu_k \end{aligned}$$

for $f = f(x,k)$, $g = g(x,k)$. For the generalization of the "x^4" system[8][9] we get

$$f_{xx} = kf - f\int_\Omega f^2 \, d\mu_k. \tag{3.4}$$

First of all we must stress that φ, Ψ, f, g are not necessarily continuous functions of k; we demand only L_2-integrability with respect to k. For example, the most interesting cases correspond to the situation, when $\varphi(x,t,k)$ has some singular components. E.g. the n-th component equations (3.1)-(3.3) correspond to the cases

$$\varphi(x,t,k) = \Sigma_{j=1}^{n} \varphi_j(x,t)\,\delta(k-j),$$

$$f(x,k) = \Sigma_{j=1}^{n} f_j(x)\;\delta(k-\lambda_j).$$

In general, φ, Ψ, f, g have smooth and singular parts.

While the system (3.2) corresponds to the spectral problem for the time-dependent Schrodinger, (3.4) corresponds to the usual Sturm-Liouville problem.

Let's present the Hamiltonian structure and conservation laws of the stationary equation (3.4).

We introduce a natural Hamiltonian

$$H = \frac{1}{2}\{\int_\Omega v^2 d\mu_k - \int_\Omega ku^2 d\mu_k + \frac{1}{2}(\int_\Omega u^2 d\mu_k)^2\}. \tag{3.5}$$

The Hamiltonian system which corresponds to (3.5) has the natural form

$$\binom{u}{v}_x = \begin{pmatrix} 0 & 1 \\ -1 & 0 \end{pmatrix} \begin{pmatrix} \delta / \delta u \\ \delta / \delta v \end{pmatrix} H \tag{3.6}$$

or, in shorter form, we have

$$u_{xx} = ku - u\int_\Omega u^2 d\mu_k \tag{3.4'}$$

for $u = u(x,k)$.

We have a very simple form of the conservation laws for (3.4). We define C(k) in the following expression

$$C(k) = 2v(x,k)^2 - 2ku(x,k)^2 + u(x,k)^2 \int_\Omega u(x,\ell)^2 d\mu_\ell$$
$$+ \int_\Omega \frac{(v(x,k)u(x,\ell) - v(x,\ell)u(x,k))^2}{k - \ell} d\mu_\ell.$$

The main result is the conservation of the C(k).

<u>Theorem</u> 3.1: <u>For the solution</u> u,v <u>of</u> (3.6) <u>and arbitrary</u> k,

$$\frac{dC(k)}{dx} = 0.$$

The conservation laws may be written, of course, in a more traditional way, e.g.

$$\int_\Omega h(k)C(k)d\mu_k = \text{const}$$

for an arbitrary function h(k), e.g.

$$H = \frac{1}{2}\int_\Omega C(k)d\mu_k = \text{const.}$$

The conservation quantities C(k) are connected with the resolvent expansion coefficients $R_k[u]$ (the Hamiltonians of the higher KdV equations). This can be explained because the infinite component system corresponds to the solution of the inverse scattering problem for a Schrodinger operator with the potential having arbitrary behavior.

In this way we have the following general <u>problem:</u> for an arbitrary potential u(x) for which the corresponding Sturm-Liouville problem has sense (e.g. in the self-adjoint case and for non-trivial spectrum) the solution of (3.6) which is naturally associated with u(x) exists in the sense that

$$u(x) = -\int_\Omega u(x,k)^2 d\mu_k + C \tag{3.7}$$

where u(x,k) is the eigenfunction associated with k, of the Schrodinger with the potential u(x) (cf. [13]).

In other words we may consider the Hamiltonian system over the set of all the potential (or, more precisely, over the set of all possible eigenfunctions).

In all cases we can prove that the representation (3.7) and even deeper results are connected with the complete integrability of (3.6) and action-angle variables (cf. early results of Borg and Levitan [13]).

However we can treat (3.4) together with the representation (3.7) rather formally as the source of information on what the inverse scattering method is. Then the complete integrability of (3.6) shows the uniqueness of the inverse scattering [where scattering data are action variables for (3.6) and angle variables correspond to the isospectral variation of u(x)]. The main problem with this approach is, of course, the proof that the representation (3.7) covers a large class of potentials.

Now inside this formal scheme we present relations between "natural" spectral data of u(x) and invariants connected with the system (3.6).

One of the main results is the following theorem, the proof of which is identical to that of lemma 3[9] or [8]:

Theorem 3.2: *For any solution u(x,k) of (3.4) let the potential u(x) be defined by (3.7)*

$$u(x) = -\int_\Omega u(x,k)^2 d\mu_k + C_1.$$

Then for an arbitrary $n \geq 0$ there is a polynomial $P_n(k)$ with constant coefficients and of degree $n - 1$ and a constant C_n such that

$$R_n[u] = \int_\Omega P_n(k) u(x,k)^2 d\mu_k + C_n: \quad n \geq 0. \tag{3.8}$$

The simplest way to prove theorem 3.2 is to look at any system $\varphi(x,k)$ of the eigenfunctions for the potential u(x):

$$-\varphi_{xx} + u(x) = \varphi k,$$

$$E(k) = \int_\Omega \frac{\varphi^2(x,\ell) d\mu_\ell}{\ell - k}.$$

Then E(k) satisfies the resolvent -like equation

(3.9) $$-E_{xxx} + 4(u-k)E_x + 2u_x E = 2\{\int_\Omega \varphi^2(x,\ell)d\mu_\ell\}_x$$

independently of the representation (3.7). You can obtain (3.8) from (3.7), (3.9), and the definition of the resolventa $R(x,\lambda)$.

Remark 3.3: The constants C_n are very important.

1) The coefficients of the polynomial $P_n(k)$ are expressed algebraically only in terms of $C_0,\ldots,C_{n-1}$ using simple recurrence formulae[8][9].

2) The quantities C_n are the first integrals of (3.4). There are expressions of C_n in terms of the momentae

$$\int_\Omega h_n(k)C(k)d\mu_k$$

of the conservation laws $C(k)$ which can be deduced from (3.1) and (3.9).

3) On the other hand C_n can be expressed in terms of the first integrals of the higher KdV, i.e. the terms of

$$\int_{-\infty}^{\infty} R_i[u]dx \quad \text{for} \quad i = 0,\ldots,n.$$

In this connection we would like to mention a very interesting duality between the integrals of $u(x,k)^2$ with respect to k and with respect to x.

While the quantities $\int k^n u(x,k)^2 d\mu_k$ lead to the higher KdV Hamiltonian $R_n[u]$, the quantities like $\int_{-\infty}^{\infty} u(x,k)^2 u_x(x)dx$ lead to the scattering coefficient $R(k)$ in the case of $u(x)$ rapidly decreasing on infinity.

§4.

We proved already[8],[9],[10] that the system (3.6) together with the representation (3.7) describes all finite band potentials. In this case $f(x,k)$ is singular and of the form

$$f(x,k) = \Sigma_{i=1}^n f_i(x)\delta(k-\lambda_i)$$

for an arbitrary λ_i. The same representation takes place for an arbitrary C^∞-periodic potential u(x). In this case we have the same representation

$$u(x) = -\Sigma_{i=1}^{\infty} f_i^2(x) + C$$

for eigenvalues λ_i being non-degenerate periodic [left ends of the zonae] and $f_i(x)$ being eigenfunctions corresponding to λ_i. Here

$$\lambda_{i_n} = n^2 \frac{\pi^2}{T^2} + 0(1) + 0(\frac{1}{n^2}) \ldots .$$

Naturally, singular solutions of the form

$$f(x,k) = \Sigma_{i=1}^{\infty} f_i(x)\delta(k-\lambda_i)$$

with arbitrary λ_i may correspond to an arbitrary quasiperiodic potential u(x). However for the general quasiperiodic potential we have no idea what the spectrum is. Probably we need some diophantine conditions for quasi-periods of u(x).

We have a complete [and interesting] reduction of the inverse scattering problem for the system (3.6)-(3.7) in the case of u(x) rapidly decreasing on infinity[13].

Let us define all spectral data in this case. Let $\int_{-\infty}^{\infty} |u(x)|(1+x^2)dx < \infty$ and f(x,k) be the solution of $-f'' + uf = k^2 f$ with

$$f(x,k) \sim e^{ikx}: \quad x \to +\infty$$

and

$$f(x,k) = a(k)e^{ikx} + b(k)e^{-ikx}: \quad x \to -\infty.$$

The reflection coefficient r(k) can be defined as

$$r(k) = -\frac{b^*(k)}{a(k)}.$$

It is known that u(x) may have only a finite number of bound states $-\beta_n^2 < \ldots < -\beta_1^2 < 0$. Then the potential is uniquely determined by the reflection coefficient r(k), n bound states $\beta_1^2, \ldots, \beta_n^2$ and n norming constants

$$c_j = \{\int_{-\infty}^{\infty} f^2(x;i\beta_j)dx\}^{-1}: \ j = 1,\dots,n.$$

It is possible to express u(x) in terms of $f(x;k)^2$ in the form of the system (3.7)-(3.7). This was done for the first time by M. Kruskal, C. Gardner, J. Greene and R. Miura[2] for purely soliton solutions, for which $r(k) \equiv 0$. In this case

$$u(x) = -4 \Sigma_{j=1}^{n} c_j \beta_j f^2(x;i\beta_j).$$

However for an arbitrary potential with $r(k) \not\equiv 0$ we have in the representation (3.7), both singular and smooth parts.

There is however one case, when we have a smooth representation only. This is the case of u(x) without bound states treated in [7]. In this case, a very simple formula exists:

$$u(x) = \frac{2i}{\pi} \int_{-\infty}^{\infty} kr(k)f^2(x,k)dk$$

However we found that a formula containing both reflectionless and bound stateless potentials exists. This corresponds to (3.7) with both singular and smooth parts:

$$u(x) = \frac{2i}{\pi} \int_{-\infty}^{\infty} kr(k)f^2(x,k)dk + \Sigma_{j=1}^{n} d_j f^2(x;i\beta_j).$$

Moreover, according to our previous results we have a more general representation for $R_K[u]$, e.g. we have

<u>Theorem</u> 4.1: a) <u>In the case</u> $r(k) = 0$ <u>we have</u>

$$R_{k+1}[u] = (-1)^k \Sigma_{j=1}^{n} \beta_j^{2K-1} c_j f(x;i\beta_j)^2$$

b) <u>In the case of the absence of the bound states</u> $(n = 0)$:

$$R_{k+1}[u] = (-1)^{k+1} \frac{i}{2\pi} \int_{-\infty}^{\infty} k^{2K-1} r(k) f^2(x;k)dk.$$

<u>In the most general case we have</u>

$$R_{k+1}[u] = \frac{i}{2\pi}\int_{-\infty}^{\infty} (-1)^{K+1} k^{2K-1} r(k) f^2(x,k)\,dk$$

$$+ \Sigma_{j=1}^{n} (-1)^K \beta_j^{2K-1} c_j f(x;i\beta_j)^2 : \quad k = 0,1,2,3,\ldots$$

We would like to mention that such a representation of $u(x)$ as (3.7) is not unique! E.g. a finite band potential $u(x)$ can be represented as a linear combination of any n from the $2n + 1$ squares of eigenfunctions which correspond to the end of the bands. An analogical assertion is true for arbitrary C^∞-potential, which is periodic.

However under a proper normalization and for a special measure, we have uniqueness and a very simple representation for a potential $u(x)$ and higher quantities $R_k[u]$.

Main Assertion 4.2: *For the case of self-adjoint* L *and the spectral measure* $d\mu_k$ *corresponding to a* $L\varphi = k\varphi$ *there exists a system* $\varphi(x,k)$ *of eigenfunctions of*

(A) $$\left(-\frac{d^2}{dx^2} + u(x)\right)\varphi = k\varphi$$

such that

(B) $$R_k[u] = (-1)^{K+1}\int k^{K-1}\varphi^2(x,k)\,d\mu_k$$

for all $K = 1,2,3,\ldots$.

In particular,

$$\frac{1}{2} = \int \varphi^2(x,k)\,d\mu_k;$$

$$\frac{u}{4} = \int k\varphi^2(x,k)\,d\mu_k.$$

Now the measure $d\mu_k$ *and the system of eigenfunctions* $\varphi(x,k)$ *is fixed by the conditions* (B).

Remark 4.3: If we demand only

$$R_K[u] = (-1)^{K+1}\int P_{K-1}(k)\varphi^2(x,k)\,d\mu_k' + C_k$$

for polynomials $P_{K-1}(k)$, then the measure $d\mu'_k$ is not unique at all (we know only that supp $d\mu'_k \subset$ supp $d\mu_k$).

Of course there are multicomponent of infinitecomponent systems corresponding to an arbitrary one-component completely integrable system. However, these general infinitecomponent systems and their relationships with the uniqueness of the inverse scattering for some linear operator are not examined in detail

We have one more example: the multicomponent modified MKdV which corresponds to the inverse scattering method of the third order operator. It is known that in general, there is no well established inverse scattering procedure for operators of the 3rd order. The absence of inverse scattering for the simplest third order operators creates difficulties in proving the complete integrability of the Boussinesque equation

$$u_{tt} + \frac{\partial}{\partial x}(6uu_x + u_{xxx}) = 0$$

even for the potential which rapidly decreases on infinity. The connection between the Boussinesq equation and the 3rd order operator is through the Lax representation of the form

$$\frac{\partial L_3}{\partial t} = [L_2, L_3],$$

where $L_3 = \frac{\partial^3}{\partial x^3} - \frac{3}{4}[u\frac{\partial}{\partial x} + \frac{\partial}{\partial x}u + w]$, $L_2 = \frac{\partial^2}{\partial x^2} - u$. This equation is equivalent to the Boussinesq, where $u_t = w_x$. The inverse scattering for the 3rd order operators is connected with the following multicomponent MKdV equation. In the finite component case it was the form

$$\varphi_{it} = -\varphi_{ixxx} + \varphi_{ix}\Sigma_{j=1}^{n}\varphi_j\Psi_j + \varphi_i\Sigma_{j=1}^{n}\varphi_{jx}\Psi_j;$$

$$\Psi_{it} = -\Psi_{ixxx} + \Psi_{ix}\Sigma_{j=1}^{n}\varphi_i\Psi_j + \Psi_i\Sigma_{j=1}^{n}\Psi_{jx}\varphi_j:$$

$i = 1,\dots,n$. E.g. the multicomponent MKdV has the form

$$\varphi_{it} = -\varphi_{ixxx} + \varphi_{ix}\Sigma_{j=1}^{n}\varphi_j^2 + \varphi_j\frac{1}{2}\Sigma_{j=1}^{n}(\varphi_j^2)_x.$$

The corresponding infinite component equations are

$$\varphi_t = -\varphi_{xxx} + \varphi_x \int_\Omega \varphi\Psi \, d\mu_k + \varphi \int_\Omega \varphi_x \Psi \, d\mu_k,$$

$$\Psi_t = -\Psi_{xxx} + \Psi_x \int_\Omega \varphi\Psi \, d\mu_k + \Psi \int_\Omega \Psi_x \varphi \, d\mu_k.$$

References

[1] Y. Nogami, C.S. Warke, Phys. Lett., 59A, 251, (1976).

[2] G. Gardner, J.M. Greene, M.D. Kruskal, R.U. Miura, Comm. Pure and Appl. Math., 27, 97, (1974).

[3] H.F. Baker, Proc. Royal Soc. London, A117, 584, (1928).

[4] I.M. Gelfand, L.A. Dikij, Russian Math Survey 30, 77, (1975) P.D. Lax, Comm Pure and Appl. Math., 28, 141, (1975), B.A. Dubrovin, V.B. Matveev, S.P. Novikov, Russian Math. Survey, 31, 59, (1976).

[5] J. Drach, C.R. Acad. Sci Paris 168, 48, (1919);
J. Drach, C.R. Acad. Sci Paris 168, 337 (1919).

[6] I.V. Cherednik, Functional Anal. Appl. 12, 45, (1978).

[7] D. Kaup. J. Math. Analys, 59 (1976), 849. P. Deift, P. Trubowitz, Comm. Pure and Appl. Math. No. 3, (1980).

[8] D.V. Chudnovsky, G.V. Chudnovsky, Lett Nuovo Cimento, 22, 31, (1978).

[9] D.V. Chudnovsky, G.V. Chudnovsky, Lett. Nuovo Cimento, 22, 47 (1978);
D.V. Chudnovsky, G.V. Chudnovsky, C.R. Acad. Sci. Paris 286A, 1075 (1978).

[10] D.V. Chudnovsky, G.V. Chudnovsky, Seminaire sur les equations non lineaires I, (1977-1978), Centre de Mathematiques, Ecole Polytechnique, 300pp. (1978).

[11] H.P. McKean, P. Van Moerbeke, Inventiones Math. 30, 217, (1975).

[12] D. Gross, A. Neveu, Phys. Rev. D10, 3235, (1974).

[13] G. Borg, Acta Math., 78 (1946), No. 1-2; B. Levitan, Doklady Acad. Sci. USSR, 83 (1952), 349.

[14] D.V. Chudnovsky, Proceedings of the Les Houches summer school on Complex Analysis, Microlocal Calculus and Quantum Field Theory, 1979, Lecture Notes in Physics, Springer, v. 126, 1980, 352-416.

The Representation of an Arbitrary, Two-dimensional Completely Integrable System as the Common Action of Two Commuting One-dimensional Hamiltonian Flows.

by

D.V. Chudnovsky

We construct a canonical, one-parameter Hamiltonian flow acting on an infinite dimensional symplectic manifold (parametrized by pairs of operators or matrices). This system is shown to be completely integrable. By fixing a certain invariant submanifold and restricting two of the commuting Hamiltonian flows associated with our system to that submanifold we obtain any given two dimensional isospectral deformation equation (in scalar, or matrix unknowns). Those isospectral deformation equations are determined uniquely by the dimension of the system, on the orbit of our symplectic manifold, and by the two prescribed commuting Hamiltonian flows.

§0. We begin first with the canonical one dimensional Hamiltonian depending on operator valued functions and then derive the associated, two dimensional isospectral deformation equations. At the second part of the paper we show how, conversely, the two-dimensional isospectral deformation equations determine the same canonical, one dimensional Hamiltonian via a matrix moment problem.

In order to exhibit the connection with isospectral deformations we fix a certain operator-valued measure $d\Sigma_\lambda$ $(\lambda \in \mathbb{C})$. In the most interesting applications, $d\Sigma_\lambda$ is either a) a measure $d\Sigma_\lambda = P(\lambda)d\lambda$ absolutely continuous with respect to Lebesgue measure; or b) a singular measure $d\Sigma_\lambda = \Sigma_{i=1}^{n} \Sigma_i^0 \delta(\lambda - \lambda_i)d\lambda$. In both cases we can, by multiplication with a suitable operator replace $d\Sigma_\lambda$ by a measure of the form a') $d\lambda$ or b') $\Sigma_{i=1}^{n} \delta(\lambda - \lambda_i)d\lambda$.

We introduce a Hilbert space H. The symbols $\varphi, \psi, U, V, A, B\ldots$ denote operators on H. We shall frequently use matrix notations, which means that in some basis of H, fixed once and forever, operators, B, can be represented as matrices $B = (b_{i,j})$. This is convenient in the finite-dimensional case. In general, the indices i,j may be even continuous variables, so that B is an integral operator with kernel $b(i,j)$. We assume that spectral measure $d\Sigma_\lambda = (d(\sigma_\lambda)_{ij})$ to have support in a region $\Omega \subseteq \mathbb{C}$. We now consider the following general system of equations in unknown operators $\varphi_\lambda, \hat{\varphi}_\lambda$, $\lambda \in \Omega$, depending on a parameter x:

$$\check{\varphi}_{\lambda,x} = \int_\mu (\lambda + \mu)\check{\varphi}_\mu d\Sigma_\mu \hat{\varphi}_\mu . \check{\varphi}_\lambda ;$$

(0.1)

$$\hat{\varphi}_{\lambda,x} = -\varphi_\lambda \int_\mu (\lambda + \mu)\check{\varphi}_\mu d\Sigma_\mu \hat{\varphi}_\mu ;$$

$\lambda \in \Omega$. The sequences of operator-valued functions $(\check{\varphi}_\lambda : \lambda \in \Omega)$, $(\varphi_\lambda : \lambda \in \Omega)$ are defined on the space $\int_\Omega^\oplus d\Sigma_\lambda H_\lambda$, $H_\lambda \cong H$ in L^2 norm.

§1. The system (0.1) can be regarded as a natural generalization of the operator Russian chain [1], [2]. In the case where dim H = 2 that system can be related to a standard Russian chain [1], [3], [5] (with finitely many unknown matrices $\check{\varphi}_\lambda, \hat{\varphi}_\lambda$, provided Ω is finite). We now rewrite the system (0,1) in Hamiltonian form. For this purpose we introduce a natural symplectic structure on a set of pairs of sequence$(\check{\varphi}_\lambda)$, $(\hat{\varphi}_\lambda)$ of operators.

We use the following notations:

(1.1) $$\check{\varphi}_\lambda = (\check{\varphi}_{\lambda,ij}); \quad \hat{\varphi}_\lambda = (\hat{\varphi}_{\lambda,ij}); \ \lambda \in \Omega.$$

Then we view $\check{\varphi}_{\lambda,ij}$, $\hat{\varphi}_{\lambda,ij}$ as a set of canonically conjugated variables; namely for each pair (i,j) the variable $\hat{\varphi}_{\lambda,ji}$ (momentum) is conjugate to the variable $\check{\varphi}_{\lambda,ij}$ (coordinate). The Poisson brackets (between conjugate variables) are given by

(1.2) $$\{\hat{\varphi}_{\lambda,ab};\check{\varphi}_{\mu,cd}\} = \delta_{\lambda\mu}\cdot\delta_{ad}\cdot\delta_{bc}.$$

Using the following abbreviated notations

(1.3) $$\check{\varphi} = (\check{\varphi}_\lambda : \lambda \in \Omega), \quad \hat{\varphi}^t = (\hat{\varphi}^t_\lambda : \lambda \in \Omega),$$

(t denotes transposition), then the general Hamiltonian equations can be represented in the form

(1.4) $$\check{\varphi}_x = \frac{\delta H}{\delta(\hat{\varphi}^t)}, \qquad (\hat{\varphi}^t)_x = -\frac{\delta H}{\delta\check{\varphi}}.$$

For $F = \langle(F_{\lambda,ij}): \lambda \in \Omega\rangle$, $\frac{\delta H}{\delta F}$ is defined by

(1.5) $$\frac{\delta H}{\delta F} = \langle(\frac{\partial H}{\partial F_{\lambda,ij}}): \lambda \in \Omega\rangle.$$

The system (0.1) can now be reinterpreted as a Hamiltonian system, with the following Hamiltonian

(1.6) $$H_{\Pi} = \mathrm{tr}_H\{\int_\Omega \check{\varphi}_\mu d\Sigma_\mu \hat{\varphi}_\mu)\cdot \int_\Omega \mu\check{\varphi}_\mu d\Sigma_\mu \hat{\varphi}_\mu)\}.$$

§2. The Hamiltonian (1.6) has many internal symmetries the existence of which can be established in various ways. We use the method of reference [2].

Theorem 2.1: For arbitrary $\lambda \in \Omega$, the following operator-valued functions are the first integrals of (0.1):

(2.2) $$K[\lambda] = \hat{\varphi}_\lambda\check{\varphi}_\lambda;$$

Moreover, if $K[\lambda]=1$, then also

(2.3) $$C[\lambda] = \hat{\varphi}_\lambda\cdot\int_\mu \frac{1}{\lambda-\mu}\check{\varphi}_\mu d\Sigma_\mu\hat{\varphi}_\mu\cdot\check{\varphi}_\lambda.$$

And

(2.4) $\{\mathrm{tr}_H C[\lambda], \mathrm{tr}_H C[\eta]\} = 0$, for arbitrary $\lambda, \eta \in \Omega$.

Because of his particular importance, we exhibit one further, similar first integral separately:

Proposition 2.5: The quantity

(2.5) $$\mathfrak{M}_0 = \int_\Omega \check{\varphi}_\lambda d\Sigma_\lambda \hat{\varphi}_\lambda$$

is a first integral of (0.1).

Other integrals of the form (2.3) have the following canonical moment-type representation. Multiply $C[\lambda]$ by λ, and look at the **expansion** in powers of λ^{-1} as $\lambda \to \infty$. This yields

Corollary 2.7: _For any_ $\lambda \in \Omega$ _and any integer_ $n \geq 0$,

(2.8) $$\hat{\varphi}_\lambda \int_\mu \mu^n \check{\varphi}_\mu d\Sigma_\mu \hat{\varphi}_\mu \check{\varphi}_\lambda$$

is a first integral of (0.1).

§3. Using the "moment notation",

(3.1) $$\mathfrak{M}_n = \int_\Omega \check{\varphi}_\lambda \lambda^n d\Sigma_\lambda \hat{\varphi}_\lambda : \quad n = 0,1,2,\ldots,$$

we can rewrite the first integrals given in (2.8) in a short form,

$$\hat{\varphi}_\lambda \cdot \mathfrak{M}_n \cdot \check{\varphi}_\lambda : \quad n = 0,1,2,\ldots; \quad \lambda \in \Omega.$$

Proposition 3.2: _Consider the following commuting Hamiltonians_

(3.3) $$H_m = \mathrm{tr}_H\{\int_\Omega \lambda^m C[\lambda] d\Sigma_\lambda\} : \quad m = 0,1,2,3,\ldots$$

Then we have

(3.3') $$H_m = \mathrm{tr}_H\{\int_\lambda \int_\Omega \lambda^m (\lambda-\mu) \check{\varphi}_\lambda d\Sigma_\lambda \hat{\varphi}_\lambda \check{\varphi}_\mu d\Sigma_\mu \hat{\varphi}_\mu\},$$

$m = 0,1,2,\ldots$; _in particular_,

(3.4)
$$H_0 = 0, \qquad H_1 = \mathrm{tr}\{(\int_\lambda \check{\varphi}_\lambda d\Sigma_\lambda \hat{\varphi}_\lambda)^2\}$$
$$H_2 = \mathrm{tr}\{(\int_\lambda \check{\varphi}_\lambda d\Sigma_\lambda \hat{\varphi}_\lambda) \cdot (\int_\lambda \lambda \check{\varphi}_\lambda d\Sigma_\lambda \hat{\varphi}_\lambda)\} \equiv H_{\Pi}.$$

If t_m denotes the time-parameter of the evolution generated by the Hamiltonian H_m $(m = 0,1,2,\ldots)$, then we have the following evolution of the "collective" potential.

(3.5) $$\mathfrak{M}_1 = \int_\lambda \lambda\check{\varphi}_\lambda d\Sigma_\lambda \hat{\varphi}_\lambda .$$

For $m = 0,1,2,\ldots$

(3.6) $$\frac{\partial}{\partial t_m} \mathfrak{M}_1 = [\int_\mu \check{\varphi}_\mu d\Sigma_\mu \hat{\varphi}_\mu , \int_\mu \mu^m \check{\varphi}_\mu d\Sigma_\mu \hat{\varphi}_\mu],$$

or in the notations of (3.1),

(3.6') $$\frac{\partial}{\partial t_m} \mathfrak{M}_1 = [\mathfrak{M}_0, \mathfrak{M}_m] : m \geq 0.$$

Let us now look at the evolution of $\mathfrak{M}_n$ with respect to $x(\equiv t_2)$, i.e. under the flow generated by the Hamiltonian H_{II} $(\equiv H_2)$ introduced in (1.6), (3.4).

Proposition 3.7: For $n = 0,1,2,\ldots$, the following linear differential equations for the moments $\mathfrak{M}_n$ hold:

(3.8) $$\frac{\partial \mathfrak{M}_n}{\partial x} = [\mathfrak{M}_n, \mathfrak{M}_1] + [\mathfrak{M}_0, \mathfrak{M}_{n+1}].$$

With the notations introduced so far, we can rewrite a generalized operator Russian chain in the form of a linear differential equation

(3.9) $$\check{\varphi}_{\lambda,x} = (\mathfrak{M}_1 + \lambda\mathfrak{M}_0)\check{\varphi}_\lambda ;$$
$$\hat{\varphi}_{\lambda,x} = -\hat{\varphi}_\lambda (\mathfrak{M}_1 + \lambda\mathfrak{M}_0).$$

By (3.8) the quantity

$$\Lambda = \Sigma_{n=0}^{\infty} \mathfrak{M}_n \lambda^{-n}$$

is, by (3.8), a formal resolvent expansion for the linear problem (3.9).

Moreover, since $\mathfrak{M}_0$ is a first integral of (0.1) (i.e. x-independent) we can find a transformation

$$\check{\varphi}_\lambda \to T\check{\varphi}_\lambda T^{-1}, \qquad \hat{\varphi}_\lambda \to T\hat{\varphi}_\lambda T^{-1}$$

(replacing $d\Sigma_\lambda$ by $T^{-1}d\Sigma_\lambda T$) that diagonalizes $\mathfrak{M}_0$ (or, in general, reduces $\mathfrak{M}_0$ into a normal form).

§4. Now we show how two dimensional completely integrable system, arising from an arbitrary isospectral deformation problem for a matrix first order differential operator can be transformed into some generalized operator Russian chain [2].

We consider linear differential operators with coefficients that are operators on a Hilbert space H. The simplex linear differential operator, and the operator conjugate to it are

$$(4.1) \qquad L_{\mathfrak{a},E} = \frac{d}{dx} + V_{\mathfrak{a}} - E\mathfrak{a};$$

$$-L^t_{\mathfrak{a},E} = \frac{d}{dx} - V^t_{\mathfrak{a}} + E\mathfrak{a}^t,$$

where E is (complex) spectral parameter. In the non-degenerate case we assume further that $V_{\mathfrak{a}}$ is of the form

$$(4.2) \qquad V_{\mathfrak{a}} = [\mathfrak{a}, V].$$

Since the operators (4.1) depend only on the orbit (normal form) of $\mathfrak{a}$, we may assume, without loss of generality, that $\mathfrak{a}$ is a diagonal operator on H, and, in some fixed basis of H, it can be written as

$$\mathfrak{a} = (a_i \cdot \delta_{ij}).$$

The resolvent expansion $\Lambda_{\mathfrak{a}}(E,x)$ of the linear problem (4.1) is determined by the following well known rule

Lemma 4.3: _The equation_

$$\frac{\partial\Lambda}{\partial x} = [\Lambda, V_{G} - EG]$$

<u>has a unique solution</u> $\Lambda = \Lambda_{G}(E,x)$ <u>in the space</u> $\mathbb{C}[[1/E]]$ <u>of formal power series in</u> $1/E$ <u>such that</u>

(4.5)
$$\Lambda_{G} = G + \lambda_{1,G}/E + \ldots + \Sigma_{n=0}^{\infty} \frac{\lambda_{n,G}}{E^{n}}$$

<u>with</u> $\lambda_{0,G} = G$, $\lambda_{1,G} = -V_{G}$, <u>and such that</u>

(4.6)
$$\lambda_{G} \equiv G \quad \text{if} \quad V_{G} = 0.$$

These quantities $\lambda_{n,G}$ are local functionals in V_{G}, $V_{G,x}, \ldots$. They give rise to the canonical series of two dimensional isospectral deformation equations associated with the operator (4.1). This series has the following canonical form:

(4.7)
$$[G, V_{t} + \lambda_{n,G}] = 0 \quad \text{or}$$
$$-[G, V_{t}] = [G, \lambda_{n,t}] : n = 1, 2, \ldots .$$

The general two dimensional isospectral deformation equation can be represented as an analytic combination of the canonical Hamiltonians (4.7), see [5].

<u>Example</u> 4.8: Let us consider the simplest case $H = H_{1} \times H_{2}$ where G is $\mathbb{I}$ (identity operator) on H_{1} and $-\mathbb{I}$ on H_{2}. Then the first two (non-trivial) two-dimensional equations are the non-linear Schrodinger equation

(4.9)
$$i\varphi_{t} = \varphi_{xx} + 2\varphi\psi\varphi;$$
$$-i\psi_{t} = \psi_{xx} + 2\psi\varphi\psi$$

and the modified Korteweg-de-Vries equation

$$\varphi_y = \varphi_{xxx} + 3(\varphi_x\psi\varphi + \varphi\psi\varphi_x);$$

(4.10)

$$\psi_y = \psi_{xxx} + 3(\psi_x\varphi\psi = \psi\varphi\psi_x)$$

for operators φ and ψ. In the matrix case one can view φ as $n \times n$ matrix and ψ as $m \times n$ matrix.

We now consider the fundamental solutions $\check{F}_E$, $\hat{F}_E$ of the operator spectral problem (4.1):

$$\frac{d}{dx}\check{F}_E = -V_{\mathfrak{a}}\check{F}_E + E\mathfrak{a}\check{F}_E;$$

(4.11)

$$\frac{d}{dx}\hat{F}_E = \hat{F}_E V_{\mathfrak{a}} - \hat{F}_E E\mathfrak{a}.$$

There is a canonical choice of eigenfunctions for (4.11), namely the Jost eigenfunctions:

<u>Lemma</u> 4.12: <u>There are fundamental solutions</u> $\check{\varphi}_E$ <u>and</u> $\hat{\varphi}_E$ <u>of</u> (4.11) <u>which can be represented as formal power series in</u> E, <u>as</u> $E \to \infty$:

(4.13) $$\check{\varphi}_E = \{\Sigma_{m=0}^{\infty} T_m E^{-m}\}\exp(E\mathfrak{a}x)$$

<u>and</u>

(4.14) $$\hat{\varphi}_E = (\check{\varphi}_E)^{-1}$$

<u>with</u> $\check{T}_0 = \mathbb{I}$. <u>Here the matrix elements of</u> $\check{T}_m$ <u>are local functionals of</u> $\mathfrak{a}$, $V_{\mathfrak{a}}$, $V_{\mathfrak{a}x}$, ... <u>Moreover, for</u>

(4.15) $$\eta_{\mathfrak{a}}(E) \stackrel{\text{def}}{=} \check{\varphi}_E \cdot \mathfrak{a} \cdot \hat{\varphi}_E,$$

<u>we have the same asymptotic expansion in</u> E <u>as</u> $E \to \infty$, <u>as for</u> $\Lambda_{\mathfrak{a}}(E,x)$:

(4.16) $$\eta_{\mathfrak{a}}(E) \cong \Sigma_{n=0}^{\infty} \frac{\lambda_{n,\mathfrak{a}}}{E^n}.$$

This lemma enables us to express the Hamiltonian densities as moments of a certain operator spectral measure $d\Sigma_E$. We adopt the terminology of the matrix (operator) moment problem introduced by Krein and Berezansky [4].

Lemma 4.17: *Let $\check{\varphi}_E$, $\hat{\varphi}_E$ be arbitrary solutions of the linear problem (4.11). Then there exists a spectral measure $d\Sigma_E(=(d\sigma_{ij}))$ which is a solution of a moment problem*

$$\int_E \check{\varphi}_E \cdot E^n d\Sigma_E \hat{\varphi}_E = \lambda_{n,\mathfrak{a}}: n = 0,1,2,\dots . \tag{4.18}$$

The canonical choice of the spectral measure $d\Sigma_E$ corresponds to the choice of the solutions $\check{\varphi}_E$, $\hat{\varphi}_E$ given in Lemma 4.12.

The existence of this spectral measure is merely a consequence of the Riesz theorem; the x independence of $d\Sigma_E$ follows from the structure of the solutions of equation (4.4).

The spectral measure $d\Sigma_E$ solving (4.18) provides the reduction of the linear problem (4.11) to a nonlinear generalized operator Russian chain. The most important ones are the first two moments in (4.18)

$$\int \check{\varphi}_E d\Sigma_E \hat{\varphi}_E = \mathfrak{a},$$

(4.19)

$$\int E\check{\varphi}_E d\Sigma_E \hat{\varphi}_E = -V_{\mathfrak{a}}.$$

Using (4.19) we rewrite the linear problem (4.11) in the natural nonlinear form of a generalized operator Russian chain (0.1):

$$\check{\varphi}_{\lambda,x} = \int_\mu (\lambda + \mu)\check{\varphi}_\mu d\Sigma_\mu \hat{\varphi}_\mu \cdot \check{\varphi}_\lambda$$

(4.20)

$$\hat{\varphi}_{\lambda,x} = -\hat{\varphi}_\lambda \cdot \int_\mu (\lambda + \mu)\check{\varphi}_\mu d\Sigma_\lambda \hat{\varphi}_\mu .$$

In particular, any two of the commuting Hamiltonians H_m introduced in (3.3),

$$H_m = \mathrm{tr}\{\int_\lambda \lambda^m C[\lambda] d\Sigma_\lambda\}: m = 0,1,2,\dots,$$

give rise to some two dimensional canonical system (4.7).

Corollary 4.21: For any spectral measure $d\Sigma_E$ satisfying (4.18) and functions $\check{\varphi}_E$, $\hat{\varphi}_E$ satisfying, in the x-direction, the generalized operator Russian chain equations (4.20) corresponding to the Hamiltonian H_2, and, in the t-direction, the equations obtained from a Hamiltonian H_m for a given m, the m-th equation (4.7) holds, i.e.:

$$-\frac{\partial}{\partial t} V_G = [G,\lambda_{m,G}].$$

Here

$$V_G = -\int_\mu \mu\check{\varphi}_\mu d\Sigma_\mu \hat{\varphi}_\mu,$$

by (4.18).

In particular, the nonlinear Schrodinger (modified KdV) equation (4.9) (respectively (4.10)) corresponds to the combination of the flows generated by H_2 in the x-direction and H_3, in the t-direction (respectively H_4, in the t-direction) for the operator Russian chain variables $\check{\varphi}_\lambda, \hat{\varphi}_\lambda$.

References

[1] D.V. Chudnovsky, Phys, Lett. 74A (1979), p. 185-188.

[2] D.V. Chudnovsky, Les Houches Lectures, August 1979, Lecture Notes in Physics, v. 126, Springer-Verlag, 1980, pp. 352-416.

[3] D.V. Chudnovsky, Lecce Lectures, June 1979, Lecture Notes in Physics, v. 120, Springer-Verlag 1980, pp. 103-150.

[4] M.G. Krein, M.A. Krasnoselsky, Uspehki, Math. Nank 2 (1947), No. 3, pp. 60-106; Yu. M. Berezansky, Trudy Moscow Math. Soc., 21 (1970), pp. 47-102.

[5] A.C. Newell, Proc. Royal. Soc. London, A365 (1979), pp. 283-311.

[6] D.V. Chudnovsky, C.R. Acad. Sci. Paris, 289A (1979), pp. A-731-A-734.

Department of Mathematics
Columbia University
New York, NY
USA

Self-duality of Yang-Mills Fields and of
Gravitational Instantons
by

Jean Pierre Bourguignon

In this talk we survey recent results establishing self-duality for certain classes of Yang-Mills fields and gravitational instantons. This involves discussing the dynamical system defined by the Yang-Mills functional on the space of connections on a vector bundle.

We shall confine our considerations to mathematical statements although the questions discussed are relevant to some physical problems and although the physics of the situation has sometimes been of some help.

§1. Some notations.

Let $\pi: E \to M$ be a G-vector bundle over a Riemannian manifold M (the Riemannian metric is part of the data) with G a compact Lie group. We shall restrict ourselves to four-dimensional manifolds in order that the notion of self-duality (see further) makes sense. This turns out to be also a class of physical interest.

In our discussion the compact Lie groups fall under three families: the abelian groups T^k (and among them the circle group U_1) for which the theory will reduce to the study of ordinary harmonic 2-forms; the groups SU_2, U_2, SU_3 and their products with abelian groups characterized by the fact that the commutator of non-trivial element does not contain a non-abelian group; SO_4 and the other compact groups which all contain SO_4 as a subgroup.

The G-structure on the fibers gives additional information. In particular it distinguishes a Lie algebra subbundle $\mathcal{G}_E \subset E^* \otimes E$ with fiber the Lie algebra $\mathcal{G}$ of G. This bundle $\mathcal{G}_E$ is isomorphic (if the representation ρ of G in the fibers is faithful as we suppose later on to the associated bundle $\mathcal{G}_P = P_{ad} \times \mathcal{G}$ where P is the principal bundle of E (i.e. so that $E = P \times_\rho \xi$). Notice that the sections of the bundle $\mathcal{G}_E$ form the Lie algebra of the gauge group.

The space $\mathcal{C}_E$ of G-connections on E will be our configuration space. Recall that $\mathcal{C}_E$ is an affine space modeled on the space

$\Omega^1(M,\mathcal{G}_E)$ of $\mathcal{G}_E$-valued 1-forms. We like to view a G-connection as a differential operator ∇ from $\Omega^0(M,E)$ into $\Omega^1(M,E)$ with the identity of $T^*M \otimes E$ as principal symbol. A connection ∇ being given, it is possible, by extending usual formulas, to define an exterior differential d^∇ for E-valued differential forms (of course d^∇ maps $\Omega^k(M,E)$ into $\Omega^{k+1}(M,E)$). The deviation of $\Omega^0(M,E)$ from being a complex is measured by the curvature R^∇ of the G-connection ∇ and, as it is well known, R^∇ belongs to $\Omega^2(M,\mathcal{G}_E)$.

Our potential will be the Yang-Mills functional defined on the configuration space $\mathcal{C}_E$ by

$$YM(\nabla) = \int_M \|R^\nabla\|^2.$$

Its definition requires the use of both a G-invariant metric on the fibers of $\mathcal{G}_E$ and the Riemannian metric on M (hence the double bars!). The integration is performed with respect to the volumn element by the metric g.

§2. Self-duality of Yang-Mills fields.

We suppose M oriented by a 4-form vol of length 1, so that we can regard the tangent bundle τ_M: $TM \to M$ as an SO_4-bundle. We insist that the metric g is part of our data: as a result we shall freely identify TM and T^*M.

The Hodge map $*$ defined on exterior 2-forms by $(*\omega,\omega')\mathrm{vol} = \omega \wedge \omega'$ is an involution on $\Lambda^2 TM$ and therefore gives rise to the decomposition

$$\Lambda^2 TM = \Lambda^+ TM \oplus \Lambda^- TM$$

into ε-self-dual subspaces (the eigenspaces for the eigenvalues $+I$ and $-I$). Notice that after identifying skew symmetric matrices with the Lie algebra of SO_4, this decomposition is precisely the decomposition of SO_4 into its two simple ideals isomorphic to SU_2.

Consequently, if we consider the curvature R^∇ as a map from $\Lambda^2 TM$ to $\mathcal{G}_E$ we can split R^∇ into its ε-self-dual pieces

$$R^\nabla = R^\nabla_+ + R^\nabla_-.$$

We call the connection ∇ ε-<u>self-dual</u> if $R^\nabla = R^\nabla_\varepsilon$.

On the other hand it follows from Chern-Weil theory that the expression in the curvature $\int_M |R^\nabla \wedge R^\nabla|$ is independant of the connection and is, up to a universal constant c, the so-called <u>Pontrjagin index</u> $p_1(E)$ of the bundle $\pi: E \to M$. If the group G is simple, this expression that we just considered is the only one having this property. Since

$$\int_M |R^\nabla \wedge R^\nabla| = \int_M (\|R_+^\nabla\|^2 - \|R_-^\nabla\|^2)$$

and since

$$YM(\nabla) = \int_M (\|R_+^\nabla\|^2 + \|R_-^\nabla\|^2),$$

we get that the Yang-Mills functional has a lower bound of a topological nature, namely

$$c|p_1(E)| \leq YM(\nabla).$$

Moreover we see that on a nontrivial bundle $(-\varepsilon)$-self-dual connections (with ε determined by $\varepsilon p_1(E) = |p_1(E)|$) cannot exist and that an ε-self-dual connection is an absolute minimum of YM and therefore a critical point.

The equation for general critical points of YM is

$$\delta^\nabla R^\nabla = 0$$

(where δ^∇ is the adjoint of the exterior differential, the so-called <u>codifferential operator</u>). Since, for any connection ∇, R^∇ satisfies $d^\nabla R^\nabla = 0$ (this is <u>the second Bianchi identity</u>), critical points of YM are connections with harmonic curvature (later R^∇ is called a <u>Yang-Mills field</u> or <u>an instanton</u>) i.e. we have to deal with some kind of Hodge theory. If the group G is abelian, the theory reduces to ordinary Hodge theory of harmonic 2-forms. If the group G is non-abelian, the problem is non-linear since the connection ∇ enters itself in the definition of the operators d^∇ and δ^∇. (Notice also that the Riemannian metric g eneters in the definition of δ^∇.)

When $M = S^4$ and G simple, one has now an almost complete description of self-dual connections (see [6]). The main question which remains is to decide whether there are some other critical points.

It is at this point that the distinction that we made earlier between the two last families of Lie groups is relevant. We postpone the discussion of the group SO_4 until a little later to present a theorem which holds true for the groups SU_2, U_2, SU_3.

We shall say that a critical connection ∇ (sometimes called a *Yang-Mills connection*) is *weakly stable* if the second variation of at ∇ is nonpositive. Typical examples of such critical points are local minima. This notion has clearly some physical bearing.

The we have

Theorem 1 (cf. [3],[4]). *Any weakly stable* SU_2-*Yang-Mills field on a 4-dimensional orientable Riemannian homogeneous manifold is* ε-*self-dual*.

Remarks. (i) The spaces under consideration turn out to be S^4, $\mathbb{C}P^2$, $S^2 \times S^2$, $S^2 \times T^2$, $S^3 \times S^1$ and T^4.

(ii) As stated Theorem 1 extend to the groups U_2 and SU_3 for those spaces having no 2-cohomology, i.e. S^4 and $S^1 \times S^3$ (notice that the invariant metrics on these spaces are both conformally flat).

(iii) For the other spaces Theorem 1 extend also with the minor modification that the Yang-Mills connection may be twisted by the connection of a line bundle whose curvature form is an ordinary harmonic 2-form (in fact a parallel 2-form).

Before discussion SO_4-bundles, a few words on how such a theorem can be proved (for details see [4]).

The main idea is to construct deformations of the connection related to the geometry of the situation. We do so by considering $_E$-valued 1-forms obtained from the $\mathfrak{g}_E$-valued 2-form R^∇ by interior product with infinitesimal isometries (these deformations can in fact be viewed as elements of the Lie algebra of the *enlarged gauge group*).

Then we use the second variation formula on these special deformations. Using that ∇ is a Yang-Mills connection, i.e., that R^∇ is harmonic, we get that a certain average of the second variations on these deformations vanishes. This is made possible by the use of Weitzenböck formulas for Lie algebra-bundle-valued 1- and 2-forms. We then use the stability assumption to strengthen the result on the average to a result valid for each deformation. This enables us finally to extract from the equation the algebraic statement that the algebras generated in each fiber of $\mathcal{G}_E$ by the $R^{\nabla}_{+X,Y}$'s and the $R^{\nabla}_{-X,Y}$'s (where X and Y are tangent vectors to M) commute with each other. This follows from an algebraic lemma saying that the tensor product $\Lambda^+TM \otimes \Lambda^-TM$ is isomorphic to the space S^2_0TM of traceless symmetric 2-tensors. Since the groups SU_2, U_2 and SU_3 are small enough, we can the conclude that R^{∇}_+ or R^{∇}_- either has to vanish or reduces to an abelian field.

The case of SO_4-bundles is slightly more complicated since a new integral constraint on the curvature appears, the so-called <u>Euler characteristic</u> $\chi(E)$.

It is convenient to split the Lie algebra SO_4 into its simple ideals (this can be done by another Hodge map $*$ in the fibers) and also the curvature R^{∇} of an SO_4-connection ∇ into

$$R = R^+_+ + R^+_- + R^-_+ + R^-_-.$$

With these notations the two integral constraints can be written

$$8\pi^2\chi(E) = \int_M \{\|R^+_+\|^2 + \|R^+_-\|^2 - \|R^-_+\|^2 + \|R^-_-\|^2\},$$

$$4\pi^2 p_1(E) = \int_M \{\|R^+_+\|^2 - \|R^+_-\|^2 + \|R^-_+\|^2 - \|R^-_-\|^2\}.$$

In this case the pertinent notion is that of <u>two-fold self-duality</u>. Indeed for a given bundle only one of these two constraints is going to give an effective lower bound to the Yang-Mills functional $\max\{4\pi^2|p_1(E)|, 8\pi^2|\chi(E)|\} \leq YM(\nabla)$. Hence the vanishing of two out of the four components of the curvature. We can then prove

Theorem 1' (cf. [3], [4]). <u>Any weakly stable SO_4-Yang-Mills field on S^4 is two-fold self-dual.</u>

This result extends to other Riemannian homogeneous orientable 4-manifolds with modifications analogous to the ones mentioned earlier.

§3. Self-duality of gravitational instantons.

We are now interested in describing some phenomena special to the tangent bundle. The new notion which can then be introduced is the torsion of a connection. Of particular interest on the tangent bundle are the torsion free connections, like the Levi-Civita connection D of the Riemannian metric g. The curvatures of such connections verify another Bianchi identity (the first to be discovered in fact).

In the Riemannian case it is interesting to decompose the curvature R^D into its SO_4-irreducible components

$$R^D = W^+ + W^- + Z + U$$

where $W = W^+ + W^-$ is the Weyl curvature conformal tensor (the Riemannian manifolds for which either one of W^+ and W^- vanishes are the so-called half conformally flat spaces which play a crucial role in Penrose's twistor theory, see [1]), Z measures the deviation of g being an Einstein metric and U the curvature tensor of the standard sphere times 1/6 the scalar curvature.

We can then ask ourselves when is the Levi Civita connection D a Yang-Mills connection. These metrics are sometimes called gravitational instantons, see [7]. One must be careful that since a metric is not a critical point of the functional $g \to \int_M \|R^D\|^2$ since we already underlined that the metric on the base space was used in the definition of YM, but supposed fixed.

The twofold self-duality that we introduced for SO_4-bundles can be applied here. We have

$$R_-^+ + R_+^- = Z,$$

$$R_+^+ = W^+ + \frac{1}{6}\mu \, Id_{\Lambda^+ TM},$$

$$R_-^- = W^- + \frac{1}{6}\mu \, Id_{\Lambda^- TM}.$$

Because of the symmetries of the connection notice that the vanishing of R_-^+ forces the vanishing of R_+^- and conversely and is equivalent to the metric being Einstein. The other cases reduce to conformally flat spaces with vanishing scalar curvature. Notice also that the decomposition for the tangent bundle is finer since the components W^+

and U (resp. W^- and U) belong to different irreducible spaces.

It follows from this discussion that the next theorem can be thought of as a self-duality statement.

Theorem 2. *Let M be an orientable 4-manifold. If the Pontrjagin index of its tangent bundle does not vanish, then any gravitational instanton is self-dual.*

The stability assumption in Theorem 1 has been replaced by a purely topological assumption on M (indeed a third of the Pontrjagin index is the signature of the intersection form in the 2-cohomology and hence a topological invariant).

The proof of Theorem 2 requires a detailed analysis of the Weitzenböck formulas for TM-valued 1-forms and Λ^2TM-valued 2-forms (see [2] for details), together with a differential geometric study of harmonic TM-valued differential I-forms, the so-called *traceless Codazzi tensors* described in [5].

Notice that Theorem 2 prevents manifolds M with $2\chi(M) < |p_1(M)|$ and $p_1(M) \neq 0$ from being gravitational instantons.

References

[1] M.F. Atiyah, N.J. Hitchin, and I.M. Singer, Self-duality in four dimensional Riemannian geometry, Proc. Royal Soc., **A** 362 (1978) 425-461.

[2] J.P. Bourguignon, Les Variétés de dimension 4 à signature non nulle et à courbure harmonique sont d'Einstein, Preprint, IAS, (Princeton).

[3] J.P. Bourguignon, and H.B. Lawson, Stability and isolation phenomena for Yang-Mills fields, Preprint.

[4] J.P. Bourguignon, H.B. Lawson, and J. Simons, Stability and gap phenomena for Yang-Mills fields, Proc. Nat. Acad. Sci. U.S.A. (1979), 1550-1553.

[5] A. Derdzinski, Classification of certain compact Riemannian manifolds with harmonic curvature and non-parallel Ricci tensor, to appear in Math. Z.

[6] V.G. Drinfeld, and Y.I. Manin, A description of instantons, Commun. Math. Phys. 63 (1978), 177-192.

[7] G.W. Gibbons, and C.N. Pope, $\mathbb{C}P^2$ as a gravitational instanton, Commun, Math. Phys., 61 (1978), 239-248.

Centre de Mathematique
Ecole Polytechnique
91128 Palaiseau
France

and

Institute for Advanced Study
Princeton University
Princeton, NJ
USA

On Proving the Nonintegrability of a Hamiltonian System

by

Richard C. Churchill

It has been evident since the days of Poincaré that among all Hamiltonian systems the integrable ones are the exception (all terminology peculiar to Hamiltonian systems will be defined), and proofs have recently been provided (e.g. see [9]). Nevertheless, it is a formidable task to rigorously establish that a specific Hamiltonian system does not admit a complete set of integrals, somewhat akin to showing that a number selected at random from the real line is irrational. Indeed, even the n-body problem has survived attack, although partial success was achieved by Bruns (1887) with regard to a complete set of algebraic integrals for that problem (e.g. see [14], p. 23).

Fortunately, with the advent of the Smale horseshoe mapping some practical methods are becoming available, but (with complete proofs) only for analytic systems with two degrees of freedom, and only when all integrals are assumed analytic and "global " i.e. defined on the same domain as the given Hamiltonian.

This note is an informal nonrigorous expository account of several of these recent methods. In particular, we show how an idea of V.K. Melnikov [10] might be used to prove nonintegrability, and then review methods developed by the author, D.L. Rod and G. Pecelli which apply, for example, to the Hamiltonian system defined by

$$H(x,y) = \frac{1}{2}|y|^2 + \frac{1}{2}|x|^2 - \frac{1}{2}x_1^2x_2^2, \qquad x = (x_1,x_2),\ y \in R^2.$$

Since we only wish to give the flavor of the techniques, in our treatment rigor will be slighted, for example in not stating smoothness assumptions, in giving domains as R^m, and in assuming all solutions of all differential equations encountered are defined for all time.

In the preparation of this note I am pleased to acknowledge helpful conversations with Martin Kummer and J.T. Montgomery.

§1. Basic Definitions and Notations.

If $F: R^m \to R$, then F_x will denote the gradient of F (which we regard, along with all elements of R^m, as a column vector), and F_{x_j} will denote the partial derivative of F with respect to x_j. Also, $J = J_n$ will denote the skew-symmetric $2n \times 2n$ matrix $\begin{pmatrix} 0 & I \\ -I & 0 \end{pmatrix}$, where $I = I_n$ is the $n \times n$ identity matrix.

A Hamiltonian system is a system of ordinary differential equations of the form

$$(1.1) \qquad \dot{x} = JH_x,$$

where $H: R^{2n} \to R$ is called the Hamiltonian or energy (function), and n is the number of degrees of freedom. Notice if we write $x = (u,v)$, $u,v \in R^n$, then (1.1) becomes

$$(1.2) \qquad \dot{u} = H_v, \qquad \dot{v} = -H_u,$$

i.e. $\dot{u}_j = H_{v_j}$, $\dot{v}_j = -H_{u_j}$, $j = 1,\ldots,n$; u_j and v_j are then called canonical variables.

An integral of (1.1) is a nonconstant function $G: R^{2n} \to R$ such that $(d/dt)G(x(t)) \equiv 0$ for any solution $x = x(t)$ of (1.1). If for functions $F,G: R^{2n} \to R$ we define the Poisson bracket $\{F,G\}$ of F and G by $\langle F_x, JG_x \rangle$, then the chain-rule $(d/dt)G(x(t)) = \langle G_x, \dot{x} \rangle = \langle G_x, JH_x \rangle$ shows that G is an integral of (1.1) if and only if $\{G,H\} \equiv 0$. Note by the skew-symmetry of J that $\{H,H\} \equiv 0$, proving H is always an integral of the associated Hamiltonian system. Geometrically, G is an integral of (1.1) if the level surfaces of G decompose into solution curves of that equation, i.e. if $x_0 \in R^{2n}$ and $G(x_0) = c$, then the solution $x = x(t)$ of (1.1) satisfying $x(0) = x_0$ must lie completely on the surface $G = c$, i.e. on $\{x: G(x) = c\}$. When $G = H$ such surfaces are called energy surfaces, and c is replaced by h.

(1.1) is called (completely) integrable if there are n functions $G_j: R^{2n} \to R$ such that (i) $\{G_j,H\} \equiv 0$ (they are integrals), (ii) $\{G_i,G_j\} \equiv 0$ (they are "in involution") and (iii) the set $\{(G_j)_x\}$ is linearly independent at some point of R^{2n} (they are "independent"). Often "in involution" means (ii) and (iii). Also, (iii) is sometimes required to hold at all points of R^{2n}, but this is rarely encountered in practice. (1.1) is of course nonintegrable if it is not integrable. To verify this property one tries to show there can be no integral independent of H, i.e. no integral $G: R^{2n} \to R$ with gradient field G_x independent of H_x.

As an example of an integrable system consider the linear harmonic oscillator $\ddot{x} = -Kx$, where $K: R^2 \to R^n$ is a positive definite symmetric linear operator. Choosing an orthogonal basis of R^n consisting of eigenvectors of K one may view this sytem as $\ddot{x}_j = -\omega_j^2 x_j$, $\omega_j > 0$, whereupon the substitution $u_j = \sqrt{\omega_j} x_j$, $v_j = (\sqrt{\omega_j})^{-1}\dot{x}_j$ converts the equations to

$$(1.3) \qquad \dot{u}_j = \omega_j v_j, \qquad \dot{v}_j = -\omega_j u_j.$$

This is of the form (1.2), with

$$(1.4) \qquad H = \Sigma_{j=1}^{n} (\omega_j/2)(u_j^2 + v_j^2),$$

and is integrable with $G_j = (\omega_j/2)(u_j^2 + v_j^2)$.

§2. Reduction to a Mapping.

Methods for proving the nonexistence of integrals usually begin by somehow reducing the study of the solutions of (1.1) to that of a mapping. Here we review one such method, but we remark that other methods are also used, and may be more appropriate to a specific problem.

In (1.2) make the substitution $u_j = \sqrt{2r_j} \sin \theta_j$, $v_j = \sqrt{2r_j} \cos \theta_j$, and the result is easily checked to be the Hamiltonian system

$$(2.1) \qquad \dot{\theta}_j = K_{r_j}, \qquad \dot{r}_j = -K_{\theta_j},$$

where $K(\theta, r) = H(\sqrt{2r} \sin \theta, \sqrt{2r} \cos \theta)$, i.e. K is obtained by direct substitution into the original Hamiltonian H.

Now assume

$$(2.2) \qquad \frac{\partial K}{\partial r_n} \neq 0$$

in a neighborhood of the surface $K = h$; by the implicit function theorem it follows that

$$r_n = L(r_1,\dots,r_{n-1},\theta_1,\dots,\theta_n,h) \tag{2.3}$$

near this surface, hence

$$K(\theta_1,\dots,\theta_n,r_1,\dots,r_{n-1},L) = h. \tag{2.4}$$

Differentiating w.r.t. $s = \theta_j$ or r_j this gives $K_s + K_{r_n} L_s \equiv 0$, i.e.

$$L_{\theta_j} = -K_{\theta_j}/K_{r_n}, \qquad L_{r_j} = -K_{r_j}/K_{r_n}. \tag{2.5}$$

But notice from (2.1) that (2.2) also implies $\dot\theta_n \neq 0$ hence we can replace the time variable t by θ_n. (2.1) then gives

$$K_{r_j} = \dot\theta_j = \frac{d\theta_j}{d\theta_n}\cdot\dot\theta_n = \frac{d\theta_j}{d\theta_n}\cdot K_{r_n}$$

$$-K_{\theta_j} = \dot r_j = \frac{dr_j}{d\theta_n}\cdot\dot\theta_n = \frac{dr_j}{d\theta_n}\cdot K_{r_n},$$

i.e.

$$\frac{d\theta_j}{d\theta_n} = K_{r_j}/K_{r_n}, \qquad \frac{dr_j}{d\theta_n} = -K_{\theta_j}/K_{r_n},$$

which in combination with (2.5) shows

$$r_j' = L_{\theta_j}, \qquad \theta_j' = -L_{r_j}, \qquad ' = \frac{d}{d\theta_n}. \tag{2.6}$$

In other words, writing $(r_1,\dots,r_{n-1},\theta_1,\dots,\theta_{n-1}) = x$, $\theta_n = t$ and $L = H$, which are not to be confused with the x, t and H of (1.1), the study of the solutions of (1.1) on $H = h$ becomes equivalent to that of

$$\dot x = JH_x(x,t), \qquad x \in R^{2n-2}, \tag{2.7}$$

where H is periodic in t.

We can view (2.7) as a special case of a (not necessarily Hamiltonian) system

$$\dot{x} = f(x,t), \qquad x \in R^m, \tag{2.8}$$

in which f is periodic in t, w.l.o.g. of minimal period 2π, an assumption usually indicated by writing $(x,t) \in R^m \times S^1$. By adjoining $\dot{t} = 1$ we can regard these equations as autonomous, and the attendant flow will then have orbits crossing t = constant planes in $R^m \times S^1$ transversely. This last fact is what we use to reduce the study of (2.8) to that of a mapping. Indeed, if for any $0 \leq t_0 < 2\pi$ we let Σ_{t_0} be the plane $R^m \times \{t_0\}$ in $R^m \times S^1$, then we can define a mapping $\varphi_{t_0}: \Sigma_{t_0} \to \Sigma_{t_0}$ by following $x \in \Sigma_{t_0}$ along the solution curve of (2.8) through x up to the point $\varphi_{t_0}(x)$ where that curve first intersects $\Sigma_{t_0+2\pi}$. By virtue of the periodicity assumption we should regard Σ_{t_0} and $\Sigma_{t_0+2\pi}$ as being the same, and this is why we have $\varphi_{t_0}: \Sigma_{t_0} \to \Sigma_{t_0}$. Notice how φ_{t_0} reflects properties of the flow of (2.8), e.g. fixed points of φ_{t_0} correspond to periodic orbits.

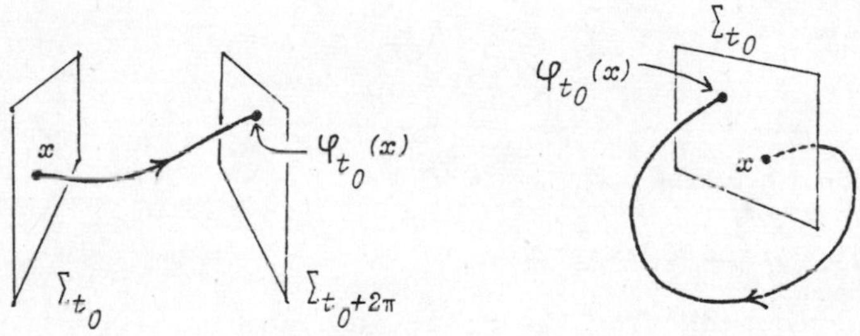

Viewing φ_{t_0} in the covering space $R^m \times R$(left) and in the quotient space $R^m \times S^1$(right).

Figure 2-1

The physics literature is replete with numerical studies of these so-called Poincare, or stroboscopic, mappings $\varphi_{t_0}: \Sigma_{t_0} \to \Sigma_{t_0}$. They suggest the existence of integrals independent of H (in (1.1)) when the iterates of points tend to remain on "nice" curves which stratify Σ_{t_0}, and the nonexistence of such integrals when such iterates seem to appear at random places in this plane.

For the example (1.4) we have K in (2.1) given by $K(\theta,r) = \omega_1 r_1 + \omega_2 r_2$ when $n = 2$, thus L in (2.3) is given by

$r_2 = (h - \omega_1 r_1)/\omega_2$, and so (2.6) becomes

$$r_1' = 0, \qquad \theta_1' = \omega_1/\omega_2,$$

which we regard as time-dependent even though t does not appear. The corresponding flow (in (r_1,θ_1,t)-space) is obviously just $r_1(t) = r_1(0)$, $\theta_1(t) = \theta_1(0) + t\omega_1/\omega_2$, $t = t$, and so $\varphi_{t_0}: (r,\theta) \to (r, +2\pi\omega_1/\omega_2)$. If we regard r and θ as polar coordinates, then φ_{t_0} is seen to be simply a planar rotation by $2\pi\theta_1/\theta_2$ radians (Figure 2-2). The iterates of points thus lie on circles, reflecting the integrability.

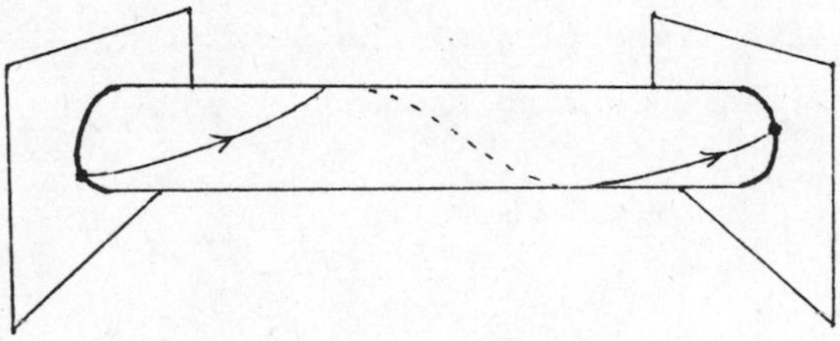

Figure 2-2

§3. Hyperbolic Periodic Orbits.

In the previous section the reduction from (2.1) to (2.7) may only be possible locally, say in a neighborhood of a periodic orbit Π, in which case φ_{t_0} will be defined only in a neighborhood D of the point $x \in R^{2n-2}$ (w.l.o.g. $x = 0$) where this orbit intersects $\Sigma_{t_0} \simeq R^{2n-2}$. More generally, suppose x in (2.8) is only allowed to vary in some open set $D \subset R^m$ containing 0, and that 0 corresponds to a periodic orbit Π of the associated flow. Writing φ for φ_0, we then see that 0 will be a fixed point of φ, and so φ must have the form $x \to Ax +$ higher order terms, where $A = \varphi'(0)$ is the Jacobian matrix of φ at 0. One speaks of Π, and also of 0, as being <u>hyperbolic</u> if none of the eigenvalues of A is on the unit circle of the complex plane.

The importance of hyperbolicity is that the behavior of φ in some neighborhood of 0 is then completely understood. Namely, this neighborhood N will contain two manifolds W^s_{loc} and W^u_{loc}, intersecting only at 0, which respectively contain all points of N approaching 0 under forward and backward iterates of φ. By taking the union of all iterates of these respective manifolds one then obtains the *stable* and *unstable manifolds* W^s and W^u of 0, which respectively contain all points in D which approach 0 under forward and backward iterates of φ. The *stable manifold theorem* asserts that these manifolds are as smooth as φ which in turn is just as smooth as f in (2.8). By following W^s and W^u along the flow, we see that the hyperbolicity of Π implies the existence of two smooth manifolds, called the *stable* and *unstable manifolds* of Π, which respectively contain all orbits positively and negatively asymptotic to Π, and which intersect Σ_0 in W^s and W^u as in Figure 3-1.

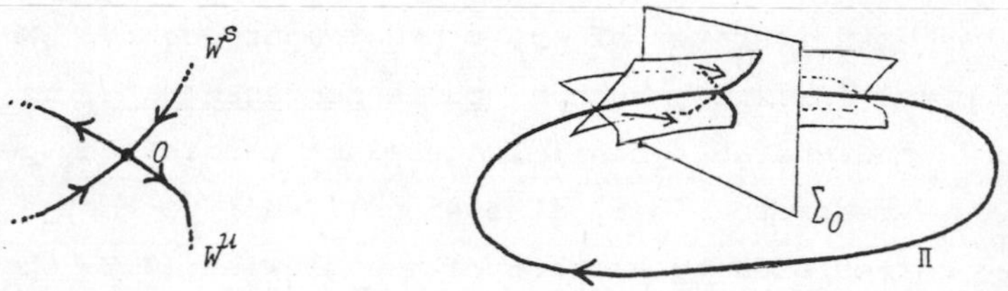

When carried along by the flow, the stable and unstable manifolds of 0 (left) "sweep out" the stable and unstable manifolds of Π (right).

Figure 3-1

The result usually quoted for verifying hyperbolicity is the following (see for example, Proposition 7.1.4, p. 214, of [1]).

3.1 Theorem: *Let $\Pi = \Pi(t)$ be a periodic solution of* (1.1) *of minimal period* 2π *and let* $y(t) = (y_{ij}(t))$ *be the unique matrix solution of the "variational equation"*

$$\dot{y} = JH_{xx}(\Pi(t))y, \qquad y(0) = I_{2n},$$

of (1.1) *along* Π. *Then* 1 *is at least a double eigenvalue of* $y(2\pi)$, *hence the set of eigenvalues of this matrix has the form* $\{1, 1, \lambda_3, \ldots, \lambda_{2n}\}$, *and* $\lambda_3, \ldots, \lambda_{2n}$, *which may also include* 1, *are the*

<u>eigenvalues of $\varphi'(0)$, where $\varphi: D \to D$ is the Poincare mapping associated with</u> Π.

The result, however, assumes $\Pi(t)$ is known in "closed form," which is often not the case and in addition that the variational equation can be solved, which is also often not the case. In fact the experience of this author is that this theorem can rarely be applied in practice, and that other means of proving hyperbolicity must be discovered (e.g. see [13]). On the other hand, for the purposes of numerical work the assertion is quite useful.

§4. <u>Homoclinic Points.</u>

Suppose 0 is a hyperbolic fixed point of $\varphi: D \to D$, where φ can arise from either (2.7) or (2.8) as one of the mappings φ_{t_0}. Then a point $0 \neq p \in W^s \cap W^u$ is called a <u>homoclinic point</u> for φ; under both forward and backward iterates of φ, p must approach 0. When $n = 2$ in (1.1), so $m = 2$ in (2.8), such a p is <u>nondegenerate</u>, or <u>transverse</u>, if these stable and unstable manifolds are not tangent at p (this concept easily extends to higher dimensions). Notice that a homoclinic point for φ corresponds to a orbit of the flow of (2.8) which is doubly asymptotic to the periodic orbit Π associated with the fixed point 0 of φ. One refers to this orbit as being a <u>homoclinic orbit</u> to Π, and as being nondegenerate if the homoclinic point is such.

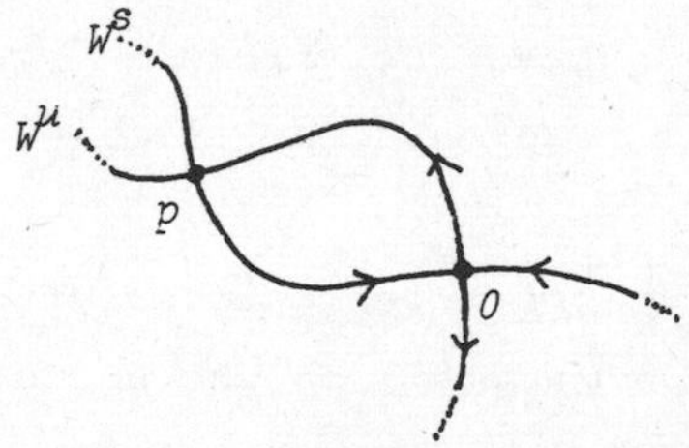

A nondegenerate homoclinic point p for φ.

Figure 4-1

The consequences of the existence of a nondegenerate homoclinic point p for φ can be seen in terms of the following pictures.

Under high forward iterates a small square I^2 abutting on W^s near 0 will intersect itself (at least) twice as shown in Figure 4-2, and we can

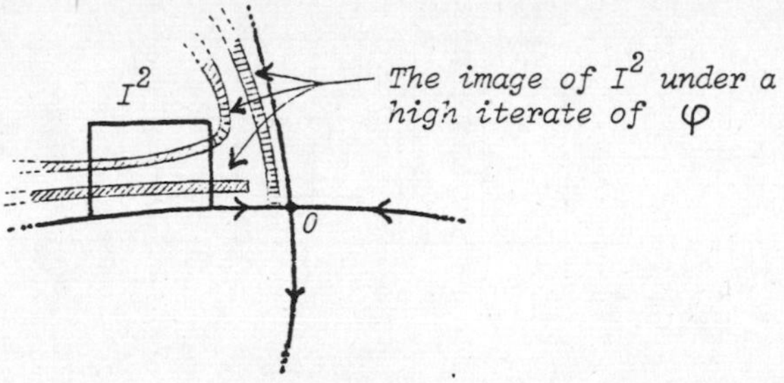

Figure 4-2

"idealize" this picture as in Figure 4-3(a), or, slightly incorrectly, as in Figure 4-3(b), which explains the name "horseshoe mapping."

Now label the horizontal strips in Figure 4-3(a) as H_0 and H_1, and their respective vertical preimages as V_0 and V_1. Then $V_i \cap H_j$ will have a vertical preimage $V_{ij} \subset V_j$ and a horizontal image $H_{ij} \subset H_i$, and we can obviously continue taking further such intersections so as to obtain infinite nested sequences of vertical and horizontal strips

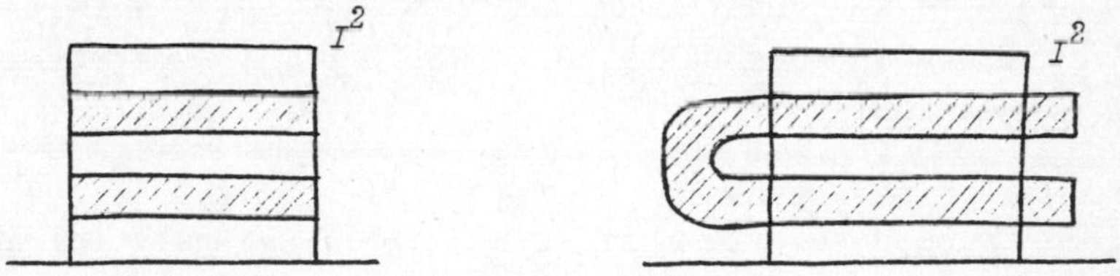

(a) (b)

Figure 4-3

in I^2 as illustrated in Figure 4-4. But then by intersecting each nest of vertical strips and each nest of horizontal strips, we see that

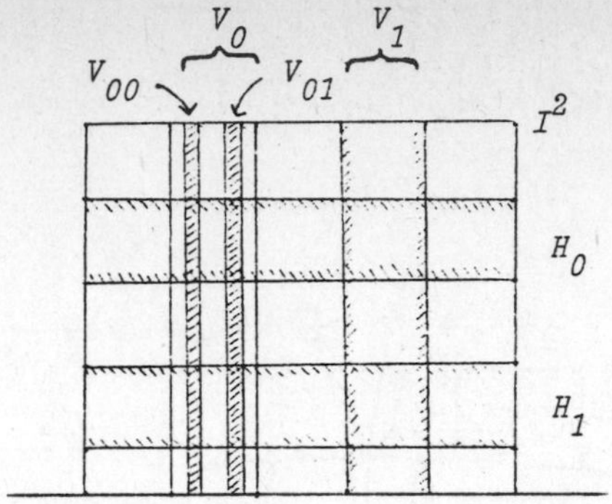

Figure 4-4

with each infinite sequence

$$(4.1) \qquad \ldots \epsilon_{-4}\epsilon_{-3}\epsilon_{-2}\epsilon_{-1}.\epsilon_0\epsilon_1\epsilon_2\epsilon_3\ldots, \qquad \epsilon_j = 0 \text{ or } 1,$$

we can associate a vertical line $V_{\epsilon_{-1}\epsilon_{-2}\ldots}$ as well as a horizontal line $H_{\epsilon_0\epsilon_1\epsilon_2}\ldots$, hence the point of intersection of these two lines. Moreover under our (high iterate) map this point will be carried to the one represented by shifting the decimal point in (4.1) one place to the left. In other words, when restricted to the "Cantor set" C which results when all vertical lines are intersected with all horizontal lines, our mapping can be described as the "shift" on the space Seq of infinite sequences of zeros and ones. One can (and does) regard Seq as a metric space, the distance from $\{\epsilon_j\}$ to $\{\epsilon'_j\}$ being $(1 + |k|)^{-1}$, where $\epsilon_k = \epsilon'_k$ but $\epsilon_j \neq \epsilon'_j$ for $0 \leq |j| < |k|$; the shift is then a homeomorphism.

4.1 Theorem: Suppose $n = 2$, that H in (1.1) is analytic, and that the flow of that equation admits a hyperbolic periodic orbit with a nondegenerate homoclinic orbit. Then there is no analytic integral of that equation with the same domain as H which is independent of H.

Outline of the Proof:

(i) First notice that the shift on Seq has a dense orbit (just string together all possible finite sequences), and so the set of iterates of some $q \in C$ under our high iterate of φ must be dense in C.

(ii) Next observe that any integral $G: R^4 \to R$ of (1.1) would result in one for (2.7), which we again call G, and so $g = G|R^2$ would be such that $g(x) = g(\varphi^k(x))$, $x \in C$, where φ^k is our fixed high iterate. Since C has a dense orbit it would follow that $g \equiv$ constant on C, and since each point of C is a limit point of C this forces $g^{(k)} \equiv 0$ on C, i.e. all derivatives vanish. But G must be constant on orbits of the flow of (2.7), and so partial derivatives of G in a direction transverse to R^2 would also vanish, i.e. all derivatives of G must vanish. Assuming analyticity it follows that $G \equiv$ constant contradicting the requirement that G should be independent of H.

Theorem 4.1 was folklore for many years, but the first complete proof was given by J. Moser ([12], Theorem 3.10, p. 107) using ideas of C. Conley.

A critical point x_0 of (1.1) is called a <u>saddle focus</u> if the eigenvalues of the Hessian matrix $JH_{xx}(x_0)$ have the form $\pm(\alpha \pm i\beta)$, $\alpha,\beta > 0$ (we again assume $n = 2$). A <u>homoclinic point</u> for x_0 is then a point p whose orbit is doubly asymptotic to x_0. Here one also has a notion of nondegeneracy, again defined in terms of a nontangential intersection of stable and unstable manifolds, and R. Devaney has shown (see [6]) that Theorem 4.1 remains true if "hyperbolic periodic orbit" is replaced by "saddle focus." On the other hand, he has also shown (see [7]) that if a critical point is not a saddle focus, but still admits a nondegenerate homoclinic point, then integrability is possible.

§5. Perturbation Techniques.

Suppose H in (1.1) depends on a parameter ε and has the form

$$H_\varepsilon = \hat{H}(x) + \varepsilon\tilde{H}(x,\varepsilon), \tag{5.1}$$

where

(5.2) $$\dot{x} = J\hat{H}_x$$

is completely understood, e.g. integrable. Here one should be able to use (5.2) to reach conclusions about the full system

(5.3) $$\dot{x} = J\hat{H}_x + \varepsilon J\tilde{H}_x,$$

at least for $|\varepsilon|$ small.

In fact near a critical point of H one can achieve the form (5.3) artificially by the technique of "stretching" (or "scaling") variables. Indeed if $H_x(x_0) = 0$, then w.l.o.g. $x_0 = 0$, and so (at least in the analytic case) H has the form $H = H^{(2)} + H^{(3)} + \ldots$, where $H^{(j)}(x)$ is a homogeneous polynomial of degree $j \geq 2$ in the variables $x_1, \ldots, x_{2n}$. The substitution $x = \varepsilon\tilde{x}$ in (1.1) then reduces that equation to (5.3) (after the tildes are dropped) with $H = H^{(2)}$. Notice that solutions of (5.3) on $H_\varepsilon = h$ correspond to solutions of (1.1) on $h = \varepsilon^2 h$ via $\tilde{x} \to \varepsilon^1 x$, i.e. in this case conclusions about (1.1) from (5.3) will only reflect behavior at energies near zero.

If H in (1.1) has the form (5.1), then the function L of (2.3) will have the form $L = \hat{L} + \varepsilon\tilde{L}$, and so here we should replace (2.7) by

(5.4) $$\dot{x} = J\hat{H}_x(x,t) + \varepsilon J\tilde{H}_x(x,t;\varepsilon).$$

However, one can usually do better, and actually achieve the form

(5.5) $$\dot{x} = J\hat{H}_x(x) + \varepsilon J\tilde{H}_x(x,t;\varepsilon).$$

For example, if $n = 2$ and H is given by $H = \Sigma_{j=1}^{2}(\omega_j/2)(u_j^2 + v_j^2) + \varepsilon\tilde{H}$, then from the last paragraph of §2 we see that L in this case will be given by $L = (h - \omega_1/r_1)/\omega_2 + \varepsilon\tilde{L}$ as in (5.5).

Mimicking the approach in §2 we view (5.5) as a special case of the (not necessarily Hamiltonian) system

(5.6) $$\dot{x} = f_0(x) + \varepsilon f_1(x,t;\varepsilon), \qquad x \in R^m,$$

and our Poincare mappings now take the form $\varphi_{t_0,\varepsilon} = \hat{\varphi}_{t_0} + \varepsilon\tilde{\varphi}_{t_0} : \Sigma_{t_0} \to \Sigma_{t_0}$. The idea thus becomes: study $\hat{\varphi}_{t_0}$ thoroughly so as to make conclusions about $\varphi_{t_0,\varepsilon}$, at least for $|\varepsilon|$ small.

Notice that when $\varepsilon = 0$, (5.5) actually defines a flow ρ^t on R^m, whereas we always consider the flow as being on $R^m \times S^1$ since we include $\dot{t} = 1$ with that equation. In fact when $\varepsilon = 0$ this "larger" flow is just (ρ^t, t), and $\hat{\varphi}_{t_0}$ thus amounts to a photograph of ρ^t in R^m after 2π units of time have elapsed.

To state the next result we need the following terminology: If $m = 2$ and the stable and unstable manifolds of hyperbolic fixed point x_0 of $\hat{\varphi}_{t_0}: \Sigma_{t_0} \to \Sigma_{t_0}$ "mesh" as in Figure 5.1, then that component of

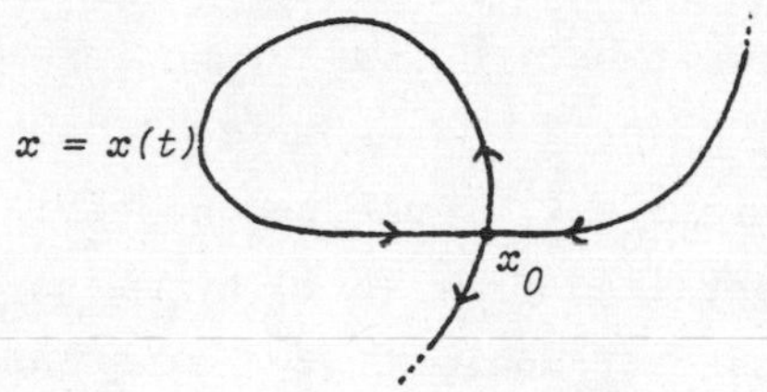

Figure 5-1

these manifolds stretching from x_0 to x_0 as in the figure (there could be two such components) will be called a <u>saddle connection</u>. Notice that a saddle connection $x = x(t)$ is actually an orbit of the flow ρ^t, hence satisfies

$$\dot{x} = f_0(x). \tag{5.7}$$

We also need the following notation: if $x = (x_1, x_2)$ and $y = (y_1, y_2)$ are vectors in R^2, then $x \wedge y = x_1 y_2 - x_2 y_1$ will denote the (signed) area of the parallelogram determined by x and y.

5.1 Theorem (Melnikov): *Assume in (5.4) that $m = 2$, and that $\hat{\varphi}_0$ admits a hyperbolic fixed point x_0 with a saddle connection $x = x(t)$, $-\infty < t < \infty$. For $0 \le t_0 < 2\pi$ define*

$$\Delta(t_0) = \Delta_\varepsilon(t_0) =$$
$$\int_{-\infty}^{\infty} \exp\left\{-\int_0^{s-t_0} \mathrm{tr} Df_0(x(u))\,du\right\} f_0(x(s-t_0)) \wedge f_1(x(s - t_0), s; \varepsilon)\,ds,$$

and suppose for some sufficiently small $\varepsilon > 0$ that $\Delta(t_0)$ has a simple zero at some $t_0 = t_0'$ (i.e. $\Delta(t_0') = 0$ but $\Delta'(t_0') \neq 0$). Then $\varphi_{t_0',\varepsilon}: \Sigma_{t_0'} \to \Sigma_{t_0'}$ has a hyperbolic fixed point with a nondegenerate homoclinic point.

Remarks:

(a) In the case of interest (5.5) the function $\Delta(t_0)$ can be written

$$\Delta(t_0) = \int_{-\infty}^{\infty}\{\frac{d}{ds}\tilde{H}(x(s - t_0),s;\varepsilon) - D_2\tilde{H}(x(s - t_0),s;\varepsilon)\}ds,$$

where in $\frac{d}{ds}\tilde{H}$ we treat $\tilde{H}$ as a function of s, whereas in $D_2\tilde{H}$ we treat H as a (parametrized) function of two variables and differentiate w.r.t. the second variable.

(b) If (5.5) is obtained from (5.3) by means of the reduction discussed in §2, if analyticity holds, and if Theorem 5.1 applied, then by Theorem 4.1 the system (5.3) can have no global analytic integral independent of H.

Outline of the Proof:

(i) First observe that x_0 corresponds to a hyperbolic orbit Π_0 of (5.6), and that the saddle connection corresponds to a "sheet" as in Figure 5-2. Notice that each plane Σ_{t_0} will

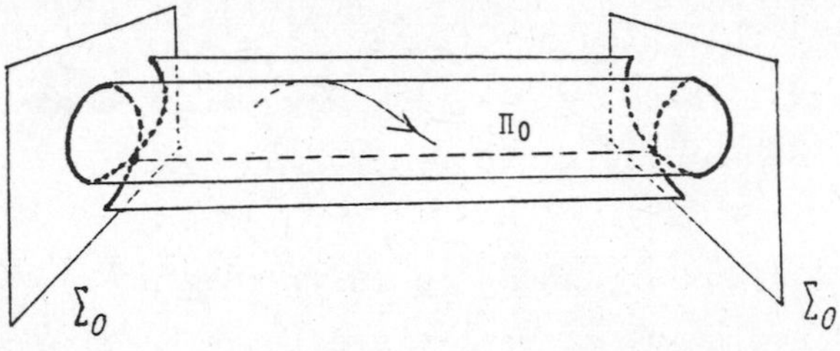

Figure 5-2

will intersect this sheet just as in Figure 5-1. Under pertubation Π_0 is replaced by a new hyperbolic periodic orbit Π_ε (this is an easy consequence of the implicit function theorem), and the sheet might be expected to "snap," now intersecting a typical plane Σ_{t_0} as in Figure 5-3. Indeed, this is precisely what we would wish!

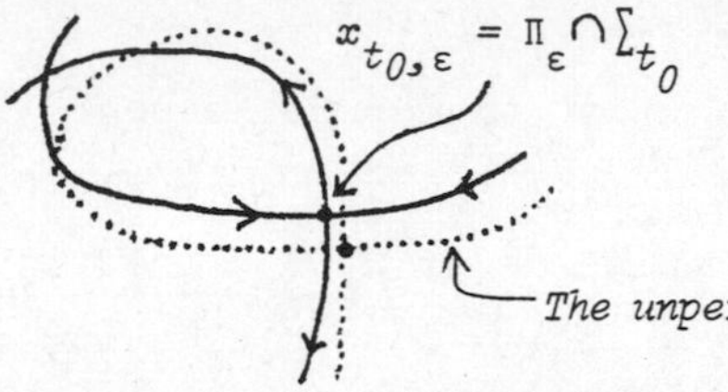

Figure 5-3

(ii) Now construct a short segment S through $x(0)$ in Σ_0 transverse to $x(t)$, let $T = S \times [0,2\Delta]$, and let $S_{t_0} = T \cap \Sigma_{t_0}$ (see Figure 5-4). Under small perturbations the stable

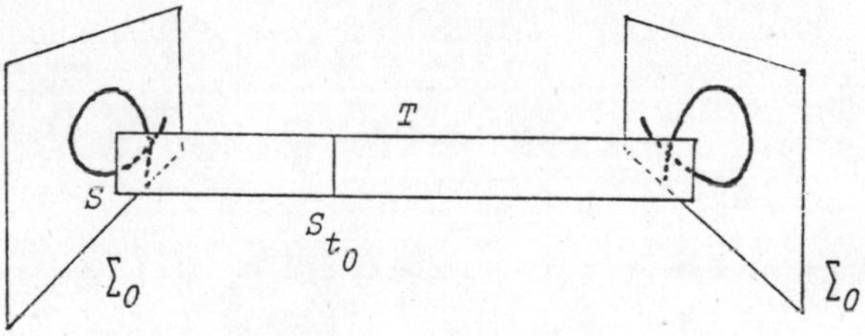

Figure 5-4

and unstable manifolds of $x_{t_0,\varepsilon} = \Pi_\varepsilon \cap \Sigma_{t_0}$ w.r.t. $\varphi_{t_0,\varepsilon}$ will "first" intersect S_{t_0} at points $x^s_{t_0,\varepsilon}$ and $x^u_{t_0,\varepsilon}$ respectively, as shown in Figure 5-5. We define the Melnikov function $m_\varepsilon(t_0)$

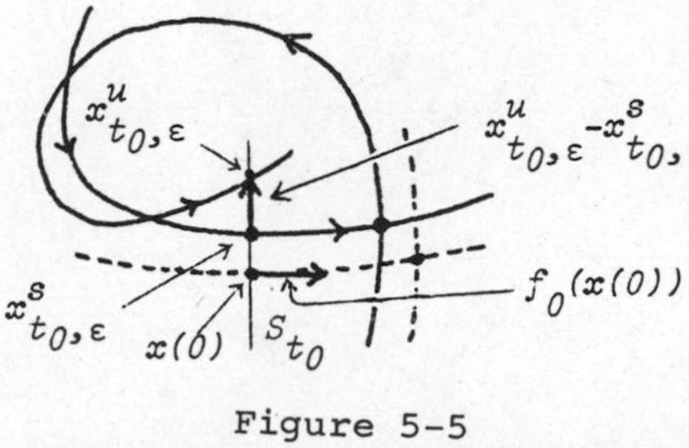

Figure 5-5

as

$$m_\varepsilon(t_0) = (x^u_{t_0,\varepsilon} - x^s_{t_0,\varepsilon}) \wedge f_0(x(0)), \tag{5.8}$$

which we note measures the "distance" from $x^u_{t_0,\varepsilon}$ to $x^s_{t_0,\varepsilon}$ along S_{t_0} (see Figure 5-5). In fact $x^u_{t_0,\varepsilon} - x^s_{t_0,\varepsilon}$ alone would do this but it is not so amenable to computations. Notice that if $m_\varepsilon(t_0) = 0$ for some $t_0' \varepsilon [0,2\pi)$, then the stable and unstable manifolds of $x_{t_0',\varepsilon}$ would intersect on $S_{t_0'}$, and if in addition $m'(t_0') = 0$ then this intersection would be transverse (look at the analogue of Figure 5-2 for the perturbed system). By proving

$$m_\varepsilon(t_0) = \varepsilon\Delta_\varepsilon(t_0) + o(\varepsilon^2), \tag{5.9}$$

we can then establish Theorem 5.1.

(iii) The trick in establishing (5.9) is to realize that the stable and unstable manifolds $x^s_{t_0,\varepsilon}$ and $x^u_{t_0,\varepsilon}$ of $x_{t_0,\varepsilon}$ can be parametrized as

$$x^\lambda_{t_0,\varepsilon}(t) = x(t - t_0) + \varepsilon x^\lambda_1(t,t_0) + o(\varepsilon^2), \qquad \lambda = s,u,$$

where

(a) x^λ_1 satisfies the variational equation of (5.6) along $x = x(t-t_0)$, and

(b) $x^\lambda_{t_0,\varepsilon}(t_0) = x_{t_0,\varepsilon}$.

Using (b) we can extend $m_\varepsilon(t_0)$ to a function of two variables by setting

$$m_\varepsilon(t,t_0) = (x^u_{t_0,\varepsilon}(t) - x^s_{t_0,\varepsilon}(t)) \wedge f_0(x(t - t_0)),$$

and using (a) and the fact that $x(t-t_0)$ satisfies (5.7) one can then easily derive linear ordinary differential equations for $m^u(t,t_0) = x^u \wedge f_0$ and $m^s(t,t_0) = x^s \wedge f_0$ (in the variable t) which when solved give (5.9). (The divergence term $\mathrm{trD}f_0$ in the statement of Melnikov's theorem is a result of the simple identity $(Mx)\wedge y + x\wedge(My) = (\mathrm{tr}M)x\wedge y$ for any 2×2 matrix M.)

Concrete applications of Theorem 5.1, as well as more detailed proofs and further references, are found in [10], [8] and [11]. The emphasis in these references, however, is not on proving nonintegrability, and they begin with (5.6). Nevertheless, the fact that this result could be used to establish nonintegrability should be clear.

§6. Nonperturbation Techniques

Reference [3] presents geometrical criteria for proving the nonintegrability of Hamiltonian systems governed by Hamiltonians of the form

(6.1) $$H(x\ y) = \frac{1}{2}|y|^2 + V(x), \qquad x,y \in R^2,$$

with $V\colon R^2 \to R$ analytic. Here we illustrate the method in the case

(6.2) $$V(x) = \frac{1}{2}|x|^2 - \frac{1}{2}x_1^2x_2^2.$$

The system associated with (6.1) is equivalent to $\ddot{x} = -V_x$, which we (not quite correctly) regard as the equations governing the motion of a bead rolling in the "potential well" described by the graph of V. For $h > \frac{1}{2}$ the level curve $V = h$ appears as in Figure 6-1, and there are four hyperbolic periodic orbits within $H = h$ [13] with projections Π_j as indicated in that illustration. For simplicity we will also refer to the periodic orbits as Π_j, $j = 1,\ldots,4$.

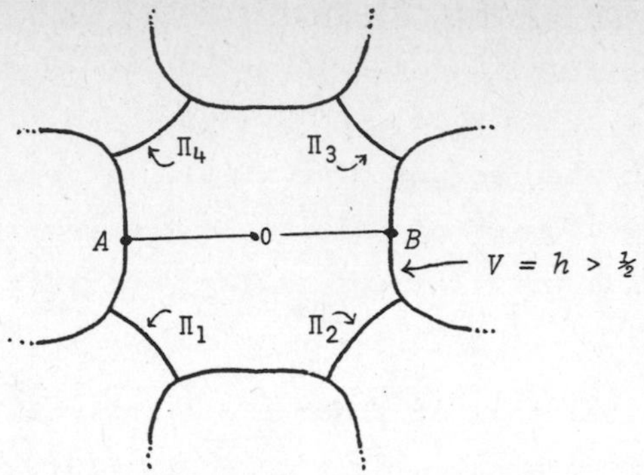

Figure 6-1

Now consider the subset of H = h which projects to the segment AB of the x_1-axis shown in Figure 6-1. From (6.1) we see that for each x on that segment there is a circle $|y|^2 = 2(V(x) - h)$ of choices for y, although at the endpoints A and B, where V(x) = h, the two circles collapse to points (all the energy is potential energy). In other words, the subset in question is a sphere, whose equator we may consider to be the periodic orbit which projects to AB, and whose northern hemisphere, a disc D, we may consider as representing solutions crossing AB in a downward direction (i.e. $y_2 < 0$).

Observe that the unstable manifold of Π_4 will intersect D in (at least) a circle, as will the stable manifold of Π_1. Moreover, if there is an orbit segment with projection connecting Π_4 to Π_1, then these two circles will intersect, resulting in an orbit γ_{41} connecting Π_4 with Π_1, a so-called <u>heteroclinic orbit from</u> Π_4 <u>to</u> Π_1. Indeed, such and orbit segment would intersect D in a point interior to both circles, and using area-preserving properties of Hamiltonian flows one can easily show that neither circle can be interior to the other; hence the intersection. (If the above orbit segment intersects D more than once this argument requires modification, but can still be carried through.) In [2] it is shown that the required orbit segment exists (at least at some energies $h > \frac{1}{2}$), and as a result the heteroclinic orbit γ_{41} exists.

Of course it is conceivable that the two circles coincide, resulting in a continuum of heteroclinic orbits from Π_4 to Π_1. This, however, is excluded by symmetry, which shows that there must also be a heteroclinic orbit from Π_4 to Π_3. On the other hand, symmetry also shows that there is a heteroclinic orbit γ_{14} from Π_1 to Π_4.

Now construct a small tube stretching from Π_1 to Π_4 in $H = h$ along γ_{14}, and then returning along γ_{41}. By carefully searching this tube one can locate a homoclinic orbit to Π_1 which, because of analyticity and the fact that the two circles do not coincide, will be nondegenerate [4],[3]. By Theorem 4.1 the system associated with (6.1) and (6.2) can therefore have no global analytic integral which is independent of H.

Actually, in this instance one can use an argument of R. Cushman to bypass the horseshoe mapping in the proof on nonintegrability [4]. Reduction to some sort of mapping, however, is still required.

References

[1] R. Abraham and J. Marsden, Foundations of Mechanics, 2nd ed., Benjamin/Cummings, Reading, Mass., 1978

[2] R.C. Churchill G. Pecelli, S. Sacolick and D.L. Rod, Coexistence of Stable and Random Motion, Rocky Mr. J. of Math. 7 (1977), 445-456.

[3] R.C. Churchill and D.L. Rod, Pathology in Dynamical Systems III: Analytic Hamiltonians, to appear in J. Differential Equations.

[4] C. Conley, Twist Mappings, Linking, Analyticity and Periodic Orbits which Pass Close to an Unstable Periodic Solution, in "Topological Dynamics," (J. Auslander, Ed.), W.A. Benjamin, New York 1968.

[5] R. Chusman, Examples of Nonintegrable Analytic Hamiltonian Vector Fields with no Small Divisors, Trans. Am. Math. Soc. 238 (1978), 45-55.

[6] R. Devaney, Homoclinic Orbits in Hamiltonian Systems, J. Differential Equations 21 (1976) 431-438.

[7] R. Devaney, Transversal Homoclinic Orbits in an Integrable System, Am. J. Math. 100 (1978), 631-642.

[8] P.J. Holmes, Averaging and Chaotic Motions in Forced Oscillations, to appear in SIAM J. Appl. Math.

[9] L. Markus and K. Meyer, Generic Hamiltonian Systems are neither Integrable nor Ergodic, Memoirs Am. Math. Soc. 144, 1974.

[10] V.K. Melnikov, On the Stability of the Center for Time Periodic Perturbations, Trans. Moscow Math. Soc. 12 (1) (1963), 1-57.

[11] J. Marsden, Geometric Methods in Mathematical Physics, to appear.

[12] J. Moser, Stable and Random Motions in Dynamical Systems, Annals of Mathematics Studies 77, Princeton University Press, 1973.

[13] D.L. Rod, G. Pecelli and R.C. Churchill, Hyperbolic Periodic Orbits, J. Differential Equations 24 (1977), 329-348, and 28 (1978), 163-165.

[14] C.L. Siegel and J.K. Moser, Lectures on Celestial Mechanics, Springer, Berlin, 1971.

Hunter College
New York, NY 10021
USA

Classical Solutions in Nonlinear Euclidean Field Theory and Complete Integrability

by

Mel S. Berger

In this note I shall present my approach to complete integrability for nonlinear elliptic partial differential equations. I then relate this approach to the classical solutions of nonlinear Euclidean field theory such as occur in contemporary Yang-Mills theories.

§1. Complete Integrability in Classical Mechanics.

For classical Hamiltonian systems of dimension 2N

$$\text{(1)} \qquad \dot{\beta}_i = \frac{\partial H}{\partial q_i}, \qquad \dot{q}_i = -\frac{\partial H}{\partial \beta_i} \quad (i = 1,\ldots,N)$$

complete integrability implies integrability of the system (1) by quadrature after appropriate coordinate transformations. The appropriate coordinate transformations here

$$\begin{cases} P_i = P_i(p_i, q_i) \\ Q_i = Q_i(p_i, q_i) \end{cases}$$

are called "canonical transformations" and have as invariants the Poisson bracket

$$[f,g] = \Sigma\left(\frac{\partial f}{\partial p_i}\frac{\partial g}{\partial q_i} - \frac{\partial f}{\partial q_i}\frac{\partial g}{\partial p_i}\right)$$

of any two functions.

The criteria for complete integrability in classical mechanics is generally formulated in terms of separable systems relative to the Hamilton-Jacobi equations or Liouville's theorem. Both situations require the existence of N independent first integrals of the motions defined by (1) $f_1,\ldots,f_N$ whose Poisson brackets vanish.

Assuming the set $M = \{(p,q) \mid F_i = c_i,\ i = 1,\ldots,N\}$ is compact, angle action variables (F_i, φ_i) can be introduced so that the system (1) can be reduced to

$$(2) \qquad \frac{dF_i}{dt} = 0, \qquad \frac{d\varphi_i}{dt} = c_i(F)$$

on M. This system's i^{th} term is basically a linear one. This approach can be extended to infinite dimensional Hamiltonian systems and partial differential equations. The celebrated Korteweg de Vries equation .

$$u_t + uu_x + u_{xxx} = 0$$

is completely integrable in this extended sense.

Two difficulties arise in this connection however. First, higher dimensional examples of nonlinear partial differential equations, integrable by this means have been very difficult to find and secondly stability arguments associated with Liouville's motion of complete integrability require a radically new conceptual framework (ex. KAM theory). Moreover, it is believed by many that the only integrable systems are those that can canonically be transformed to linear ones.

Furthermore infinite dimensional systems, (integrable in the Liouville sense) require an infinite number of independent conservation laws. Higher dimensional systems of partial differential equations rarely possess this property. In the next section we shall describe a different approach to complete integrability, that is strong enough to overcome these difficulties.

§2. Key Ideas on Complete Integrability--Euclidean Field Theory.

The classical ideas just mentioned need a complete rethinking for nonlinear elliptic partial differential equations and consequently for classical solutions of Euclidean field theory, since the Hamiltonian formulation isolates the time variable and the necessary first integrals of motion are simply not available.

We now list the desiratum for each new nonlinear theory:

(1) complete integrability should be formulated in terms of coordinate transformations and their invariants

(2) such a theory should be capable of describing "nonperturbative effects" and "bifurcation phenomena"

(3) complete integrability should be capable of discussing nonlinear problems whose linearization is nonself-adjoint (e.g.

instantons in Yang-Mills theory) and nonlinear systems independent of dimension

(4) inclusion of effects of external sources or fields (e.g. Jackiw and Adler's study of monopoles) [1], [9]

(5) involve simple examples of a "truly" nonlinear effect (and not be reducible to a linear system)

(6) inclusion of stability arguments in an intrinsic, systematic fashion.

§3. The Notion of Complete Integrability.

We are given a nonlinear operator A (supplemented by boundary conditions) defined by partial differential equations of elliptic type. We suppose A is a mapping between two Banach spaces $A: H \to Y$. Then we say A is completely integrable (of class k) if there are C^k diffeomorphisms h_1 h_2 (i.e. C^k changes of variables $h_1: H \to H$ and $h_2: Y \to Y$) such that

$$h_1^{-1} A h_2 = D$$

where D is a diagonal map.

Symbolically

$$\begin{array}{ccc} H & \xrightarrow{A} & Y \\ {\scriptstyle h_1}\downarrow & & \downarrow{\scriptstyle h_2} \\ H & \xrightarrow{D} & Y \end{array} .$$

In other words, this means that A can be diagonalized (after coordinate transformations). So A is completely integrable if it is (C^k) conjugate to a diagonal map.

Standard result from linear algebra and functional analysis yields the following facts:

(A) Every $N \times N$ matrix A of rank k is (linearly conjugate) to a diagonal map D whose diagonal consists of k ones and $(N - k)$ zeros.

(B) Every Fredholm operator of index zero $A: H \to Y$ is completely integrable since it is conjugate to a linear diagonal map.

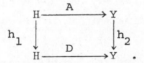

This notion of complete integrability has as <u>invariants</u> the "<u>singular points</u>" and "<u>singular values</u>" of the operator A and this involves "bifurcation phenomena" in an intrinsic way. Recall the singular points S(A) of an operator A, the points $\{x \mid x \in X, A'(x)$ is not surfective linear operator$\}$, while the singular values of A are the images of S(A) under A. This notion is fundamental for complete integrability in our sense.

<u>Remark</u>: We have chosen the term "complete integrability" in this situation purposefully. Indeed, as in the classical setting, the solutions of a completely integrable problem (in our sense) are obtained by "quadrature" after canonical coordinative transformation. This quadrature of course involves the diagonal mapping D and the solutions of the analogous diagonalized system

$$Dv = g, \qquad g \in Y \text{ fixed.}$$

Pictorially we have

$$\begin{array}{ccc} S(A) & \xrightarrow{\quad A \quad} & A[S(A)] \\ h_1 \downarrow & & \downarrow h_2 \\ S(D) & \xrightarrow{\quad D \quad} & D[S(D)]. \end{array}$$

This means that the singular points and singular values of A and D differ by changes of coordinates. This is the desired clue to construct the changes of coordinates h_1 and h_2.

The simplest cases occur for proper operators A with $S(A) = \emptyset$, i.e. the equation

$$(4) \qquad Au = f$$

is well posed in the sense that (4) has exactly one solution that depends continuously on the "external source" f. In this case, A is conjugate to the identity mapping.

However, in general the solutions of (4) are finite in number, nonunique and vary with the "size" of f. We may discuss the simplest nonlinear example of this situation.

§4. A Completely Integrable Nonlinear Elliptic Problem (in Arbitrary Dimension N).

The Elliptic Log Cosh Gordon Equation

Let Ω be a bounded domain in $\mathbb{R}^N$ (N arbitrary) with boundary $\partial\Omega$.

Let $Au = \Delta u + \alpha u + \beta \log \cosh u$ be supplemented by the Dirichlet boundary condition $u|_{2\Omega} = 0$. Here the constants α, β are positive and restricted by the lowest two eigenvalues λ_1, and λ_2 of the Laplace operator Δ

$$0 < \alpha - \beta < \lambda_1 < \alpha + \beta < \lambda_2.$$

We then consider the nonlinear Dirichlet problem

$$(5) \qquad \begin{cases} Au = f, & f \in L_2(\Omega) \\ u|_{\partial\Omega} = 0 \end{cases}$$

Theorem: [6] *The equation* (5) *has either* 0, 1, *or* 2 *solutions depending on whether the size of the projection of* f *on* $\mathrm{Ker}(\Delta + \lambda_1)$ *is less than, equal to, or greater than a certain computable critical number.*

To understand this result it is useful to investigate its complete integrability and, in fact we can prove:

Theorem 1: [8] *The operator* A *is completely integrable as a mapping between the Holder space* $C^{2,\alpha}(\bar{\Omega})$ *and* $C^{0,\alpha}(\bar{\Omega})$ $(0 < \alpha < 1)$. *In fact,* A *is* C^2 *conjugate to the diagonal mapping*

$$(6) \qquad (t,w) \longrightarrow (t^2, w)$$

where a general element u *of* $C^{2,\alpha}(\bar{\Omega})$ *is written* $u = tu_1 + w$ *with* u_1 *a simple positive eigenvector of* Δ *associated with* λ_1.

Theorem 2: [8] *The global normal form* (6) *is "stable" for* A *in the sense that if the log cosh Gordon operator*

$$Au = \Delta u + \alpha u + \beta \log \cosh u, \qquad u|_{\partial\Omega} = 0$$

is perturbed to $\widetilde{A}(u) = \Delta u + f(u), u|_{\partial\Omega} = 0$ *with*

$$\|f(u) - (\alpha u + \beta \log \cosh u)\|_{C^2(\mathbb{R}^1)} \leq \varepsilon, \quad \text{with } \varepsilon > 0$$

sufficiently small, then $\widetilde{A}u$ *is also* C^2 *conjugate to* (6).

The proof of these results can be divided into two parts. First, an analytic part in which one first finds explicit cartesian coordinates for the singular points and singular values of A in terms of the eigenfunctions of the Laplacean as basic and prove a priori estimates that enable the local normal forms to hold outside of small neighborhoods of the singular points. Secondly, the geometric part, in which one constructs smooth diffeomorphic mappings of the set $S(A)$ of singular points of A onto the set $S(D)$ of singular points of the diagonal map D (and similarly for the singular values of A and D).

The construction turns out not to depend on the strict analytic expression for the elliptic log cosh Gordon operator, but rather on qualitative properties of the function $f(u) = \alpha u + \beta \log \cosh u$ namely, its linear growth convexity and limiting behavior of $f'(u)$ as $u \to \pm\infty$. This fact is crucial for the validity of Theorem 2. Indeed one need only retrace the steps given below to deduce the stability result of Theorem 2.

Additional remark. If we suppose $f \in C^k$ $(k \geq 2)$ we may show that A is $C^{(k-2)}$ equivalent to B provided we work with Banach spaces of Holder continuous functions.

To this end we first use our results coupled with Nirenberg [3].

§5. *Idea of the Proof of Theorem 1*.

The proof divides into 2 distinct parts:

Part I -- An *analytical part* consisting of 4 steps:

Step 1: Reduction to a finite dimensional problem;

Step 2: Explicit cartesian representation for the singular points of A;

Step 3: Explicit cartesian representation for the singular values of A;

Step 4: Coerciveness estimates for the mapping A.

We sketch the main ideas of this part.

We write the mapping $A: H \to H$ in the form associated with the orthogonal decomposition $H = \text{Ker}(\Delta + \lambda_1) \oplus H_1$, i.e. we write an element $u \in H$ is the form $u = tu_1 + \omega$ (where u_1 is a normalized eigenfunction of Δ on Ω associated with λ_1) and so $u_1 > 0$ in Ω, with $\omega \in H_1$. Then we show for fixed t that the mapping A_1 defined by

(7) $$(A_1(t,\omega),\varphi) = \int_\Omega \{\nabla\omega\cdot\nabla\varphi - f(tu_1 + \omega)\varphi\}$$

(for $\varphi \in H_1$) is a global homeomorphism of H_1 into itself. This is achieved by using the Lax-Milgram theorem to prove that the inequality

$$(A_1'(t,\omega)\varphi,\varphi) \geq \frac{\epsilon}{\lambda_2} \|\varphi\|_H^2$$

implies $\|[A_1'(t,\omega)]^{-1}\| \leq \frac{\lambda_2}{\epsilon}$ for fixed $\epsilon > 0$. The global result follows now from Hadamard's theorem [1]. Then we find that the singular points and values can be determined by the coordinate representation

(8) $$A(tu_1 + \omega) = h(t)u_1 + g_1.$$

Or more explicitly set $u(t) = tu_1 + \omega(t)$

(9) $$\Delta u(t) + f(u(t)) = h(t)u_1 + g_1$$

Let's examine what happens at a singular value, so that $h'(t) = 0$.

Lemma: At a singular value with $t = t_0$

(10) $$h''(t_0) = \int_\Omega f''(u(t))[u'(t_0)]^3$$

so that by our assumptions $h''(t_0) > 0$.

Sketch of Proof: Consider (9) and differentiate twice with respect to t assuming $h'(t_0) = 0$

(11) $$\Delta u'(t) + f'(u(t))u'(t) = h'(t)u_1.$$

Since $h'(t_0) = 0$, $u'(t_0)$ is a nontrivial solution of (10) and by the asymptotic conditions (3) we may suppose $u'(t_0) > 0$ in Ω (see [2]).

(12) $$\Delta u''(t) + f'(u(t))u''(t) + f''(u(t))[u'(t)]^2 = h''(t_0)u_1.$$

Since $u''(t_0)$ is a nontrivial solution of (11) for this inhomogeneous equation

(13)
$$\{f''(u(t_0))[u'(t_0)]^2 - h''(t_0)u_1\} \perp \mathrm{Ker}[\Delta + f'(u(t_0)]$$
$$\text{i.e., } \int_\Omega [f''(u(t_0)[u'(t_0)]^2 - h''(t_0)u_1]u'(t_0) = 0.$$

This relation shows (10). This result and the convexity of $f(u)$ yield the lemma.

Another important fact is that $h(t) \to \infty$ as $t \to \infty$. This fact follows from the representation

$$h(t) = -\lambda_1 t + \underset{\Omega}{} f(tu_1 + \omega(t))u_1$$

the asymptotic relation (3) and the fact that as $t \to \infty$ the contribution due to $\omega(t)$ is negligible via the a priori estimate

$$\|\omega'(t,g_1)\|_H \leq c \quad \text{(independent of } t \text{ and } g_1).$$

The following picture illustrates the behaviour of the function $h(t)$.

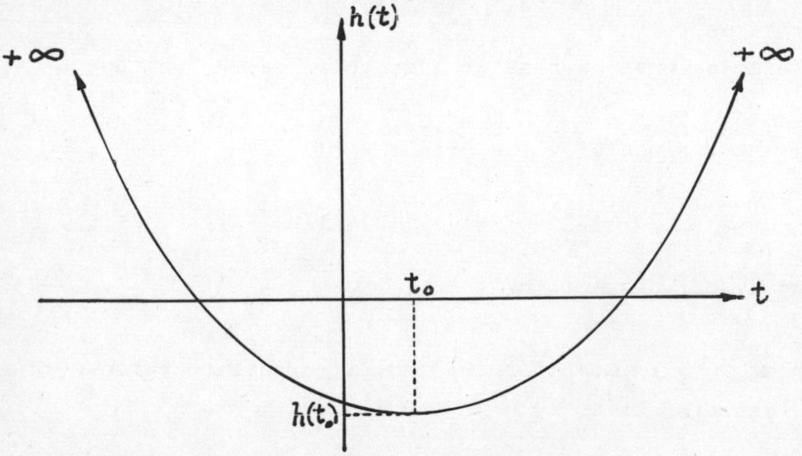

From this picture we read off the cartesian representation of the singular points and singular values of A.

Part II -- The second part of the proof is geometric namely construction of the diffeomorphisms α and β using the fact of Part I.

This part consists of 4 steps also.

Step 1: Layering of the mapping A in accord with Step 1 of Part I by a diffeomorphism α_1.

Step 2: "Translation" of the Singular Points of the Mapping A to those of B by a diffeomorphism α_2;

Step 3: Translation of the Singular Value of A to those of B by a diffeomorphism α_3;

Step 4: The final homeomorphism.

Indeed after Step 3 we find

$$\alpha_3 A \alpha_1 \alpha_2 = (\alpha(t,\omega),\omega). \tag{14}$$

Using Step 4 of Part I we represent the right hand side of (14) by the composition $B\varphi$ where φ is a diffeomorphism $H \to H$.

Thus

$$\alpha_3 A \alpha_1 \alpha_2 = B\varphi \tag{15}$$

which is the desired equation.

§6. Relation with Instantons for Yang-Mills Theory.

By definition, instantons are absolute minima of the "classical" Yang-Mills action functional with given topological charge N. It turns out that as consequence of work of Polyakov, Schwartz, Atiyah, Singer, Hitchin, Drinfeld and Manin [3] [4], [5] that for a given Lie group G (= SU(2) say) that these solutions of the associated nonlinear elliptic system of partial differential equations can be determined by explicit formulae. This system (the so-called self-dual equations) can be written $A(a) = 0$ where A is a nonlinear Fredholm operator of index $8N - 3$ between appropriately defined Sobolev spaces. The usual property of the operator A that is crucial for the explicit formulae calculations is the fact that A has no singular points!

Thus, there arises that important question as to the sense in which the Yang-Mills equations are completely integrable? Although it has been discussed at great length, it has become clear that the self-dual equations do not possess infinitely many conservation laws and thus, cannot be integrable in the sense of Liouville. However, with appropriate guage fixing, it is clear that A is conjugate (globally) to a linear surjection. Thus our definition of complete integrability is relevant in this four dimensional context.

§7. SU(2) Gauge Theory for Monopoles.

A virtue of our method of studying complete integrability for an operator $A: X \to Y$ is its relevance to an equation $Ax = y$ as y varies over Y. Physically the element $y \in Y$ represents an external field. The number of solutions of $Ax = y$ generally changes as y changes (i.e. as the external field is altered) since this change is invariant under our notion of conjugacy. These changes can be read off from the diagonal form of A if A is completely integrable.

These ideas are very significant in the current theory of S. Adler [1] [9] and R. Jackiw on the effect of external field on SU(2) monopoles. Although it is too early at present to state definite mathematical conclusions, S. Adler finds interesting nonperturbative effects in trying to extend the radially symmetric Prasad-Sommerfeld monopole solution to an axially symmetric family of solutions depending on a parameter k. This parameter k is a measure ofthe size of the external source for the problem and for $k = 0$, the solution coincides with the explicit solution of Prasad-Sommerfeld. This process, I call "nonlinear desingularization" and I have worked out the details in a number of classical problems of mathematical physics. See M.S. Berger and L.E. Fraenkel, On Nonlinear Desingularization, Bull. Amer. Math. Soc. Vol. 2, p. 165-167 (1980).

References

[1] S. Adler, Global structure of Static SU(2) Solutions. Physical Review D 19 (1979).

[2] A. Ambrosetti and G. Prodi, Annali di Math. vol. 93 (1972), pp. 231-246.

[3] M.F. Atiyah, V.G. Drinfeld, N.J. Hitchin and Yu.I. Manin, Construction of Instantons, Phys. Lett. A 65 (1978), pp. 185-187.

[4] M.F. Atiyah, N.J. Hitchin, I.M. Singer, Self-duality in Four Dimensional Riemannian Geometry, Proc. Roy. Soc. London (to appear).

[5] A.A. Belavin, A.M. Polyakov, A.S. Schwartz and Yu.S. Tyupkin, Pseudoparticle Solutions of the Yang-Mills Equations, Phys. Lett., B 59 (1975), pp. 85-87.

[6] M. Berger and E. Podolak, On the Solutions of a Nonlinear Dirichlet Problem, Ind. J. of Math., vol 24 (1975), pp. 827-846.

[7] M. Berger, Nonlinearity and Functional Analysis, Academic Press, 1977.

[8] M. Berger and P. Church, Complete Integrability and Stability of a Nonlinear Dirichlet Problem, (Indiana J. of Math., to appear).

[9] R. Jackiw, L. Jacobs and C. Rebbi, Static-Yang Mills Fields with Sources. Physical Review.

[10] Nirenberg, Nonlinear Functional Analysis, NYU Lecture Notes (1974).

University of Massachusetts
Amherst, MA 01003
USA

Hamiltonian Structure of Isospectral Deformation Equations. Elliptic Curve Case

by

D.V. Chudnovsky

G.V. Chudnovsky

Abstract

We continue the investigation of semiclassical limits of factorized S-matrices and Hamiltonian structure they induce on two dimensional isospectral deformation equations. Before [7] we dealt with rational Riemann surfaces. Here we propose a new class of semiclassical factorized S-matrices associated with an arbitrary elliptic curve and torsion subgroup of it. New two-dimensional field theories associated with them generalize both sin-Gordon and Baxter's systems.

In the previous paper [7] we considered the relationship between semiclassical approximations to the solutions of unitarity and factorization equation for S-matrices and local Hamiltonian structure on isospectral deformation equations. In [7] we have seen that the simplest semiclassical factorized s-matrix

$$r(\lambda,\mu)_{ij,k\ell} = (\delta_{ik}\delta_{j\ell} - \delta_{i\ell}\delta_{jk}) \frac{1}{\lambda - \mu}$$

gives rise to a local Hamiltonian structure for a spectral problem

$$\frac{d}{dx}\psi(x;\lambda) = U(x;\lambda)\psi(x;\lambda), \qquad (0.1)$$

with an arbitrary potential $U(x,\lambda)$ being a rational function of λ on $\mathbb{CP}^1$. However, we claim that there is a large class of completely integrable isospectral deformation equations that arise from a linear problem of the type (0.1) not on $\mathbb{CP}^1$ and but on a Riemann surface Γ. These new equations have different Hamiltonian structure which is related to a semiclassical factorized s-matrix being a rational function on Γ. Up to now there are only two examples of such completely integrable systems. Both are connected with an elliptic curve E. The first of them is the discrete, Baxter 8-vertex model [1] and another its continuous counterpart considered by Sklyanin [9] under the name of Landau-Lifchitz model. In both cases the Hamiltonian structure is induced by the S-matrix [1],[2], which is the famous Baxter factorized S-matrix corresponding to an elliptic curve E and 2-torsion subgroup of it. In [4] we presented a quantum S-matrix corresponding to an arbitrary torsion subgroup of an Abelian variety. We apply the construction of [4] to an elliptic curve E and obtain a semiclassical factorized s-matrix corresponding to an arbitrary n-torsion subgroup of E. In this note we present the Hamiltonian structure corresponding to this semiclassical s-matrices and new two dimensional completely integrable systems associated with them.

1. First of all we recall the notion of factorized S and s-matrices [2],[3],[5],[6]. According to its geometric sense quantum S-matrix $R(\theta_1,\theta_2)$ defines an isomorphism between tensor products of vector spaces of dimensions n

(1.1) $$R(\theta_1,\theta_2) : V(\theta_1) \otimes V(\theta_2) \to V(\theta_2) \otimes V(\theta_1),$$

where $\{V(\theta)\}$ form a category C of vector spaces. Practically speaking, $R(\theta_1,\theta_2)$ is an $n^2 \times n^2$ matrix written in some basis $A_1(\theta_1),\ldots,A_n(\theta)$ of $V(\theta)$. The unitarity and factorization equations are the condition for the category C with operation (1.1) to be monoidal. The unitarity equations are

(1.2) $$R(\theta_1,\theta_2).R(\theta_2,\theta_1) = \mathbb{I}_{n^2}$$

for identity $n^2 \times n^2$ matrix $\mathbb{I}_{n^2}$. The factorization equations are simply the consistency condition between two isomorphisms $V(\theta_1) \otimes V(\theta_2) \otimes V(\theta_3) \to V(\theta_3) \otimes V(\theta_2) \otimes V(\theta_1)$ that one obtains by three successful applications of the isomorphism $R(\theta_i,\theta_j) : V(\theta_i) \otimes V(\theta_j) \to V(\theta_j) \otimes V(\theta_i)$. The factorization equations are

$$(1.3) \qquad (R(\theta_2,\theta_3) \otimes \mathbb{I})(\mathbb{I} \otimes R(\theta_1,\theta_3))(R(\theta_1,\theta_2) \otimes \mathbb{I}$$

$$= (\mathbb{I} \otimes R(\theta_1,\theta_2))(R(\theta_1,\theta_3) \otimes \mathbb{I})(\mathbb{I} \otimes R(\theta_2,\theta_3)),$$

where $\mathbb{I}$ is an $n \times n$ identity matrix.

Semiclassical s-matrix is the approximation to the quantum S-matrix up to the first order. Namely, if $R(\theta_1,\theta_2)$ depends on a parameter $\hbar$, which is usually denoted by η, then we can assume $R(\theta_1,\theta_2)$ turns into the identity S-matrix $\mathbb{I}^\tau$:

$$(1.4) \qquad R(\theta_1,\theta_2) = \mathbb{I}^\tau + \hbar\mathbb{I}^\tau r(\theta_1,\theta_2) + O(\hbar^2)$$

as $\hbar \to 0$. Here the identity S-matrix $\mathbb{I}^\tau$ is defined as

$$(1.5) \qquad \mathbb{I}^\tau_{ij,k\ell} = \delta_{i\ell}\delta_{jk}.$$

The $n^2 \times n^2$ s-matrix $R(\theta_1,\theta_2)$ is called a factorized s-matrix if it satisfies the following equations that are approximations of (1.2), (1.3) according to the approximation (1.4):

$$(1.6) \qquad \mathbb{I}^\tau_r(\theta_1,\theta_2) + r(\theta_2,\theta_1)\mathbb{I}^\tau = 0,$$

$$(1.7) \quad (\mathbb{I}^\tau \otimes I)(I \otimes \mathbb{I}^\tau r(\theta_{13}))(\mathbb{I}^\tau r(\theta_{12}) \otimes I) - (I \otimes \mathbb{I}^\tau r(\theta_{12}))(\mathbb{I}^\tau r(\theta_{13}) \otimes I)(I \otimes \mathbb{I}^\tau)$$

$$+ (\mathbb{I}^\tau r(\theta_{23}) \otimes I)(I \otimes \mathbb{I}^\tau)(\mathbb{I}^\tau r(\theta_{12}) \otimes I) - (I \otimes \mathbb{I}^\tau r(\theta_{12}))(\mathbb{I}^\tau \otimes I)(I \otimes \mathbb{I}^\tau r(\theta_{23}))$$

$$+ (\mathbb{I}^\tau r(\theta_{23}) \otimes I)(I \otimes \mathbb{I}^\tau r(\theta_{13}))(\mathbb{I}^\tau \otimes I) - (I \otimes \mathbb{I}^\tau)(\mathbb{I}^\tau r(\theta_{13}) \otimes I)(I \otimes \mathbb{I}^\tau r(\theta_{23}))$$

$$= 0,$$

identical in θ_1, θ_2, θ_3.

Direct analysis of the equation (1.6), (1.7) can be performed. In particular, it is possible to show that a non trivial factorized s-matrix $r(\theta_1,\theta_2)$ that depends only on the difference $\theta_1 - \theta_2$ is expressed in terms of either rational, exponential, elliptic or Abelian functions. In each of these cases $r(\theta_1 - \theta_2)$ is a meromorphic function of $\theta_1 - \theta_2$ in $\mathbb{C}$. The equations (1.7) can be rewritten in terms of pole expansions of $r(\theta)$. For example, in the particular case when we do know that $r(\theta)$ is an elliptic function, one can claim that the poles of this function form a torsion subgroup of E. Similar arguments may be applied for an arbitrary Abelian variety. From our paper [4] we take the expression for the factorized completely X-symmetric quantum S-matrix corresponding to an arbitrary g-dimensional complex torus. Here is the coordinate expression for the S-matrix in the case, when $n = e_1 \ldots e_g$ and one identifies $\{0,1,\ldots,n-1\}$ with the Abelian group $\oplus_{i=1}^{g} \mathbb{Z}/\mathbb{Z}e_i$. The formulae involve g-dimensional θ-functions.

We take an arbitrary Riemann matrix Ω in a reduced form, when 2g periods are the following ones:

$$T_i = (0,\ldots, \underset{i-th}{1} ,\ldots,0),\ T_i^* = \{\tau_{1,i},\ldots,\tau_{g,i}\} : i = 1,\ldots,g.$$

The generalized θ-functions corresponding to the Abelian variety defined by Ω can be written as

$$(1.8) \qquad \theta\left[\frac{\vec{A}_1}{\vec{A}_2}\right](x) = \sum_{n\in\mathbb{Z}^g} \exp\{\pi i((\vec{n}+\vec{A}_2)^t B(\vec{u}+\vec{A}_2) + 2(\vec{n}+\vec{A}_2)(\vec{x}+\vec{A}_1))\},$$

for $\vec{A}_1,\vec{A}_2 \in \mathbb{C}^g$ and $g \times g$ matrix B corresponding to T_i^*: $i = 1,\ldots,g$. The S-matrix $R(\theta_1-\theta_2)$ corresponds to certain polarization of the Abelian variety and is defined by a system of positive integers $e_1,\ldots,e_g$ such that $e_i | e_{i+1}$: $i = 1,\ldots,g-1$. Then the index set $\{0,1,\ldots,n-1\} = X$ for $n = e_1 \ldots e_g$ is identified with $\oplus_{i=1}^{g} \mathbb{Z}/\mathbb{Z}ei$. The elements $R(\vec{\lambda})_{xy,zv}$ of $|X|^2 \times |X|^2$-matrix $R(\vec{\lambda})$ are represented in terms of generalized θ-functions in the following way:

$$(1.9) \quad R(\vec{\lambda})_{xy,zv} = \sum_{A\in\oplus_{i=1}^{g}\mathbb{Z}/\mathbb{Z}e_i} \frac{\theta\left[\frac{\vec{A}}{(x,y)}\right](\vec{\lambda}+\vec{\eta})}{\theta\left[\frac{\vec{A}}{\overline{(x-y)}}\right](\vec{\eta})} \times \exp\{2\pi\sqrt{-1}\ \Sigma_{i=1}^{g} \frac{(x_i-y_i)A_i}{e_i}\},$$

with $\overline{(x-y)} = \left(\frac{x_1-y_1}{e_1},\ldots,\frac{x_g-y_g}{e_g}\right)$. In the elliptic case when $g = 1$ this expression gives us a factorized unitary quantum S-matrix. In particular, one can derive from this expression the corresponding semi-classical factorized s matrix $r(\lambda)$, if one takes the limit $\eta \to T_1 + T_1^*$.

The paper [4] provides even more insight into the structure of completely X-symmetric factorized S-matrixes, which makes more explicit the relation between X and polarization of Abelian variety. We consider $\mathbb{C}^g$ as a 2g dimensional over $\mathbb{R}$ vector space with an antisymmetric bilinear form $B(x,y)$. We take a lattice L in V corresponding to θ-function (1.8) such that $B(x,y)$ is integer valued on $L \times L$. The polarization of V/L is naturally associated with complementary lattice L' of V such that $[L':L] = (e_1 \ldots e_g)^2 = n^2$. In an appropriate symplectic basis of V we have:

$$L = \{\Sigma_{i=1}^g n_i P_i + m_i Q_i : n_i \in \mathbb{Z},\ m_i \in e_i \mathbb{Z};\ i = 1,\ldots,g\}$$

$$L' = \{\Sigma_{i=1}^g n_i P_i + m_i Q_i : e_i n_i \in \mathbb{Z},\ m_i \in \mathbb{Z};\ i = 1,\ldots,g\}.$$

The bilinear form $B(x,y)$ can be represented as

$$B(x,y) = k(x,y) - k(y,x)$$

for

$$k(x,y) = -\Sigma_{i=1}^g x_i y_i'$$

with $x = \Sigma_{i=1}^g x_i P_i + y_i Q_i$, $y = \Sigma_{i=1}^g x_i' P_i + y_i' Q_i$.

There is a canonical system of $n \times n$ matrices $F_\lambda : \lambda \in L'$ generating $M_n(\mathbb{C})$ and satisfying Weyl commutation relations:

$$(1.10) \qquad F_\lambda \cdot F_\mu = \exp\{2\pi_i k(\lambda,\mu)\} F_{\lambda+\mu} \quad \text{for} \quad \lambda,\mu \in L'$$

and $F_\lambda = I$ for $\lambda \in L$. In particular, one gets the commutation relations of the Weyl type

$$F_\lambda \cdot F_\mu = \exp(2\pi i B(\lambda,\mu)\} F_\mu \cdot F_\lambda \quad \text{for} \quad \lambda,\mu \in L'.$$

The explicit description of matrices $F_\lambda : \lambda \in L'/L$ is given in [4]. In terms of matrices F_λ (and there are $|X|^2$ of them) it is very convenient to give an expression of a factorized completely X-symmetric matrix R. Following [4] we have:

$$R(\vec{x}) = \Sigma_{\lambda\in L'/L}\, w_\lambda(\vec{x}) F_\lambda \times F_{-\lambda}, \tag{1.11}$$

where $w_\lambda(\vec{x})$ are ratios of θ-functions of the form (1.8):

$$w_\lambda(\vec{x}) = \frac{\theta\left[{(\lambda)_1 \atop (\lambda)_2}\right](\vec{x}+\vec{\eta})}{\theta\left[{(\lambda)_1 \atop (\lambda)_2}\right](\vec{\eta})} \exp\{2\pi i k(\lambda,\lambda)\} \tag{1.12}$$

and generalized θ-functions $\theta\left[{\vec{A}_1 \atop \vec{A}_2}\right](\vec{x})$ corresponding to a lattice L in V. We choose here the notation $w_\lambda(\vec{x})$ in order to be in line with Baxter eight-vertex model [1]. In that case $g = 1$, $n = 2$ and the matrices F_λ are Pauli matrices. Then w_λ from [1] correspond to Baxter expression for Boltzmann weights in eight-vertex model. Naturally, our expressions for factorized S-matrices give rise to interesting multi-component generalizations of both eight-vertex and XYZ-models. Below we consider the corresponding elliptic generalizations, when $g = 1$.

2. We present now a large family of completely integrable two dimensional systems associated with semiclassical factorized s-matrix, arising from the elliptic curve E and its n-torsion subgroup. We take an elliptic curve E over $\mathbb{C}$ in its Weierstrass representation given in affine form by a cubic $y^2 = 4x^3 - g_2x - g_3$ and satisfied by $(\wp'(u),\wp(u))$. The elliptic curve E is represented as $\mathbb{C}/L$ where the lattice L is $\mathbb{Z}\omega_1 + \mathbb{Z}\omega_2$ with $\operatorname{Im}{}^{\omega_2}/\omega_1 > 0$. The n-torsion subgroup E_n of E is the subgroup of n^2 points

$$\left(\left(\wp'\left(\frac{i\omega_1}{n} + \frac{j\omega}{n}\right), \wp\left(\frac{i\omega_1}{n} + \frac{j\omega}{n}\right)\right) : 0 \le i,j \le n-1.$$

The corresponding larger lattice $L_n = \mathbb{Z}\omega_1/n + \mathbb{Z}\omega_2/n$ gives rise to cosets L_n/L playing, as before, the role of $\{0,1,\ldots,n-1\} \times \{0,1,\ldots,n-1\}$. Corresponding to L_n/L there is a set of operators F_λ: $\lambda \in L_n/L$, arising from an induced representation of the Lie algebra g [4], [8]. These operators obey the rule (1.10) for $\lambda,\mu \in L_n$:

(2.1) $$F_\lambda \cdot F_\mu = \exp\{2\pi i k(\lambda,\mu)\} F_{\lambda+\mu},$$

where for $\lambda = \lambda_1\omega_1 + \lambda_2\omega_2$, $\mu = \mu_1\omega_1 + \mu_2\omega_2 \in L_n$,

(2.2) $$k(\lambda,\mu) = -n\,\lambda_2\mu_1; \quad B(\lambda,\mu) = n(\lambda_1\mu_2 - \lambda_2\mu_1).$$

E.g.

(2.3) $$F_{\lambda'+\lambda} = F_{\lambda'} \text{ for } \lambda' \in L_n,\ \lambda \in L.$$

The operators $F_\lambda : \lambda \in L_n$ generate an algebra isomorphic to $M_n(\mathbb{C})$ and can be realized as $n \times n$ matrices following [4]:

(2.4) $$F_\lambda = A^{\lambda_1 \cdot n} \cdot B^{\lambda_2 \cdot n} \quad \text{for } \lambda = \lambda_1\omega_1 + \lambda_2\omega_2 \in L_n$$

with

(2.5) $$(A)_{ij} = \delta_{ij} \exp\left(\frac{2\pi\sqrt{-1}\, i}{n}\right); \quad (B_{ij}) = \delta_{i+1,j} \pmod n.$$

We are going to present the spectral problem that is induced by the semi-classical s-matrix generated by L and L_n. This spectral problem is defined for the values of spectral parameter on the elliptic curve E with poles of the first order at points of E_n. For the description of rational functions on E we use Weierstrass ζ-function $\zeta(x)$, $\zeta'(x) = -\wp(x)$, and derivatives $\wp^{(j)}(x)$ of $\wp(x)$. Any function $f(\theta)$ on E having its poles only at points of E_n and all of the order k, can be written as

(2.6) $$f(\theta) = \Sigma_{\lambda\in L_n/L}\, c_\lambda^{(o)} \zeta(\theta+\lambda) + \Sigma_{j=1}^k \Sigma_{\lambda\in L_n/L}\, c_\lambda^{(j)} \wp^{(j-1)}(\theta+\lambda),$$

with $\Sigma_{\lambda\in L_n/L}\, c_\lambda^{(o)} = 0$.

This decomposition (2.6) is used in order to define a spectral problem that has on E poles of arbitrary order k. Following the structure of semiclassical s-matrix associated with L_n and L we present the initial spectral problem with poles only of the first order on E in the form

$$(2.7)\qquad (\frac{d}{dx} - U_0(x,\theta))\Phi(x,\theta) = 0,$$

where $n \times n$ matrix $U_0(x,\theta)$ is expressed in terms of the matrices F_λ (2.4), (2.5) as

$$(2.8)\quad U(x,\theta) = \Sigma_{\mu\in L_n/L,\mu\neq 0} F_\mu u_\mu(x)\, \Sigma_{\omega\in L_n/L}\zeta(\theta+\omega)\exp\{2\pi\sqrt{-1}\, B(\omega,\mu)\}.$$

In other words the linear problem (2.7) depends on n^2-1 scalar functions $u_\mu(x)$. Naturally, in the quantum case $u_\mu(x)$ are field operators. In order to stress the generic relationship with the Baxter model [1] we denote elliptic functions of the form involved in (2.8) by w's. In general, for $j = 0,1,2,\dots$ and $\mu \in L_n/L$, $\mu \neq 0$:

$$(2.9)\qquad w_\mu^{(j)}(\theta) = \Sigma_{\omega\in L_n/L}\, \wp^{(j-1)}(\theta+\omega)\exp\{2\pi\sqrt{-1}B(\omega,\mu)\}$$

where, formally, $\wp^{(-1)}(u) = \zeta(u)$. From the definition (2.2) it follows that the functions $w_\mu^{(o)}(\theta)$ are elliptic for $\mu \neq 0$, $\mu \in L_n/L$, since the sum of their residues in the fundamental domain of E is zero. The higher functions $w_\mu^{(j)}(\theta)$ by definition elliptic, as $j \geq 1$.

The semiclassical s-matrix determines at the same time Poisson brackets between functions $u_\mu(x)$, $u_\lambda(y)$ for $\mu,\lambda \in L_n/L$. Namely, in the notations of (2.2) we put

$$(2.10)\qquad \{u_\mu(x),u_\lambda(y)\} = \delta(x-y)(e^{2\pi ik(\mu,\lambda)}-e^{2\pi ik(\lambda,\mu)})u_{\mu+\lambda}(x),$$

for $\mu,\lambda \in L_n/L$ (i.e. for μ,λ being elements of L_n (mod L) and $\mu \neq 0$, $\lambda \neq 0$). The initial spectral problem (2.7) gives rise to a large class of two dimensional completely integrable systems with $n^2 - 1$ variables $u_\mu(x)$. The Hamiltonians of these systems belong to a family of commuting Hamiltonians generated by the monodromy matrix $\Phi(x,y,\theta)$ of (2.7):

(2.11) $$\mathrm{tr}\ (\Phi(x,y,\theta).$$

Indeed, according to our result about semiclassical factorized s-matrices [7] all Hamiltonians (2.11) for fixed x,y and different θ's commute, if Poisson brackets are defined as in (2.10). We decided to call this family of two dimensional system elliptic Kortewed de Vries equations or simply elliptic KdV. Also there is an option to call them elliptic sin-Gordon equation. The reason for this is the following. When E degenerates into a rational, unicursal curve, i.e. the Abelian variety E is substituted by $\mathbb{A}^1$ (additive group), then the spectral problem (2.7) turns out to be a matrix linear differential operator of the first order with a single pole in $\mathbb{P}\,\mathbb{C}^1$. If this pole is at ∞ we come to the situation familiar from KdV or non linear Schrodinger equations [10], [11]. However if pole is at zero or at any other finite point, then this gives rise to sin-Gordon equation (n = 2) or different non linear σ-models.

The most interesting class of elliptic KdV or sin-Gordon equations arises when these equations can be written as a commutativity condition of two linear problems, one of which is the problem (2.7) itself:

(2.12)
$$\frac{d\Phi(x,\theta)}{dx} = U(x,\theta)\Phi(x,\theta);$$
$$\frac{d\Phi(x,\theta)}{dt} = V(x,\theta)\Phi(x,\theta).$$

Then the two-dimensional equations have the form

(2.13) $$\frac{d}{dt}U(x,\theta) - \frac{d}{dx}V(x,\theta) + [U(x,\theta),V(x,\theta)] = 0.$$

The class of equations (2.13) is indeed a rich one if one takes $V(x,\theta)$ in (2.12) as a rational function of θ on E with poles of the order k. It is most natural to take the set of poles to be a translation of E_n. E.g. we can present an equation we call an elliptic principal chiral field. This equation corresponds to the case (2.12), (2.13), when $V(x,\theta)$ has the same form as $U(x,\theta+\alpha)$ for $\alpha \in E$, $\alpha \neq 0$ and with different scalar coefficients.

In other words we take $V(x,\theta)$ in the following form

$$(2.14)\quad V(x,\theta) = \Sigma_{\mu\in L_n/L,\mu=0} F_\mu v_\mu (x).\ \Sigma_{\omega\in L_n/L} \zeta(\theta+\omega+\alpha)\exp\{2\pi\sqrt{-1}B(\omega,\mu)\},$$

for $\alpha \neq 0$. If the potential $U(x,\theta)$ is defined as in (2.8) and $V(x,\theta)$ is as in (2.14) then the equation (2.13) can be written as $2n^2-2$ equations on $2n^2-2$ variables $u_\mu(x,t)$, $v_\mu(x,t)$:

$$(2.15)\quad \begin{aligned} &\frac{\partial}{\partial t} u_\nu(x,t) + \Sigma_{\eta\neq 0,\nu} K(\alpha,\nu-\eta)\{e^{2\pi ik(\eta,\nu-\eta)}-e^{2\pi ik(\nu-\eta,\eta)}\}u_\eta v_{\nu-\eta} = 0;\\ &\frac{\partial}{\partial x} v_\nu(x,t) + \Sigma_{\eta\neq 0,\nu} K(-\alpha,\nu-\eta)\{e^{2\pi ik(\eta,\nu-\eta)}-e^{2\pi ik(\nu-\eta,\eta)}\}v_\eta u_{\nu-\eta} = 0. \end{aligned}$$

Here we denote

$$(2.16)\qquad K(\alpha,\lambda) = \Sigma_{\omega\in L_n/L} \zeta(\alpha+\omega)e^{-2\pi iB(\omega,\lambda)}.$$

The system (2.15)-(2.16) is the elliptic principal chiral field model and one should, perhaps, indicate briefly why such name is given. The first reason for this is a deep geometric one connected with Kählerian manifolds that will be examined in a separate paper. However there is also an immediate formal explanation for this name. One can consider degeneration of an elliptic curve, e.g. when the module k of E tends to zero. In this case the functions $w_\mu^{(o)}(\theta)$ all tend to $1/\theta$ on $\mathbb{C}$. Then the linear problems (2.12) for $U(x,\theta)$ and $V(x,\theta)$ defined as in (2.8) and (2.14) turns into a linear spectral problem on $\mathbb{C}$:

$$(2.17)\quad \begin{aligned} &\frac{d\Phi_\theta}{dx} = \frac{U}{\theta}\Phi_\theta;\\ &\frac{d\Phi_\theta}{dt} = \frac{V}{\theta+\alpha}\Phi_\theta, \end{aligned}$$

where U and V are $n\times n$ matrices that are linear combination of F_μ $\mu\neq 0$, $\mu\in L_n/L$ with scalar function coefficients. In other words, U and V are arbitrary traceless matrices. The consistency condition for linear problem (2.17) is called principal ciral field equation (principal chiral field equation for an algebra g), if U and V are arbitrary traceless matrices (belonging to an algebra g). In other words equations (2.15) are natural generalizations of the principal chiral field equations (2.17), if one considers the corresponding spectral problem (2.12) over E instead of (2.17) over $\mathbb{C}$. In particular, one can propose invariant restrictions on u_μ, v_μ in (2.15) in order to generate different σ-models corresponding to Grassmanian manifolds over elliptic curves.

The most natural object for elliptic generalizations is matrix Heisenberg spin system. IT should be noted that the 8-vertex model [1], [9] is an elliptic generalization of the Heisenberg ferromagnet. In the same way we now generalize an arbitrary matrix Heisenberg system. The general matrix Heisenberg chain had been introduced in our paper [10], example 1.2. This system arises as the consistency condition of two linear problems of the following sort

$$\frac{d}{dx}\Phi_\lambda = \frac{S}{\lambda}\Phi_\lambda;$$

(2.18)

$$\frac{d}{dt}\Phi_\lambda = \left(\frac{T}{\lambda} + \frac{2iS}{\lambda^2}\right)\Phi_\lambda.$$

The two dimensional equations corresponding to (2.18) have the form:

(2.19) $$-2S_x + [S,T] = 0, \quad iS_t = T_x,$$

One obtains from (2.19) a matrix Heisenberg spin system, if one imposes an invariant restriction on S: $S^2 = \mathbb{I}$. In this case the equations (2.19) take the familiar form

(2.20) $$S_t = \frac{1}{2i}[S,S_{xx}], \quad S^2 = \mathbb{I}.$$

We had noted in [10] that the system (2.20) is gauge equivalent to the matrix nonlinear Schrodinger equation. In order to obtain natural generalizations of (2.20) in the same way as the Baxter model [1] or [9] is a generalization of Heisenberg ferromagnet, it is necessary to consider linear problem (2.12) with $U(x,\theta)$ as in (2.8) and $V(x,\theta)$ with poles of the second order at E_n. This way we imitate the structure of linear problem (2.18). Consequently, $V_1(x,\theta)$ has the following form

$$(2.21)\qquad V_1(x,\theta) = \Sigma_{\mu\in L_n/L,\mu\neq 0}\, v_\mu^0(x)F_\mu w_\mu^{(0)}(\theta) + \Sigma_{\mu\in L_n/L,\mu\neq 0}\, v_\mu^1(x)F_\mu w_\mu^{(1)}(\theta).$$

The system (2.13) is of the form:

$$(2.22)\qquad \frac{d}{dt}U(x,t,\theta) - \frac{d}{dx}V_1(x,t,\theta) + [U(x,t,\theta),V_1(x,t,\theta)] = 0.$$

The structure of v_μ^1 resembles that of u_μ:

$$(2.23)\qquad v_\mu^1 = au_\mu : \mu\in L_n/L,\ \mu\neq 0,$$

for some scalar $a\neq 0$. The system (2.22) - (2.23) is the precise elliptic generalization of nonreduced system (2.19). E.g., there is an equation determining $v_\mu^0(x,t)$

$$(2.24)\qquad \Sigma_{\eta\neq 0,\nu}\{e^{2\pi ik(\eta,\nu-\eta)}-e^{2\pi ik(\nu-\eta,\eta)}\}u_\eta u_{\nu-\eta} \times (\Sigma_{\omega\in L_n/L,\omega\neq 0}\,\zeta(\omega)e^{2\pi iB(\omega,\eta)}) + a^{-1}\Sigma_{\eta\neq 0,\nu}\{e^{2\pi ik(\eta,\nu-\eta)}-e^{2\pi ik(\nu-\eta,\eta)}\}u_\eta v_{\nu-\eta}^0 = \frac{\partial}{\partial x}u_\nu$$

for any $\nu\in L_n/L$, $\nu\neq 0$. Imposing on system (2.22)-(2.23) the same kind as on (2.19) one obtains an elliptic generalization of Heisenberg spin system. E.g. for n = 2 these restrictions are the following

$$u_{(0,1)}^2 + u_{(1,0)}^2 - u_{(1,1)}^2 \equiv 1$$

where $(i,j)\in \mathbb{Z}/\mathbb{Z}2 + \mathbb{Z}/\mathbb{Z}2$ is identified with $i\frac{\omega_1}{2} + j\frac{\omega_2}{2}$ from L_2/L. For general n the number of these restrictions is much larger and they may be taken in the following form

(2.25) $$\Sigma_{\mu,\eta\in L_n/L;\mu+\eta=\nu,\mu\neq 0,\eta\neq 0}\ e^{2\pi ik(\mu,\eta)}u_\mu u_\eta = 0,$$

for $\nu \in L_n/L$, $\nu \neq 0$ and

(2.26) $$\Sigma_{\mu\neq 0}\ e^{-2\pi ik(\mu,\mu)}u_\mu u_{-\mu} = 1.$$

Moreover further restrictions can be added to (2.25) - (2.26). They correspond to the restrictions on (2.20) of the form $S = \mathbb{I} - 2P$, where P is an one-dimensional projector. This way one gets certain system being elliptic generalizations of nonlinear Schrödinger equation.

References

[1] R. Baxter, Ann. Phys. 76, 1, 25, 48 (1973).

[2] D.V. Chudnovsky, G.V. Chudnovsky, Phys. Lett. A, 79A (1980), 36-38.

[3] A.B. Zamolodchikov, Comm. Math. Phys. 69, 165 (1979).

[4] D.V. Chudnovsky, G.V. Chudnovsky, Phys. Lett. A. 81A (1981), 105-110.

[5] D. Iagonitzer, Lecture Notes Physics, Springer 126, 1 (1980).

[6] D.V. Chudnovsky, G.V. Chudnovsky, Phys. Lett. B, 98B (1981), 83-88.

[7] D.V. Chudnovsky, G.V. Chudnovsky, Lett. Math. Phys. 4 (1980), 485-493.

[8] P. Cartier, Proc. Symp. Pure Math. 9, Providence, 1965, 361-387.

[9] E.K. Sklanin, LOMI-preprint E-3, 1979, Leningrad, 1979.

[10] D.V. Chudnovsky, G.V. Chudnovsky, Z. Phys. D5, 55 (1980).

[11] V.E. Zakhorov, A.V. Mikhailov, JETP 74, 1953 (1978).

Department of Mathematics
Columbia University
New York, NY
USA

Quantum Hamiltonians associated with finite-dimensional Lie algebras and factorized S-matrices.

by

D.V. Chudnovsky

G.V. Chudnovsky

Abstract. We consider quantum Hamiltonian systems of lattice type which are associated with classical finite dimensional Lie algebras (like MH(2), SO(3) etc.), and possess complete integrability properties. These Hamiltonian systems are written in canonical Heisenberg p_n, q_n variables. They have factorized S-matrices and are generalizations of the quantum Toda lattice Hamiltonian as well as XYZ-models of statistical mechanics.

1. In this paper we consider Hamiltonian systems which are associated with a given symmetry group, and of course, the most typical example will be SO(3). Many classical mechanical systems, field theory models and models of statistical mechanics possess a given group of G symmetries and are written in symplectic coordinates different from those of Darboux [1], [2]. Though we know that locally any symplectic coordinates can be always reduced to Darboux form [1], the quantization of those systems becomes ambiguous. We propose an alternative approach making use of symplectic coordinates corresponding to a Lie algebra. This allows us to write directly the quantized system associated to group G. Moreover the Hamiltonian can be represented in Heisenberg variables p_n, q_n which are quantized counterparts of Darboux symplectic coordinates. This is not accomplished using the Darboux theory, but by parametrizing directly the elements of a Lie algebra by the elements of a corresponding Weyl algebra [3] A_n which is generated by p_i, q_i $(i = 1,\dots,n)$, or by representation of G in A_n. This representation is achieved by representing G in various function spaces and writing down infinitesimal operators corresponding to generators of G in form of differential operators.

We start with generators e^i: $i = 1,\dots,m$ of G and basic commutation relations $[e^i,e^j] = \Sigma_{k=1}^m C_{ij}^k e^k$ for structural constants C_{ij}^k of the group G. The Lie group G is associated with a manifold of local variables v^i: $i = 1,\dots,m$ and Poisson brackets between functions of these variables $f(v)$, $g(v)$ defined as

$\{f,g\}_G = \Sigma_{i,j=1}^n \frac{\partial f}{\partial v^i}\frac{\partial g}{\partial v^j}\{v^i,v^j\}_G$ with $\{v^i,v^j\}_G = \Sigma_{k=1}^m C_{ij}^k v^k$. This definition of Poisson brackets induces a symplectic structure on the symplectic manifolds which are defined by constant values of Casimir operators of G. Namely, one gets an orbit T of G defined by a system of equations $\Sigma\, a_{i_1,\dots,i_k} v^{i_1}\dots v^{i_k}$ = const for every Casimir operator $\Sigma\, a_{i_1,\dots,i_k} e^{i_1}\dots e^{i_k}$ of G (i.e. the elements of the center of universal enveloping algebra of G). Every orbit T is a symplectic even-dimensional manifold with Poisson brackets $\{\cdot,\cdot\}_G$. These Poisson brackets can be reduced to Poisson brackets in Darboux form using the representation of infinitesimal operators corresponding to one parametric subgroups of G generated by e^i. Using the description of irreducible representations of G one obtains parametrization of G in the form $v^i = V_i\,(p_1,\dots,p_\ell,q_1,\dots,q_\ell)$, depending on values of Casimir operators defining an orbit T, and $V_i(p_1,\dots,p_\ell,q_1,\dots,q_\ell)$ being an element of $A_\ell = \mathbb{C}[[p_1,\dots,p_\ell,q_1,\dots,q_\ell]]$-- the ring of formal power series in $p_1,\dots,p_\ell$ and $q_1,\dots,q_\ell$,--for the number ℓ equal to half real dimension of T and p_i, q_i being the Darboux coordinates $\{p_i,q_j\} = \delta_{ij}$ of Weyl algebra A_ℓ.

2. Now, inspired by lattice models [4], we consider the quantum one-dimensional (or classical two dimensional) lattice systems associated with the group G. These systems have local variables denoted by v_n^i $(i = 1,\dots,m)$, $n = 0, \pm 1, \pm 2,\dots$ and represented by operators e_n^i satisfying the fundamental commutation relations:

$$[e_n^i,e_r^j] = \delta_{nr}\, \Sigma_{k=1}^m C_{ij}^k e_n^k.$$

In terms of the Poisson brackets $\{\cdot,\cdot\}_G$, the corresponding symplectic manifold is $T \times T \times T \times \dots$, with local variables v_n^i obeying $\{v_n^i,v_r^j\}_G = 0$ if $n \neq r$ and $\{v_n^i,v_n^j\}_G = \Sigma_{k=1}^m C_{ij}^k v_n^k$. According to the lattice interpretation, models written in these variables represent a chain of equations with subscript n, which label the rows of a two dimensional lattice. We use this lattice-theoretic interpretation and construct completely integrable quantum lattice Hamiltonians with the help of the Onsager-Baxter method of local transfer matrices (Onsager [4], Baxter [5]). This method is based on so called Baxter lemma [5], [6] given below. In order to have a completely integrable quantum Hamiltonian, one actually needs a large family of commuting Hamiltonians. This family of Hamiltonians is given by $H(\theta) = \mathrm{Tr}\{\prod_{n=1}^{N} \mathcal{L}_n(\theta)\}$, where $\mathcal{L}_n(\theta)$ (a local transfer matrix) is an $e \times e$ matrix with elements from G. The trace $\mathrm{tr}\{\cdot\}$ is taken over G (i.e. as a sum of e elements of G on the diagonal of an $e \times e$ matrix). If different Hamiltonians $H(\theta_i)$ are to commute, then $\mathrm{tr}\{\mathcal{L}_n(\theta_1) \otimes \mathcal{L}_n(\theta_2)\} =$ $= \mathrm{tr}\{\mathcal{L}_n(\theta_2) \otimes \mathcal{L}_n(\theta_1)\}$. According to Baxter, one can demand for $e^2 \times e^2$ matrices that $\mathcal{L}_n(\theta_1) \otimes \mathcal{L}_n(\theta_2)$ and $\mathcal{L}_n(\theta_2) \otimes \mathcal{L}_n(\theta_1)$ be similar. This will guarantee the commutativity of $H(\theta_i)$ according to the commutation in G:

<u>Lemma</u> 1. Let $\mathcal{L}_n(\theta)$ be an $e \times e$ matrix with elements from $G[[\theta]]$ [6] satisfying the following commutation relations

$$(1) \qquad R(\theta_1-\theta_2)(\mathcal{L}_n(\theta_1) \otimes \mathcal{L}_n(\theta_2)) = (\mathcal{L}_n(\theta_2) \otimes \mathcal{L}_n(\theta_1))R(\theta_1-\theta_2)$$

for a scalar nonsingular $e^2 \times e^2$ matrix $R(\theta_1-\theta_2)$. Let us assume also that elements of the matrices $\mathcal{L}_n(\theta_1)$ and $\mathcal{L}_r(\theta_2)$ are commuting in G, if $n \neq r$. Then one obtains a family of commuting Hamiltonians written in coordinates e_n^i: the Hamiltonians $H_N(\theta) = \mathrm{tr}\{\mathcal{L}_N(\theta)\ldots\mathcal{L}_1(\theta)\}$ are commuting in G for all θ.

The Baxter lemma 1 is an algebraic statement, which covers the essence of the quantum and the classical inverse scattering method and isospectral deformations. In the similar, but different circumstance of the Kostant-Kirillov coadjoint representation method, one can find a similar statement, known as the Symes-Kostant-Adler lemma. However the most general combinatorial statement is, undoubtedly, the Baxter lemma 1. The scalar, $e^2 \times e^2$ matrix $R(\lambda - \mu)$ in (1) is called a factorized S-matrix [6], [8], [11]. The matrix $R(\lambda - \mu)$ is not an arbitrary one, as its elements satisfy a nonlinear system of equations, known as the factorization and unitarity equations (Yang-Karowsky-Zamolodchikov equations [11], [6]). It is known that a matrix $R(\lambda - \mu)$, which satisfies the factorization and unitarity equations, is indeed the scattering matrix of a certain quantum one-dimensional completely integrable mechanical system (see the discussion in [11a]).

The system of commuting local Hamiltonians can be obtained by expanding $\log H(\theta)$ in powers of θ. Examples of such systems can be found in [7]. Models of statistical mechanics arising from lemma 1 corresponds to the lowest dimensional representations of G. For example: Ising model, eight-vertex and XYZ models (see [5], [9]) all correspond to e_n^i coming from two dimensional representation of $SO(3)$ by Pauli matrices. Hence more general models of similar type which correspond to the same S-matrix $R(\theta_1 - \theta_2)$ as for XYZ model but for arbitrary representations of $SO(3)$, are called generalized lattice XYZ-models.

3. In order to present concrete examples of lattice XYZ-models we consider parametrizations of classical groups by elements of Weyl algebra. In the examples below we have $\ell = 1$ and the two dimensional symplectic manifold T is determined by the value of a single Casimir operator [3] (which is constant on irreducible representations of G according to the Schur lemma). Below are listed commutation relations between generators of the corresponding groups and parametrizations of these generators in terms of operators p and q. To simplify the notations the subscript n is everywhere suppressed. However it should be remembered that the value of the Casimir operator depends on n.

<u>Examples</u>. 1. <u>$M(2)$</u>: $[e^1,e^2] = 0$, $[e^2,e^3] = e^1$, $[e^3,e^1] = e^2$ and $v^1 = R\cos q$, $v^2 = R\sin q$, $v^3 = -p$.

2. <u>$MH(2)$</u>: $[e^-,e^+] = 0$, $[e^3,e^+] = e^+$, $[e^3,e^-] = -e^-$; $v^3 = p$, $v^+ = Re^q$, $v^- = Re^{-q}$.

3. <u>$SO(3)$</u>: $[e^1,e^2] = e^3$, $[e^2,e^3] = e^1$, $[e^3,e^1] = e^2$, $v^1 = i\{\nu q + 1/2(1 - q^2)p\}$, $v^2 = -\nu q + 1/2(1 + q^2)p$; $v^3 = i\{-\nu + qp\}$.

4. QU(2): $[e^1,e^2] = -e^3$, $[e^2,e^3] = e^1$, $[e^3,e^1] = e^2$;
$v^1 = 1/2[(\nu + \varepsilon)e^{iq} + (\nu - \varepsilon)e^{-iq} - 2 \sin q\, p]$;
$v^2 = i/2[(\nu + \varepsilon)e^{iq} - (\nu - \varepsilon)e^{-iq} + 2i \cos qp]$; $v^3 = -i\varepsilon + p$.

5. $SL(2,\mathbb{R})$: $[e^1,e^2] = -2e^3$, $[e^2,e^3] = -2e^1$, $[e^3,e^1] = 2e^2$;
$v^1 = 2\nu q + (1 - q^2)p$; $v^2 = -2\nu q + (1 + q^2)p$; $v^3 = 2\nu -2qp$.

Formulas for the parametrization of an arbitrary finite dimensional semi-simple Lie algebra in terms of differential operators can be presented similarly using their Dynkin diagrams in terms of classical roots systems [3], [10]. The formulas 1-5 or their generalizations can be used to represent commuting quantum Hamiltonians $H(\theta)$ in terms of differential operators in $q_1,\ldots,q_N$ and $\partial/\partial q_1,\ldots,\partial/\partial q_N$.

4. In order to define these completely integrable quantum Hamiltonians $H(\theta)$ one has to define only local transfer matrices $\mathscr{L}_n(\theta)$ obeying equations (1) for some $e^2 \times e^2$ matrix $R(\theta)$ called an S-matrix [6], [11]. Following the methods of [6], [8], it is enough to specify the analytic properties of $\mathscr{L}_n(\theta)$ as a function of θ on a finite Riemann surface Γ. $\mathscr{L}_n(\theta)$ is rational, corresponding to vector bundles over Γ of rank e. In the examples below Γ has genus $g = 0$ or 1. Also we restrict ourself to the case when $\mathscr{L}_n(\theta)$ has a minimal allowed degree, equal to $g + 1$ in our case. For a rational Riemann surface Γ we consider local transfer matrices having a single pole on Γ. The precise form of $\mathscr{L}_n(\lambda)$ can be then determined taking into account: a) symmetry conditions imposed by the group G, which is SO(3) for rational Γ and b) an additional symmetry induced by the structure of algebraic group associated with Γ and having the form of $\mathscr{L}_n(\lambda - \eta)$ being similar to $\mathscr{L}_n(\lambda)^{-1}$.

In the case of $\Gamma = \mathbb{P}^1$ one has $\mathscr{L}_n(\theta)$ corresponding to groups MH(2) (or M(2)) and SO(3) (or QU(2)). The group MH(2) gives rise to Toda lattice Hamiltonians [2]. The corresponding $\mathscr{L}_n(\theta)$ is determined by the condition (1) of the Baxter lemma and by the condition that $\mathscr{L}_n(\theta)$ has only a single simple pole at $\theta = \infty$. This form of local transfer 2×2 matrix $\mathscr{L}_n(\theta)$ is the one prescribed by the inverse scattering method for difference Sturm-Liouville problem [12].

$$\mathcal{L}_n^0(\theta) = \begin{pmatrix} \theta+e_n^3 & -e_n^+ \\ e_n^- & 0 \end{pmatrix}$$

with e_n^3, e_n^+, e_n^- satisfying the relations of MH(2). One gets quantum Toda lattice Hamiltonians if one expands

$$\mathrm{Log}[\theta^{-N} H_N^0(\theta)] = \log[\mathrm{tr}\{\theta^{-N} \mathcal{L}_N^0(\theta) \ldots \mathcal{L}_1^0(\theta)\}]$$

in power of θ and uses the parametrization of MH(2):

$$e_n^3 = p_n, \quad e_n^+ = e^{q_n}, \quad e_n^- = e^{-q_n}$$

and $[p_n, q_m] = \delta_{nm}$. The first Hamiltonians given by this expansion are quantum Toda lattice Hamiltonian and its first integrals

$$\log[\theta^{-N} H_N^0(\theta)] = \Sigma_{j=1}^{\infty} H_j \theta^{-j}, \quad H_1 = \Sigma\, p_n, \quad H_2 = \frac{1}{2}\Sigma\, p_n^2 +$$

$$+ \Sigma\, e^{q_{n+1}-q_n}, \quad H_3 = \frac{1}{3}\Sigma\, p_n^3 + \Sigma\, p_n e^{q_{n+1}-q_n} + \Sigma\, p_n e^{q_n - q_{n-1}}, \ldots .$$

Let us consider the case SO(3) and $\Gamma = \mathbb{P}^1$. We take the most general form of the local transfer 2×2 matrix as having a single simple pole at $\theta = \mu$:

$$\mathcal{L}_n^1(\theta) = \begin{pmatrix} I + \dfrac{U_{11}^{(n)}}{\theta-\mu} & \dfrac{U_{12}^{(n)}}{\theta-\mu} \\ \dfrac{U_{21}^{(n)}}{\theta-\mu} & I - \dfrac{U_{11}^{(n)}}{\theta-\mu} \end{pmatrix},$$

The conditions (1) of lemma 1 imply that the elements $U_{ij}^{(n)}$ of $\mathcal{L}_n(\theta)$ satisfy the relations

$$[U_{11}^{(n)},U_{12}^{(n)}] = -\eta U_{12}^{(n)}, [U_{11}^{(n)},U_{21}^{(n)}] = \eta U_{21}^{(n)}, [U_{12}^{(n)},U_{21}^{(n)}] = -2\eta U_{11}^{(n)}.$$

One of the sequences of quantum Hamiltonians is given by an expansion of

$$H_N^1(\theta) = (\theta-\mu)^N \operatorname{tr}[\mathcal{L}_N^1(\theta)\dots\mathcal{L}_1^1(\theta)]$$

and by the parametrization of $U_{ij}^{(n)}$ in terms of p_n, q_n. Then we have

$$H_N^1(\theta) = 2(\lambda-\mu)^{N}I + \Sigma_{j=N-2}^{0} (\lambda-\mu)^j H_j^1$$

and the Hamiltonian H_{N-2}^1 can be represented in a simple form:

$$H_{N-2}^1 = \eta^2[2\Sigma_{n\neq m}\{p_n p_m - \epsilon_m p_n - \epsilon_n p_m\} + \Sigma_{n\neq m}\{-e^{q_n-q_m} p_n p_m -$$

$$- (\ell_m-\epsilon_m)e^{q_n-q_m} p_n + (\ell_n+\epsilon_n)e^{q_n-q_m} p_m + (\ell_n+\epsilon_n)(\ell_m-\epsilon_m)e^{q_n-q_m}\}]$$

$$+ \text{const}_N.$$

This quantum Hamiltonian is also completely integrable in the sense that it possesses N - 1 additional algebraic first integrals commuting between themselves N - 2 of them are given by H_j^1 and the additional one is given by the integral of the center of mass $P = \Sigma_{n=1}^N p_n$.

Hamiltonian H_{N-2}^1 takes the simplest form, if the constants ℓ_n and ϵ_n, that are defined in terms of values of Casimir operator, are all zero and then the Hamiltonian is

$$H'_{N-2} = 2\,\Sigma_{\substack{n,m=1\\ n\neq m}}^{N}\{1 + \operatorname{ch}(q_n-q_m)\}p_n p_m.$$

Hamiltonians which are generated by $\mathcal{L}_n^1(\theta)$, as given in the expansion of $H_N^1(\theta)$, involve nonoinear interactions between all the neighbors in the lattice model, while Toda lattice Hamiltonians are degenerations of the above Hamiltonians and involve only nearest neighbor interactions. It is possible, however, to find Hamiltonians which commute with those of $H^1_N(\theta)$ and involve only nearest neighbor interactions. For this to happen one considers an expansion of $\log H^1_N(\theta)$ in the neighborhood of $\theta = \eta$. Surprisingly, the first non-trivial Hamiltonian in the new sequence has the same form as H^1_{N-2} but with summation $\Sigma_{n\neq m}$ extended only over nearest neighbors. For example, the Hamiltonian H'_{N-2} has the following commuting counterpart:

$$H''_{N-2} = 2\,\Sigma^N_{n=1}\{1 + ch(q_n - q_{n+1})\}p_n p_{n+1},$$

with $p_{n+N} \equiv p_n$, $q_{n+N} \equiv q_n$.

A general statement describing the relationship between lattice models with the nearest neighbor interaction and nonlocal one is given at the end of this note.

The group SO(3) generates another local transfer matrix which is this time connected with a cylinder instead of the complex plane. In this case after parametrization of SO(3), in the Weyl group A_1, we obtain a local transfer matrix which is expressed in terms of trigonometric functions of p_n:

$$\mathcal{L}_n(\lambda) = \begin{pmatrix} ch(\lambda+p_n) & e^{-\eta q_n} chp_n \\ e^{\eta q_n} chp_n & ch(\lambda-p_n) \end{pmatrix}$$

with $[p_n, q_m] = \delta_{nm}$. Commuting Hamiltonians are given by an expansion of $H^2(\lambda) = tr[\mathcal{L}^2_N(\lambda)\ldots\mathcal{L}^2_1(\lambda)]$.

Similarly SO(3) gives rise to local transfer matrices which are expressed in terms of Jacobi elliptic θ-functions and have as elements $\theta_1(\lambda + p_n)$, $\theta_4(\lambda + p_n)$ for Jacobi θ-functions θ_1, θ_4 [13].

These expressions of local transfer matrices are generalized to arbitrary Abelian varieties. Instead of writing down the corresponding local transfer matrices we present directly the correct Hamiltonians which are generalizations of XYZ-Hamiltonians (or Heisenberg models) of statistical mechanics [9]. We refer to [8], where multispin versions of the XYZ-Hamiltonians are presented. The group G and commutation relations in this case are induced by a Heisenberg group over a finite Abelian group X. Here we restrict ourselves to the case of cyclic group $X = \mathbb{Z}/\mathbb{Z}m$ when generators of G are $F_{(i,j)}$ $(i,j = 0,1,\ldots,m-1)$ and commutation relations are

$$[F_{(i_1,j_1)}, F_{(i_2,j_2)}] = \{\zeta_m^{-j_1 i_2} - \zeta_m^{-j_2 i_1}\} F_{(i_1+i_2, j_1+j_2)}$$

with $\zeta_m = \exp(2\pi\sqrt{-1}/m)$ and addition (mod m). The operator variables are $F_{\alpha,n}: \alpha \in X^2 = (\mathbb{Z}/\mathbb{Z}m)^2$, $n = 0,\pm 1,\pm 2,\ldots$ and the Hamiltonian H has the form

$$H_{XYZ} = \sum_{\substack{\alpha\in X^2, \\ \alpha\neq 0}} J_\alpha \sum_{\substack{n=1 \\ r=1 \\ n\neq r}}^{N} F_{\alpha,n} F_{-\alpha,r}.$$

The coupling constants J_α are not arbitrary and relations between them are presented in [8]. However for generalizes XXX-model (or Heisenberg model) when all J_α are equal, the corresponding Hamiltonian is commuting with a local one. Namely, the Hamiltonian H_{XYZ} for $J_\alpha = 1$ commutes with the following local Hamiltonian

$$H'_{XXX} = \sum_{\substack{\alpha\in X^2, \\ \alpha\neq 0}} \sum_{n-1}^{N} F_{\alpha,n} F_{-\alpha,n+1}.$$

It should be noted that the Hamiltonian H'_{XXX} for $m = 2$ is equivalent to the Hamiltonian H''_{N-2}.

This research was supported by the Office of Naval Research under constracts N00014-78-C-0138 and NR041-529.

References

[1] E.T. Whittaker, A treatise on the analytical dynamics of particles and rigid bodies, Cambridge, 1927.

[2] M. Gutzwiller, Ann. of Physics 124, 347 (1980); Ann. of Physics (1981) (to appear).

[3] A.A. Kirillov, Representation Theory, Springer, 1972. R. Gilmore, Lie Groups, Lie Algebras and some of their applications, John Wiley, 1974.

[4] L. Onsager, Phys. Rev. 65, 117 (1944). C.J. Thompson, Mathematical statistical mechanics, Princeton Univ. Press, 1972. J.M. Drouffe, C. Itzykson, Phys. Repts. 38C, 133 (1975).

[5] R. Baxter, Ann. Phys. 76, 1, 25, 48 (1973).

[6] D.V. Chudnovsky, G.V. Chudnovsky, Phys. Lett., 98B, 83 (1981), Phys. Lett., 79A, 36 (1980).

[7] Here $G[[\theta]]$ denotes the formal power series in θ with the coefficients from G.

[8] D.V. Chudnovsky, G.V. Chudnovsky, Lett. Math. Phys. 5, 43 (1981). (EN-Saclay Prepring DPh-T/80/131, September 1980 (to appear)).

[9] R.J. Baxter, Trans. Royal Soc. London, A289, 315 (1978); R.J. Baxter, Academic Press (to appear).

[10] N. Bourbaki, Groupes et algèbres de Lie, Herman, Paris, 1968.

[11] a) A.B. Zamolodchikov, Comm. Math. Phys. 69, 165 (1979).

b) E.K. Sklanin, L.A. Takhtadjan, L.D. Faddeev, Theor. Math. Phys. 40, 688 (1980). L. Takhtadjan, L.D. Faddeev, Usp. Mat. Nauk. 34, 13, 1979.

[12] R.M. Case, J. Math. Phys., 15, 2166 (1974). K.M. Case, M. Kac, J. Math. Phys., 14, 594 (1973).

[13] E.T. Whittaker, Watson, A course of modern analysis, v. 2, Cambridge Univ. Press, 1927.

Department of Mathematics
Columbia University
New York, NY
USA

Classical and Quantum Operator Nonlinear Schrödinger Equation. I

by

D.V. Chudnovsky, G.V. Chudnovsky,
A. Neveu

Abstract: We consider generalizations of the classical nonlinear Schrödinger equation, $i\psi_t = \psi_{xx} + 2\psi\psi^+\psi$, to operator functions $\psi = \psi(x,t)$ and their solvability via the inverse scattering method. This provides a new class of soluble field theories in one-space, one-time dimensions, which, after quantization, are equivalent to a system of many, nonidentical, particles with δ-function interactions and a spectrum of bound states richer than in the usual model.

1. Introduction.

There exists by now a rather large class of exactly soluble models in two dimensions. These models, which have accumulated slowly over the years have recently been shown all to come from some simple and very deep fundamental mathematical structure, which unifies, and simplifies, their solutions; these solutions had initially been obtained by clever tricks which at first seemed unrelated to each other, and, conceptually as well as in practice, it is very pleasant to have this unified point of view.

Physically, however, one cannot remain happy very long with the present situation. The reason is that these models are highly simplified, and only a very limited set of physical statistical mechanical systems in two-dimensions can thus be solved. From a quantum field theory point of view, the particle spectrum of the available soluble two-dimensional models is usually relatively simple, in the sense that it contains a small set of particles and/or a small set of internal quantum numbers, the only exception being the $(\bar{\psi}\psi)^2$ models. Most of these soluble models may not have much in common with more interesting higher dimensional theories (in particular full Yang-Mills in four dimensions); more importantly perhaps, there remain two-dimensional models unsolved which are believed to be soluble, like the principal chiral field, and whose fundamental fields have the same internal space structure as the gauge fields of four-dimensional Yang-Mills.

In this pqper, we propose a generalization of the classical non-linear Schrödinger equation to operator functions. This would include in particular finite-dimensional square matrices. Sending the dimension of the matrix to infinity can give us a soluble planar field theory in one space-one time dimension, which is equivalent to the quantum mechanics of a linear chain of atoms with nearest neighbour δ potentials, a hitherto unsolved problem. In section II, we present the mathematics of the operator generalization of the non-linear Schrödinger equation, and of the related inverse scattering method, showing in particular the existence of an infinite set of conservation laws. In section III, we explain how the quantization of this equation can lead to the solution of the problem of non-relativistic particles interacting via nearest neighbour potentials on a line.

2. Operator Generalization of the non-linear Schrödinger Equation.

Let H be an arbitrary Hilbert space. We consider operator equations for operators on H. Usually, we restrict operator variables to be bounded operators on H, satisfying some additional properties (e.g. a compactness is quite natural). Symbols ψ, φ, U, V, ... will denote operator variables, being functions of space x and time t, restricted by the conditions above. In previous papers [1-3], a class of operator completely integrable systems was defined. Here, we shall consider the most interesting one, the operator non-linear Schrödinger equation. One first gest coupled operator non-linear Schrödinger equations

$$i\varphi_t = \varphi_{xx} + 2\varphi\psi\varphi$$

(2.1)

$$-i\psi_t = \psi_{xx} + 2\psi\varphi\psi$$

in which one can consistently choose $\varphi = \psi^+$, obtaining the operator non-linear Schrödinger equation on H:

$$i\varphi_t = \varphi_{xx} + 2\varphi\varphi^+\varphi \tag{2.2}$$

These equations are derived as a particular case of isospectral deformation equations: they appear in the isospectral deformation of the Dirac equation on H or of the Schrodinger equation on $H \otimes H$; the Dirac equation is

$$\frac{dF}{dx} = -VF + i\zeta\sigma_3 F$$
$$\frac{d\tilde{F}}{dx} = \tilde{F}V - i\zeta\tilde{F}\sigma_3 \tag{2.3}$$

with

$$\sigma_3 = \begin{pmatrix} I & 0 \\ 0 & -I \end{pmatrix}, \quad V = \begin{pmatrix} 0 & \varphi \\ \psi & 0 \end{pmatrix}$$

the corresponding Schrodinger equation is

$$\frac{d^2F}{dx^2} = UF - \zeta^2 F$$
$$\frac{d^2\tilde{F}}{dx^2} = \tilde{F}U - \zeta^2\tilde{F} \tag{2.4}$$

with

$$U = \begin{pmatrix} \varphi\psi & \psi_x \\ \psi_2 & \psi\varphi \end{pmatrix} \tag{2.5}$$

Complete integrability of equations (2.1-2.2) means that they are linearized by means of the inverse scattering transformation. More precisely, we consider operator Jost solutions of (2.4) with the following asymptotic behaviors:

$$F(x,\zeta) \sim T(\zeta)\exp(-i\zeta x) \qquad \text{for } x \to -\infty$$
$$F(x,\zeta) \sim \exp(-i\zeta x) + R(\zeta)\exp(i\zeta x) \quad \text{for } x \to +\infty \tag{2.6}$$

(2.6) cont. $\tilde{F}(x,\zeta) \sim \tilde{T}(\zeta)\exp(-i\zeta x)$ for $x \to -\infty$

$$\tilde{F}(x,\zeta) \sim \exp(-i\zeta x) + R(\zeta)\exp(i\zeta x) \quad \text{for } x \to +\infty.$$

Then $R(\zeta)$ is called the scattering coefficient, and basically, the potential $U(x)$ in (2.4) is reconstructed in terms of $R(\zeta)$. For this, the discrete spectrum of (2.4) is needed. We consider only the case when $U(x)$ decreases at infinity exponentially or faster. In this case, $R(\zeta)$ can be analytically continued to the whole ζ plant, and may have n poles at the points ζ_j, $j = 1,\dots,n$. The points ζ_j^2 correspond to discrete eigenvalues of the problem (2.4). We introduce the scattering data corresponding to the discrete spectrum:

$$\lim_{\zeta\to\zeta_j} (\zeta - \zeta_j)R(\zeta) = P_j, \quad j = 1,\dots,n.$$

Then, the potential $U(x)$ is in one-to-one correspondence with the scattering data

$$s = \{R(\zeta);\zeta_j,P_j: j = 1,\dots,n\}.$$

For inverse scattering, we can use the Gelfand-Levitan equation. Starting from s, we build the following operator kernel

$$F(y) = \frac{1}{2\pi}\int_{-\infty}^{+\infty} R(\zeta)\exp(i\zeta y)d\zeta + \Sigma_{i=1}^{n} P_i\exp(-i\zeta_i y).$$

The operator Gelfand-Levitan equation with the kernel $F(y)$ has the form

$$(2.7) \qquad K(x,x_1) + F(x + x_1) + \int_x^\infty K(x,z)f(z + x_1)dz = 0, \; x_1 \le x.$$

The potential $U(x)$ is reconstructed from the scattering data s by using (2.7) and

$$(2.8) \qquad U(x) = -2\frac{d}{dx}K(x,x).$$

Now, equations (2.1-2.2) as any other isospectral deformation equations become linear equations for the scattering data. We already formulated in refs. 2 and 3 a general assertion of this type, and we repeat it here:

Main theorem: We introduce the following linear operator $\mathcal{L}$ connected with $U(x)$:

$$(2.9) \qquad 4\mathcal{L}\cdot H(x) = H_{xx}(x) - 2\{U(x),H(x)\} + G\cdot\int_x^\infty dx_1\, H(x_1)$$

with

$$G\cdot H(x) = \{U_x(x),H(x)\} + [U(x),\int_x^\infty dx_1 [U(x_1),H(x_1)]]$$

(as usual, $\{A,B\} = AB + BA$; $[A,B] = AB - BA$).

For fixed constant operators M and N on H and entire functions $\alpha(z)$ and $\beta(z)$, the following non-linear operator evolutionary equation of $U(x,t)$

$$(2.10) \qquad U_t(x,t) = \alpha(\mathcal{L})[N,U(x,t)] + \beta(\mathcal{L})G\cdot M$$

is equivalent to a linear differential equation for the scattering operator coefficient $R(k,t)$ of the Schrödinger operator (2.4) with the potential $U(x,t)$:

$$(2.11) \qquad R_t(k,t) = \alpha(-k^2)[N,R(k,t)] + 2ik\beta(-k^2)\{M,R(k,t)\}.$$

In particular, the main theorem contains the linearization of the operator non-linear Schrodinger equations (2.1-2.2):

Corollary: For the potential U in (2.5)

$$U = \begin{pmatrix} \varphi\psi & \varphi_x \\ \psi_x & \psi\varphi \end{pmatrix}$$

the operator non-linear Schrödinger equation (2.1) has the form

(2.12) $$U_t = i\mathcal{L}\cdot[\sigma_3, U(x,t)]$$

and is equivalent to a linear differential equation for the reflection coefficient R(k,t) of (2.4-2.5):

(2.13) $$R_t(k,t) = -ik^2[\sigma_3, R(k,t)].$$

Of course we have a n-th order operator non-linear Schrödinger equation if we change (2.12) into

(2.14) $$\frac{\partial}{\partial t_n} U = (i\mathcal{L})^n\cdot[\sigma_3, U]$$

for the same U(x,t) as in (2.5). Linearization of equation (2.14) gives us

(2.15) $$\frac{\partial}{\partial t_n} R(k,t) = (-ih^2)^n[\sigma_3, R(k,t)].$$

Complete integrability of (2.1-2.2) means also:

a) the existence of infinitely many conservation laws [3];

b) the existence of Bäcklund transformations;

c) complete integrability of a quantized version.

In the case dim H = 1, we obtain the usual non-linear Schrödinger equation $i\varphi_t = -\varphi_{xx} + 2|\varphi|^2\varphi$.

In many physical problems, there appears so-called multidimensional Schrödinger equations, having the form

$$i\varphi_t = -\Sigma_{j=1}^n \frac{\partial^2\varphi}{\partial x_j^2} + \varphi\varphi^+\varphi$$

for $\varphi = \varphi(x_1,\ldots,x_n,t)$. However, the equation is not completely integrable for $n \geq 2$ even if dim $H \geq 1$. First of all, there are unstable solutions even for n = 2, and soliton interactions are not elastic, as seen on computer calculations. Also, it for $n \geq 2$ does not have non-trivial higher conservation laws. In other words, we cannot expect any immediate multidimensional generalization of the non-linear Schrödinger equation to be completely integrable. It is an intriguing problem to consider the complete integrability in the stationary case (no t-dependence); in order to stress its non-triviality, we mention that the stationary system for n = 2 does not have non-trivial polynomial local conservation laws.

The class of n-th order non-linear Schrödinger equations generated by evolution equations (2.14) provides us with an infinite family of local Hamiltonian flows commuting with the flow of coupled operator non-linear Schrödinger equation (2.1). The fact that flows (2.14) commute with each other for $n \geq 1$ follows from the linear evolution equations (2.15) on scattering coefficients $R(k,t_1,t_2,\ldots,t_n)$. In particular, as the second canonical flow commuting with the coupled operator nonlinear Schrödinger equation one obtains operator coupled modified Korteweg-de Vries equation. In the notations (2.5) and (2.14) with $t_2 = y$ these equations are written in the canonical form

(2.16)
$$\varphi_y = \varphi_{xxx} - 3(\varphi_x\psi\varphi + \varphi\psi\varphi_x);$$
$$\psi_y = \psi_{xxx} - 3(\psi_x\varphi\psi + \psi\varphi\psi_x).$$

One should notice proper normalization of equations (2.1) and (2.16).

Higher order operator non-linear Schrödinger equations (2.14) also provide us with the description of local polynomial higher conservation laws [3]. Indeed the fact that dynamical flows (2.14) commute, means simply that any equation

(2.17)
$$\frac{\partial}{\partial t_p}U = p(i\mathcal{L})\cdot[\sigma_3,U]$$

for polynomial $p(i\mathcal{L})$ in $i\mathcal{L}$ describes an algebraic local conservation law of the non-linear Schrödinger equations (2.1) and, simultaneously of any of the higher equations (2.14). To prove this we rewrite the condition of commutativity:

$$\frac{\partial}{\partial t_p}\cdot\frac{\partial}{\partial t_q}\,U = \frac{\partial}{\partial t_q}\cdot\frac{\partial}{\partial t_p}\,U$$

for two polynomials $p(i\mathcal{L})$, $q(i\mathcal{L})$. In particular,

$$(2.18)\qquad \frac{\partial}{\partial t_p}\{i\mathcal{L}\cdot[\sigma_3,U]\} = \frac{\partial}{\partial t}\{p(i\mathcal{L})\cdot[\sigma_3,U]\}.$$

Now we use the explicit form of $\mathcal{L}$ (2.9). The left hand side of the equation (2.18) can be written as

$$\frac{\partial}{\partial x}\begin{pmatrix} \frac{\partial}{\partial t_p}(\varphi\psi_x - \varphi_x\psi) & ; & \frac{\partial}{\partial t_p}\varphi \\ \frac{\partial}{\partial t_p}\psi & & \frac{\partial}{\partial t_p}(\psi\varphi_x - \psi_x\varphi) \end{pmatrix}.$$

Since, by (2.17), $\partial/\partial t_p\ \varphi$ and $\partial/\partial t_p\ \psi$ are expressed as polynomials in $\varphi,\psi,\varphi_x,\psi_x,\dots,\varphi_{x\dots x},\psi_{x\dots x}$; the equation (2.18) takes the form of conservation law

$$(2.19)\qquad \frac{\partial}{\partial x}\{\mathfrak{M}\cdot U\} = \frac{\partial}{\partial t}\{p(i\mathcal{L})\cdot[\sigma_3,U]\}.$$

Local polynomial conservation laws (2.19) are consistent with the Hamiltonian structure for the coupled operator nonlinear Schrödinger equation (2.1), and coupled operator MKdV equation (2.16). The corresponding symplectic structure is generated by the following definition of Poisson brackets $\{F,G\}$ between functionals of φ and ψ:

$$(2.20)\qquad \{F,G\} = \int \frac{\delta F}{\delta(\varphi,\psi)}\begin{pmatrix} 0 & I \\ -I & 0\end{pmatrix}\frac{\delta G}{\delta(\varphi,\psi)}^t dx$$

where $\frac{\delta K}{\delta(f_1,\dots,f_r)} = (\frac{\delta K}{\delta f_1},\dots,\frac{\delta K}{\delta f_r})$ and $\frac{\delta K}{\delta(f_1,\dots,f_r)}^t$ $= (\frac{\delta K}{\delta f_1},\dots,\frac{\delta K}{\delta f_r})^t$.

With the definition (2.20) of Poisson brackets all Hamiltonian flows (2.17) are in involution. Consequently conservation laws (2.19) are also involutive. Symplectic structure (2.20) still applies for the reduction (2.2) of (2.1), when $\psi = \varphi^+$. All conservation laws (2.19) do not degenerate under this reduction.

As we see, it is easy to describe local conservation laws that are differential polynomials in φ and ψ and lead to an infinite sequence of local commuting Hamiltonians. However, the complete description of the hidden symmetry structure of equations like the matrix nonlinear Schrödinger equation reveals more conservation laws. Those laws are nonlocal in ϕ and ψ. Their existence for equations of the nonlinear Schrödinger type seems to go unnoticed, though similar nonlocal conservation laws were found for nonlinear σ-models by Lüscher and Pohlmeyer (Nucl. Phys. B137 (1978), 46). These nonlocal conservation laws do commute (are involutive) with local ones, but do not commute with each other and generate quite nontrivial infinite dimensional Lie algebra of conservation laws. This Lie algebra is one of the Generalized Cartan matrix (GCM) Lie algebras that belongs to a series of affine Lie algebras (called Kac-Moody algebras). These algebras currently are investigated in connection with completely integrable systems, but in a different way, when the symplectic manifold for a completely integrable system is taken as an orbit under coadjoint action of an affine Lie algebra. We are not going to consider here the representation theory part of the problem, important for the proper understanding of the quantized nonlocal conservation laws, but simply describe the algebra obeyed by nonlocal conservation laws.

Nonlocal conservation laws arise from eigenfunction expansion of an auxiliary linear problem (2.3) for the Dirac equation associated as a linear spectral problem with nonlinear Schrödinger equation. For this one takes normalized fundamental solutions F of (2.3) in the form

$$F = \Sigma_{n=0}^{\infty} T_n(i\zeta)^{-n}\cdot\exp(\sigma_3 i\zeta x),$$

$T_0=\begin{pmatrix} I & 0 \\ 0 & I\end{pmatrix}$ [3] and $T_n=0$ for $x=-\infty$ and $n>0$. If

$$T_n\Big|_{x=+\infty} = \begin{pmatrix} Q_n & 0 \\ 0 & R_n \end{pmatrix}$$

then nonlocal conservation laws (matrix or operator ones) have the form

$$Q_n \text{ and } R_n,$$

which is the proper form of nontrivial (nonzero) nonlocal conservation laws provided that $|\phi| \to 0$, $|\psi| \to 0$ as $|x| \to \infty$.

To write a complete list of commutation relations between elements of Q_n and R_n one introduces $\phi(\zeta) = \Sigma_{n=0}^{\infty} T_n|_{\bar{x}=+\infty} \cdot \zeta^{-n}$ and then in the symplectic structure (2.20):

$$(\zeta-\zeta')\{\phi(\zeta)_{ac}, \phi(\zeta')_{bd}\} = \phi(\zeta)_{bc}\, \phi(\zeta')_{ad} - \phi(\zeta)_{ad}\, \phi(\zeta')_{bd}.$$

Moreoever, in order to present non-local conservation laws in a more convenient way we make one more transformation:

$$\ln \phi(\zeta) = \Sigma_{n=0}^{\infty} \begin{pmatrix} Q^{(n)} & 0 \\ 0 & R^{(n)} \end{pmatrix} \zeta^{-n-1}$$

The algebra of commutation relations beween elements of the matrices $Q^{(n)}$ and $R^{(n)}$ in symplectic structure (2.20) seems to correspond to that of appropriate affine algebra $SU(N)[[t, t^{-1}]]$ for N equal to the dimension of H. Let ϕ be $N\times M$ and ψ be $M\times N$ matrices so $Q^{(n)}$ are $N\times N$ and $R^{(n)}$ are $M\times M$ matrices. The set of commutation relations between matrix elements of conservation laws $Q^{(n)}$ and $R^{(n)}$ has the following form

$$\{Q_{ij}^{(n)}, Q_{kl}^{(m)}\} = \delta_{kj} Q_{il}^{(n+m)} - \delta_{il} Q_{kj}^{(n+m)} + \tilde{Q}_{n,m};$$

(*)
$$\{R_{ij}^{(n)}, R_{kl}^{(m)}\} = \delta_{il} R_{kj}^{(n+m)} - \delta_{kj} R_{il}^{(n+m)} + \tilde{R}_{n+m};$$

$$\{Q_{ij}^{(n)}, R_{kl}^{(m)}\} = 0.$$

Here $Q_{\alpha\beta}^{(n)}$ and $R_{\alpha\beta}^{(n)}$ are corresponding matrix elements of $Q^{(n)}$ and $R^{(n)}$ and $\tilde{Q}_{n,m}$ and $\tilde{R}_{n,m}$ are polynomials in the matrix elements of $Q^{(k)}$ and $R^{(k)}$ for $k< \min\{n,m\}$. However the commutation relations (*) take a form of generating relations for GCM algebras $SU(N)[[\lambda, \lambda^{-1}]]$ and $SU(M)[[\lambda, \lambda^{-1}]]$, if one looks not on the Poisson brackets but on the commutators between vector fields. Let, e.g. $\partial_{ij,n}$ denote vector field acting on elements of ϕ and ψ, defined by a (nonlocal) Hamiltonian $Q_{ij}^{(n)}$. Then

(**)
$$[\partial_{ij,n}, \partial_{kl,m}] \doteq \delta_{kj}\, \partial_{il,n+m} - \delta_{il}\partial_{kj,n+m}.$$

Here we used gauge transformations on ϕ and ψ, which are of the form $(\phi,\psi) \mapsto (A\phi D^{-1}, D\phi A^{-1})$ for invertible A and B. The sign $\doteq$ in (**) means equal up to a gauge transformation. The Hamiltonians $Q_{ij}^{(n)}$ and $R_{ij}^{(n)}$ are also connected with equations of the form (2.10) for

$\alpha(\lambda) = \lambda^{n'}$ or $\beta(\lambda) = \lambda^{n'}$ and appropriate N and M. In particular, the sequence of matrix (operator) conservation laws $Q^{(u)}$ and $R^{(u)}$ can be extended to negative integers n, if $\alpha(\lambda)$ and $\beta(\lambda)$ have poles at $\lambda=0$.

Traces of $Q^{(n)}$ and $R^{(m)}$ mutually commute for all n and m and commute with all other Hamiltonians $Q_{ij}^{(n')}$ and $R_{ij}^{(m')}$. Moreover tr $Q^{(n)}$ and tr $R^{(n)}$ are exactly local polynomial conservation laws that were derived above from (2.10). The terms of the lowest degree of $Q^{(n)}$ and $R^{(n)}$ are the following

$$Q^{(n)} = - \int_{-\infty}^{\infty} dx\{\phi \frac{\partial^n}{\partial x^n}\psi + \ldots\},$$

$$R^{(n)} = - \int_{-\infty}^{\infty} dx\{\psi \frac{\partial^n}{\partial x^n}\phi + \ldots\},$$

The first element in the sequence are the following

$$Q^{(0)} = -\int_{-\infty}^{\infty} dx\ \phi\psi,\quad R^{(0)} = -\int_{-\infty}^{\infty} dx\ \psi\phi,$$

$$Q^{(1)} = -\int_{-\infty}^{\infty} dx[\phi\psi_x - \phi\psi\int_{-\infty} dx'\ \phi\psi] - \tfrac{1}{2}Q^{(0)2}$$

$$R^{(1)} = -\int_{-\infty}^{\infty} dx[\psi\phi_x - \psi\phi\int_{-\infty}^{x} dx'\ \psi\phi] - \tfrac{1}{2}R^{(0)2}.$$

Relations (*) are to be modified in order to obtain generators of the affine Lie algebra $SU(N)[[\lambda,\lambda^{-1}]]$.

The quantization of systems (2.1), (2.16) will be discussed in the next section. Now we consider solutions of the corresponding classical equations. The transformation (2.7), (2.8) of nonlinear equations to linear integral equations determines in general reduction of any system (2.10) to the linear evolution (2.11). An alternative approach to explicit solution of completely integrable system of the type (2.1) is based on the Bäcklund transformation. Namely, Bäcklund transformation changes the scattering data s into a new set of data s', where a) reflection operator $R(\zeta)$ is multiplied by a Blaschke factor; b) new eigenvalue ζ_{n+1} is added or c) one eigenvalue ζ_j is omitted. Under the transformation from s to s' the potential $U(x,t)$ is changed to a new potential $U'(x,t)$ so that if $U(x,t)$ satisfy any of equations (2.10) then $U'(x,t)$ satisfies the same equation. Usually Bäcklund transformation is written in a complicated form but we propose below a very simple formula for the Bäcklund transformation valid for a general class of linear spectral problems including that of (2.3) and (2.4).

Theorem 2.21: Let $U_j(\lambda)$ be matrix functions with coefficients depending on z_1 (= x), z_2 (= y), z_3 (= t),... and depending rationally on λ: $j = 1,2,3,\ldots$. Let μ_1, μ_2 be two complex numbers, different from poles of $U_j(\lambda)$ and let $\Phi(\lambda)$ be a (fundamental) solution of linear spectral problems:

$$\frac{\partial}{\partial z_j}\Phi(\lambda) = U_j(\lambda)\Phi(\lambda) \tag{2.22}$$

$j = 1,2,3,\ldots$. We consider two particular eigenfunctions $\vec{\psi}^{\,t}_{\mu_2}$ and $\vec{\psi}_{\mu_1}$ of a direct and inverse spectral problem (2.22):

$$\vec{\varphi}^{\,t}_{\mu_2} = \Phi(\mu_2)c_2, \quad \vec{\psi}_{\mu_1} = c_1\Phi(\mu_1)^{-1}. \tag{2.23}$$

Let us define

$$\mathcal{L}(\lambda) = I + \frac{\mu_1-\mu_2}{\lambda-\mu_1}\{\vec{\varphi}^{\,t}_{\mu_2}\cdot\vec{\psi}_{\mu_1}\cdot\frac{1}{(\vec{\varphi}_{\mu_2},\vec{\psi}_{\mu_1})}\} \tag{2.24}$$

where $(\vec{\varphi}_{\mu_2},\vec{\psi}_{\mu_1}) = \vec{\psi}_{\mu_1}\cdot\vec{\varphi}^{\,t}_{\mu_2}$ is a scalar product of $\vec{\varphi}_{\mu_2}$ and $\vec{\psi}_{\mu_1}$. Then for new potentials $U'_j(\lambda)$ and new eigenfunctions $\Phi'(\lambda)$ defined below, equations of the type (2.22) are satisfied. If we set

$$\begin{aligned} U'_j(\lambda) &= \frac{\partial}{\partial z_j}\mathcal{L}(\lambda)\cdot\mathcal{L}(\lambda)^{-1} + \mathcal{L}(\lambda)U_j(\lambda)\mathcal{L}(\lambda)^{-1}; \\ \Phi'(\lambda) &= \mathcal{L}(\lambda)\Phi(\lambda), \end{aligned} \tag{2.25}$$

then $U'_j(\lambda)$ has poles at the same points in λ-plane as $U_j(\lambda)$ (of the same orders) and one has

$$\frac{\partial}{\partial z_j}\Phi'(\lambda) = U'_j(\lambda)\,\Phi'(\lambda): \tag{2.26}$$

$j = 1,2,3,\ldots$. In particular, if $U_1(\lambda)$, $U_2(\lambda)$ satisfy nonlinear completely integrable equations

$$\left[\frac{\partial}{\partial z_1} - U_1(\lambda), \frac{\partial}{\partial z_2} - U_2(\lambda)\right] = 0. \tag{2.27}$$

then new potentials $U'_j(\lambda)$ (2.25) satisfy the same nonlinear equations

$$\left[\frac{\partial}{\partial z_1} - U'_1(\lambda), \frac{\partial}{\partial z_2} - U'_2(\lambda)\right] = 0.$$

In particular, theorem 2.22 gives the description of soliton solutions of all equations of type (2.10). Equations of the type (2.10) are characterized by the condition that $U_1(\lambda)$ (written in (2.3)) has a single pole of the first order at infinity in the λ plane, while $U_2(\lambda)$ has the only pole at infinity in the λ plane of the order equal to the degree of $\alpha(\mathcal{L})$. For example, we can present one soliton solution for equations (2.1) or (2.16). Let us take two finite points $i\mu_1$ and $i\mu_2$ with $\mu_1 \neq \mu_2$ and $\Phi(\mu) = \exp\{\sigma_3\mu x + \sigma_3\mu^2 t + \sigma_3\mu^3 y\}$. Let $\vec{\varphi}_2^{\,t} = \Phi(i\mu_2)\vec{c}_2^{\,t}$ and $\vec{\varphi}_1 = \vec{c}_1\Phi(i\mu_1)^{-1}$, where $\vec{c}_1$, $\vec{c}_2$ are elements of $H \oplus H$. We put, as before $P = \vec{\varphi}_2^{\,t}\vec{\varphi}_1 \cdot \frac{1}{(\vec{\varphi}_1,\vec{\varphi}_2)}$, so that P is a projector in $H \oplus H$. Then the new, transformed, potential V' takes the form

$$V' = U_0 + i(\mu_1-\mu_2)[P,\sigma_3].$$

Here the starting potential $U_0 = 0$ as $\frac{\partial}{\partial x}\Phi(\mu) = \sigma_3\mu\Phi(\mu)$, $\frac{\partial}{\partial t}\Phi(\mu) = \sigma_3\mu^2\Phi(\mu)$, $\frac{\partial}{\partial y}\Phi(\mu) = \sigma_3\mu^3\Phi(\mu)$. The result is the following:

$$\varphi = 2\sqrt{-1}(\mu_2-\mu_1)\vec{\varphi}^{\,t}_{1\mu_2}\vec{\varphi}_{1\mu_1}\frac{1}{w};$$

$$\psi = 2\sqrt{-1}(\mu_1-\mu_2)\vec{\varphi}^{\,t}_{2\mu_2}\vec{\varphi}_{2\mu_1}\frac{1}{w}$$

for $w = \vec{\varphi}^t_{2\mu_1}\vec{\varphi}_{1\mu_2} - \vec{\varphi}_{1\mu_1}\vec{\varphi}^t_{2\mu_2}$ with $\vec{\varphi}^t_{1\mu_i} = \vec{c}^t_i \exp\{i\mu xI+(i\mu)^2 tI+(i\mu)^3 yI\}$ and $\vec{\varphi}^t_{2\mu_i} = \vec{c}^t_i \times \exp\{-(i\mu)xI - (i\mu)^2 It - (i\mu)^3 yI\}$ and $\vec{c}_i$ elements of H: $i = 1,2$.

Then φ, ψ satisfy a) the operator coupled nonlinear Schrodinger equation (2.1); b) the operator coupled modified KdV equation (2.16). In order to get the solution of nonlinear Schrödinger equation (2.2) one must assume that $\mu_2 = -\mu_1$.

3. The quantized field theories.

The operator systems considered in section 2 give rise to quantum field theories. The relationship with quantized field theories generalizing quantum nonlinear Schrödinger are twofold. First of all operator systems (2.1) or (2.2) can be considered itself as a generalization of quantum nonlinear Schrödinger system. In order to get quantum nonlinear Schrödinger system from (2.2) one should impose an invariant restriction in the form of canonical commutation relations

$$[\varphi^+(x),\varphi(y)] = \varkappa\delta(x - y) \tag{3.1}$$

for $\varkappa \neq 0$. Another possible quantization connected with operator systems of §2 consists in the quantization of operator system itself. In other words we take symplectic form defined by (2.20) and quantize it. In other words instead of one component field theory we consider multicomponent field theory with components φ_{ij}, ψ_{ij} where $\varphi = (\varphi_{ij})$ and $\psi = (\psi_{ji})$. In the case of φ being $N \times M$ matrix and ψ being $M \times N$ matrix, we have 2NM component field theory with components $\varphi_{ij},\varphi^*_{ij}$: $i = 1,\dots,N$: $j = 1,\dots,M$ where $\psi_{ji} = \varphi^*_{ij}$ $(i=1,\dots,N{:}j=1,\dots,M)$. The canonical commutation relations are

$$[\varphi^*_{ij}(x),\varphi_{k\ell}(y)] = \varkappa\delta_{ik}\delta_{j\ell}\delta(x - y) \tag{3.2}$$

for $\varkappa \neq 0$. Again, one can imbed the quantization (3.2) of matrix nonlinear Schrödinger equation (2.2) into a large operator nonlinear Schrödinger equation of the same type (2.2), where now φ and ψ are operators having block structure $M \times N$ and $N \times M$, respectively, with blocks corresponding to field operators φ_{ij} and φ^*_{ji}, correspondingly. This way we obtain a hierarchy of successful quantizations of field theories starting from quantum nonlinear Schrödinger equation. All equations in this hierarchy are completely integrable as quantum field theories. Namely, they possess infinitely many commuting quantum first integrals where commutation is understood in the sense of commutation rules (3.1) (for the ordinary nonlinear Schrödinger equation) and (3.2) (for matrix nonlinear Schrödinger case). The amazing property of these commuting quantum Hamiltonians is the fact that they describe certain polymers [4], [5] with potential of interaction build from δ-functions. Namely, the second quantization rules for the ordinary nonlinear Schrodinger equation (2.2) can be written on an n-particle sector [13] as a Hamiltonian describing n particles on the line with δ-potential of interaction

$$H_n^{(2)} = \frac{1}{2}\Sigma_{j=1}^{n} \frac{\partial^2}{\partial x_j^2} + \varkappa\Sigma_{i<j}\ \delta(x_i - x_j). \tag{3.3}$$

The famous Bethe Ansatz states that eigenfunctions of (3.3) can be obtained in the form of the combination of plane waves. The form of the Bethe Ansatz shows that there are n - 1 additional quantum Hamiltonians, commuting with $H_n^{(2)}$.

Results of §2 show easily how to construct these commuting Hamiltonians. Namely, one considers operator commuting Hamiltonian flow (2.14) and adds invariant restrictions (3.1). This way we obtain n canonical commuting Hamiltonians $H_n^{(i)}$: i = 1,...,n built from $\frac{\partial}{\partial x_j}$ and $\delta(x_i - x_j)$ $(i \neq j)$. These Hamiltonians are

$$\begin{aligned} H_n^{(i)} &= \frac{1}{i}\Sigma_{k=1}^{n} \frac{\partial^i}{\partial x_k^i} + \varkappa\Sigma_{k<\ell}\ \delta(x_k - x_\ell) \\ &\quad\times \left(\frac{\partial^{i-2}}{\partial x_k^{i-2}} + \frac{\partial^{i-2}}{\partial x_k^{i-3}\partial x_\ell} + \ldots + \frac{\partial^{i-2}}{\partial x_\ell^{i-2}}\right) + \ldots \end{aligned} \tag{3.4}$$

i = 1,2,...,n, where $H_n^{(i)}$ is homogeneous of the weight i, where $\frac{\partial}{\partial x_k}$ is counted with the weight 1 and $\delta(x_k - x_\ell)$ with the weight 2.

These conditions and the structure of $H_n^{(i)}$ in (3.4) determine uniquely $H_n^{(i)}$. As a consequence of the involutivity of (2.14), all Hamiltonians $H_n^{(i)}$ are commuting:

$$(3.5) \qquad [H_n^{(i)}, H_n^{(j)}] = 0: \quad i,j = 1,\dots,n.$$

Hence they possess common eigenfunctions numerated by a sequence of eigenvalues $(k_1,\dots,k_n)$. These normalized eigenfunctions naturally satisfy

$$(3.6) \qquad H_n^{(i)} \psi(k_1,\dots,k_n) = (\Sigma_{\ell=1}^n k_\ell^i)\psi(k_1,\dots,k_n).$$

The usual Bethe Ansatz [7] states that eigenfunctions $\psi(k_1,\dots,k_n)$ can be represented in terms of exponentials:

$$(3.7) \qquad \psi(k_1,\dots,k_n) = \Sigma_{\pi\in S_n} A(\pi)\exp\{\Sigma_{\ell=1}^n x_\ell k_{\pi(\ell)}\},$$

with summation over all permutation π of $\{1,\dots,n\}$. The corresponding formulae for $\psi(k_1,\dots,k_n)$ can be found in [4].

An alternative and more general approach to determination of eigenfunctions of quantized Hamiltonians is S-matrix approach. This approach is based on the Baxter method for models of statistical mechanics [8] and coincides essentially with the Yang method [7] for the case of δ-potential of interaction. The method of S-matrix is essentially quantum and determines commutation relations on coefficients of isospectral deformation equations, under which higher conservation laws survive quantization and preserve commutation.

4. S-matrix approach.

We present a description of the S-matrix approach to the determination of local quantum conservation laws and Bethe Ansatz for isospectral deformation equations. This method was essentially initiated by Yang [9] and studied in detail by Baxter for the eight-vertex model [8]. We refer the reader to [10] for an exposition of the modern developments. In the case of arbitrary isospectral deformation equations (2.27) generated by linear problems (2.22) the S-matrix method is based on the following lemma belonging to Baxter [8].

Lemma 4.1 [12]: Let $\Phi_{x_0}(\lambda;x)$ denote the fundamental solution of the linear spectral problem:

$$\frac{\partial}{\partial x}\Phi_{x_0}(\lambda;x) = U(\lambda;x)\Phi_{x_0}(\lambda;x) \tag{4.2}$$

with the initial condition $\Phi_{x_0}(\lambda;x_0) = I$. We assume that $U(\lambda;x)$ is $N \times N$ matrix with elements belonging to an associative algebra $\mathfrak{B}_0$ of operators, rationally depending on λ. We assume that commutation relations in $\mathfrak{B}_0$ on elements of $U(\lambda)$ imply the following relations for fundamental solutions $\Phi(\lambda;x)$ of (4.2):

$$R(\lambda,\mu)(\Phi_{x_0}(\lambda;x) \otimes \Phi_{x_0}(\mu;y)) = (\Phi_{x_0}(\mu;y) \otimes \Phi_{x_0}(\lambda;x))R(\lambda;\mu), \tag{4.3}$$

where $R(\lambda,\mu)$ is a nonsingular $N^2 \times N^2$ matrix. Then

$$[\mathrm{tr}\,\Phi_{x_0}(\lambda;x), \mathrm{tr}\,\Phi_{x_0}(\mu;y)] = 0. \tag{4.4}$$

In particular, $\log(\mathrm{tr}\{\Phi_{x_0}(\lambda,x)\})$ gives rise to commuting Hamiltonians determining canonical quantum isospectral deformation flows.

The $n^2 \times n^2$ matrix $R(\lambda,\mu)$ which is assumed to depend only on $\lambda - \mu$ is called quantum S-matrix of the spectral problem (4.2). The fundamental Baxter relation (4.3) determines commutation rules in the algebra $\mathfrak{B}_0$, generated by elements of $U(\lambda)$. The matrix $R(\lambda - \mu)$ itself is not arbitrary, because according to [12] it satisfies factorization equations for 1 + 1 dimensional S-matrices.

Baxter himself considered equations of the form (4.3) for difference spectral problem

$$\Phi_{n+1}(\lambda) = V_n(\lambda)\Phi_n(\lambda) \tag{4.5}$$

instead of the continuous problem (4.2). In this case equations (4.3) substituted by local equations

(4.6) $\quad R_{n,m}(\lambda,\mu)(V_n(\lambda) \otimes V_m(\mu)) = (V_m(\mu) \otimes V_n(\lambda))R_{n,m}(\lambda,\mu),$

where $R_{n,m}(\lambda,\mu) = R(\lambda,\mu)$ for $n = m$ and $R_{n,m} = \mathbb{I}^{\top}$ for $n \neq m$ [12], where $(\mathbb{I}^{\top})_{ab,cd} = \delta_{ad}\delta_{bc}$. This means e.g. that for $n \neq m$, elements of $\mathcal{L}_n(\lambda)$ and of $\mathcal{L}_m(\mu)$ commutes between each other. It is comparatively easy to verify conditions (4.6) but rather difficult to check conditions (4.3) since relations between fundamental solutions $\Phi_{x_0}(\lambda;x)$ and potential $U(\lambda;x)$ is rather complicated involving path integration. It was proposed to approximate continuous problem (4.2) by a discrete problem (4.5) in some concrete cases [11]. This approsch is not rigorous since approximation is quite arbitrary and second of all relations (4.6) for approximation (4.5) holds only approximately; limits are unlikely to be justified. However there is a rigorous procedure of proving (4.3) based on the Bäcklund transformation of the theorem 2.21 Namely, together with the continuous problem (4.2) one considers Bäcklund transformations given by (2.25) Applying successfully Bäcklund transformation n-times one gets discrete transformation equations (4.5) (see the second of the equations in (2.25)), then the S-matrix $R(\lambda,\mu)$ is chosen to satisfy exact equations (4.6). Under this condition, commutation relations (4.4) hold. Hence, we have a rigorous way of finding the S- matrix $R(\lambda,\mu)$ for any spectral problem (4.2) using Bäcklund transformation formula.

Let us present S-matrix expressions for spectral problem (2.3) giving us S-matrix of quantum operator nonlinear Schrödinger equation (2.2), (3.2).

Theorem 4.7: Let us consider quantum matrix nonlinear Schrödinger equation (2.1), where $\varphi = (\varphi_{ij})$, $\psi = (\psi_{ji})$ $(i = 1,\dots,N;\ j = 1,\dots,M)$, with commutation relations (3.2): $[\psi_{ji}(x),\varphi_{k\ell}(y)] = \varkappa\delta_{j\ell}\delta_{ik}\delta(x-y)$ $(i,k = 1,\dots,N; j,\ell = 1,\dots,M)$. Then the S-matrix $R(\lambda - \mu)$ of the spectral problem (2.3) with $V = \begin{pmatrix} 0 & \varphi \\ \psi & 0 \end{pmatrix}$, has the following nonzero elements only:

(4.8)
$$R(\lambda - \mu)_{aa,bb} = \delta_{ab};$$

$$R(\lambda - \mu)_{ab,ab} = \frac{\lambda - \mu}{\lambda-\mu+\varkappa}.$$

(4.8) cont. $$R(\lambda-\mu)_{ab,ba} = \frac{\varkappa}{\lambda-\mu+\varkappa}$$

Here $R(\lambda-\mu)$ is $(N+M)^2 \times (N+M)^2$ matrix with $\mathbb{Z}/\mathbb{Z}(N+M)$-symmetry.

Proof. In order to show the relations (4.3) we, following the program outline supra, use Bäcklund transformation in order to reduce differential equations (2.3) to the form of difference system (4.5) by Bäcklund transformation 2.21 at point $\mu_1 = \mu_2 = \infty$. Let us write down explicitly the formulae (2.24), (2.25) in this case. The application of Bäcklund transformation at points $\mu_1 = \mu_2 = \infty$ n times is denoted formally as

$$F_n(\lambda) = (B_{\mathfrak{B}_\infty})^n (F(\lambda)).$$

The recurrence formulae relation Bäcklund transformations can be written as

$$F_{n+1}(\lambda) = \mathcal{L}_n(\lambda) F_n(\lambda);$$

where

$$\mathcal{L}_n(\lambda) = \begin{pmatrix} \lambda + P_n & -U_n \\ U_n^{-1} & 0 \end{pmatrix}.$$

The matrix functions $F_n(\lambda)$ satisfy linear differential Dirac equation (2.3):

$$\frac{\partial}{\partial x} F_n(\lambda) = A_n(\lambda) F_n(\lambda) = (\lambda\sigma_3 + V_n) F_n(\lambda),$$

where

$$V_n = \begin{pmatrix} 0 & \varphi_n \\ \psi_n & 0 \end{pmatrix}$$

(and $\varphi_0 = \varphi$, $\psi_0 = \psi$). The recurrence formulae for φ_n, ψ_n are the following ones

$$\psi_{n+1} = -4\psi_n^{-1},$$

$$\varphi_{n+1} = \frac{1}{4}(\varphi_n^{-1}\varphi_{nx})_x\varphi_n - \frac{1}{4}\varphi_n\psi_n\varphi_n.$$

The relations between P_n, U_n are the following:

$$U_n = -\frac{1}{2}\varphi_n, \quad P_n = -\frac{1}{2}\varphi_n^{-1}\varphi_{nx}.$$

The commutation relations between elements of φ and ψ imply commutation relations between elements of matrices $\mathcal{L}_n(\lambda)$ and $\mathcal{L}_m(\mu)$ which take the form of the equation (4.6). Considering $R(\lambda - \mu)$ in (4.6) as a $(N + M)^2 \times (N + M)^2$ matrix rationally depending on $\lambda - \mu$ one gets an expression for its elements given in (4.8).

From the Baxter formula (4.2) one gets an algorithm for the Bethe Ansatz to the solution of the quantized Hamiltonian of system (2.1). We are not giving an expression for the corresponding eigenfunctions and eigenvalues, since the S-matrix given by (4.8) repeats the block structure of Yang's operators Y [7] and, hence the iterative formulae for Bethe Ansatz are those given by Yang [9].

References

[1] D.V. Chudnovsky and G.V. Chudnovsky, Phys. Lett. 73A (1979) 292.

[2] D.V. Chudnovsky, Phys. Lett. 74A (1979) 185.

[3] D.V. Chudnovsky, "One and multidimensional completely integrable systems arising from the isospectral deformation". Proceedings of the Les Houches International colloquium on complex analysis and relativistic quantum theory, Lecture notes in Physics, vol. 120, Springer-Verlag (1980).pp.352.

[4] E.Lieb,D.Mattis "Mathematical Physics in one dimension",AP, N.Y.

[5] G.'t Hooft, Nucl. Phys. B72 (1974) 461.

[6] E. Brézin, G. Parisi, G. Itzykson and J.B. Zuber, Commun. Math. Phys. 59 (1978) 35.

[7] M. Gaudin, Modeles exacts en mechanique satistique: la méthode de Bethe et sen généralisations, Note CEN-S1559 (1), 1559(2), 1972, 1973.

[8] R.J. Baxter, Trans. Royal Soc. London A289, 315 (1978).

[9] C.N. Yang, Phys. Rev. Lett. 19, 1312 (1967).

[10] J. Honerkam, Lecture Notes inPhys., 126, 417-428, Springer, 1980.
M. Karowsky, Lecture Notes in Phys., 126, 346,1980.

[11] E. Sklanin, L.A. Tachtadjan, L.D. Faddeev, Theor. Math. Phys. 40 (1979), no 3.

[12] D.V. Chudnovsky, G.V. Chudnovsky, Phys. Lett 79A, 36 (1980). Phys. Lett. 98 B, 83 (1981).

[13] F. Berezin, Method of second quantization

Department of Mathematics
Columbia Univeristy
New York, NY
USA

Laboratoire de Physique Théorique de l'École Normale Supérieure
24, rue Lhomond
75231 Paris
France

TRACE IDENTITIES FOR THE SCHROEDINGER OPERATOR AND THE WKB METHOD

G. Parisi

Abstract

Trace identities are derived for the operator $p^2 + x^q$, q positive even. For q=4 these identities can be used to estimate the ground state energy within an accuracy of about 10^{-4}.

1. Introduction

Trace identities for differential operators play an important role in many domains of mathematical physics[1-5]. In this paper, we consider the differential operator

$$H = p^2/2 + x^q = -\frac{\hbar^2}{2}\left(\frac{d}{dx}\right)^2 + x^q \tag{1}$$

i.e. the Schroedinger operator with potential x^q, q positive even.

This case differs from the one studied by Faddeev[5] in that the potential does not go to zero at infinity. In Section 2 of this paper we present a formal proof of the trace identities for H and we point out the relation between trace identities and the WKB expansion for the eigenvalues of H. In Section 3, we show how trace identities can be used to improve the WKB method for computing the ground state energy.

2. The trace identities

Let us consider the operator (1) where p and x satisfy the canonical commutation relations

$$[p, x] = i\hbar \tag{2}$$

(In the following $\hbar$ will be used as an expansion parameter which will finally be set to 1). H is a positive self-adjoint operator with discrete spectrum $\{\lambda_n\}$.

We define the following functions: the trace of the resolvent,

$$R(E) = \mathrm{Tr}(1/(H+E)) = \sum_{0}^{\infty} {}_n\, 1/(E+\lambda_n) = \int_0^\infty dt\, G(t)\exp(-Et)$$
$$G(t) = \mathrm{Tr}\left[\exp(-tH)\right] \qquad (3)$$

the Mellin transform of R(E),

$$M(s) = \int_0^\infty E^{-s} R(E)\, dE \qquad (4)$$

and the Zeta function of the operator H,

$$Z(s) = \mathrm{Tr}(H^{-s}) = \sum_{0}^{\infty} {}_n\, \lambda_n^{-s} . \qquad (5)$$

After simple manipulations we obtain:

$$Z(s) = \frac{1}{\pi} \sin(\pi s)\, M(s) . \qquad (6)$$

The analytic structure of the functions M(s) and Z(s) is quite interesting. Notice that if M(s) is not singular for integer s, (s = m), Z(m) must be zero.

Using the WKB method we find the asymptotic expansion[6, 7]:

$$\lambda_n \xrightarrow[n\to\infty]{} N^{2q/(q+2)}(C_o + C_1/N^2 + C_2/N^4 \cdots) \qquad N = n + \frac{1}{2} \qquad (7)$$

Eq. (7) implies that Z(s) is a meromorphic function with simple poles at the points:

$$s_k = \left(\frac{1}{2} + \frac{1}{q}\right)(1 - 2k) \qquad k = 0, 1, 2, \ldots . \qquad (8)$$

The residuum of Z(s) (Z_k) at $s = s_k$ is an algebraic function of $C_o, C_1, C_2 \ldots C_k$.

We want to derive a similar result for the function M(s). We need to know the large E behaviour of the function R(E), which can be obtained by studying the small t behaviour of G(t) using semiclassical methods. Indeed we know that

$$\mathrm{Tr}\, f(p,x) = \frac{\hbar}{2\pi} \int dp\, dx\, f(p,x) \tag{9}$$

if all p operators are on the left of the x operator. Using the Baker-Hausdorff formula for $\exp - tA - tB$, $A = p^2/2$, $B = x^q$, all x operators may be carried to the right of the p operators, thus producing an expansion in powers of $\hbar$ (small t):

$$G(t) \underset{t \to 0}{\simeq} \sum_0^\infty {}_k B_k \hbar^{2k-1} t^{-s_k}/\Gamma(-s_k + 1). \tag{10}$$

Equivalently:

$$R(E) \underset{E \to \infty}{\simeq} \sum_0^\infty {}_k B_k \hbar^{2k-1} E^{-1+s_k}. \tag{11}$$

Therefore, the only poles of $M(s)$ for $s < 1$ are at the points $s = s_k$; their residua being B_k. Of course the following relation holds:

$$Z_k = \sin(\pi s_k)/\pi B_k . \tag{12}$$

We now have at our disposal all the results needed to find the trace identities; they are

$$\begin{aligned} Z(m) &= 0 && \text{if } m \notin \{s_k\}, \\ Z(m) &= (-1)^m B_k && \text{if } m \in \{s_k\}, \end{aligned} \tag{13}$$

where m is a negative integer. If the second case is realized, the pole of $Z(s)$ at $s = s_k$ is absent ($C_k = 0$).

As an example, let us consider the harmonic oscillator $q = 2$. In this case, we find:

$$\begin{aligned} \lambda_n &= n + \frac{1}{2}, \qquad s_k = 1 - 2k, \\ Z(s) &= \sum_0^\infty {}_n 1/(n+\tfrac{1}{2})^s = (2^s - 1)\, \zeta(s), \end{aligned} \tag{14}$$

where $\zeta(s)$ is the Riemann ζ function.

Our results imply that $Z(n) = 0$, n being a negative even integer, i.e. we have recovered the trivial zeros of the $\zeta(s)$ function.

3. A numerical application

As stressed by Dikii[3], trace identities are useful for numerical purposes also if they involve the value of Z(s) in a region where the representation (5) does not converge.

Let us define the WKB approximant of order k,

$$W_n^{(k)} = N^{2q/(q+2)} \left[\sum_0^k {}_i C_i N^{-2i} \right] , \qquad N = n + \frac{1}{2} . \tag{15}$$

We can write

$$Z(s) = \sum_0^\infty {}_n (\lambda_n^{-s} - (W_n^{(j)})^{-s}) + \sum_0^\infty {}_n (W_n^{(j)})^{-s} . \tag{16}$$

The first sum is convergent for $s > s_{j+1}$ while the second sum may be calculated by explicit analytic continuation in m. When m is a negative integer, the trace identities (if $m \notin \{s_k\}$) can be written as:

$$\sum_0^\infty {}_n \left[\lambda_n^{-m} - (W_n^{(j)})^{-m} \right] = - \sum_0^\infty {}_n (W_n^{(j)})^{-m} \tag{17}$$

provided that $m > s_{j+1}$.

The identities (17) can be used to evaluate the energies of the lower states by approximating the convergent sum in the l. h. s. of (17) with a finite number of terms.

Let us consider the case q = 4. The simplest approximation is

$$\lambda_o - W_o^{(1)} = - \sum_0^\infty {}_n W_n^{(1)} . \tag{18}$$

A more refined approximation is

$$\begin{aligned} \lambda_o + \lambda_1 - W_o^{(2)} - W_o^{(2)} &= - \sum_0^\infty {}_n W_n^{(2)} , \\ \lambda_o^2 + \lambda_1^2 - \left[W_o^{(2)}\right]^2 - \left[W_o^{(2)}\right]^2 &= - \sum_0^\infty {}_n \left[W_n^{(2)}\right]^2 . \end{aligned} \tag{19}$$

We use the values[6, 7]

$$C_o = 3^{4/3}\pi^2\left[\Gamma(1/4)\right]^{-8/3}, \qquad C_1 = C_o/9\pi, \tag{20}$$

$$C_2 = -C_o\left[5/(2^3 3^4\pi^2) + 11/(2^9 3^5\pi^6)(\Gamma(1/4))^8\right].$$

Solving for the λ's, we find in the first and second approximations: $\lambda_o = 0.6603$ and $\lambda_o = 0.6677$, $\lambda_1 = 2.3941$ respectively; the exact values are $\lambda_o = 0.6680$ and $\lambda_1 = 2.3936$ (Ref. 8).

For comparison, we report on Table I, the values of $W_n^{(o)}$ $W_n^{(1)}$ $W_n^{(2)}$ for n = 0, 3. The values of λ_o and λ_1 - using two trace identities - are more accurate than the second order WKB approximants $W^{(2)}$ of a factor 300 and 15 respectively.

Table I - $W_n^{(0)}$, $W_n^{(1)}$ and $W_n^{(2)}$ are the zeroth, the first and the second order WKB approximants respectively. $\overline{W}_0^{(1)}$ has been computed using the first trace identity as indicated in the text, $\overline{W}_n^{(1)} = W_n^{(1)}$ for $n \neq 0$. Similarly, $\overline{W}_0^{(2)}$ and $\overline{W}_1^{(2)}$ have been computed using the first two trace identities, $\overline{W}_n^{(2)} = W_n^{(2)}$ for $n > 1$. λ_n are the exact values.

n	$W_n^{(0)}$	$W_n^{(1)}$	$\overline{W}_n^{(1)}$	$W_n^{(2)}$	$\overline{W}_n^{(2)}$	λ_n
0	0.5463	0.6235	0.6603	0.5960	0.6677	0.6680
1	2.3636	2.4007	2.4007	2.3993	2.3941	2.3936
2	4.6705	4.6970	4.6970	4.6966	4.6966	4.6968
3	7.3148	7.3359	7.3359	7.3357	7.3357	7.3357

The reader interested in high energy physics has quite likely remarked the strong similarity of the method here described with the use of superconvergence sum rules in the Regge theory[9)].

Acknowledgements

The author is grateful to C. Itzykson, J. Lascoux and A. Voros for stimulating discussions. He also acknowledges S. Graffi for useful suggestions.

References

1) I. M. Gelfand, Uspekhi Mat. Nauk 11, 191 (1956), MR 18;129.
2) L. A. Dikii, Izv. Akad. Nauk SSSR Ser. Mat. 19, 187 (1955), MR 17;619.
3) L. A. Dikii, Uspekhi Mat. Nauk 13, 111 (1958), MR 20;6655.
4) I. M. Gelfand and B. M. Levitan, Dokl. Akad. Nauk SSSR 88, 593 (1953), MR 13;588.
5) L. D. Faddeev, Dokl. Akad. Nauk SSSR 115, 878 (1957), MR 20;1029.
6) A. Voros, Doctorat Thesis, Saclay preprint.
7) C. M. Bender, K. Olaussen and P. S. Wang, Phys. Rev. D16, 1740 (1977).
8) F. Hioe and E. Montroll, J. Math. Phys. 16, 1945 (1975).
9) V. De Alfaro, S. Fubini, G. Furlan and C. Rossetti, Phys. Rev. Letters 8, 576 (1966).

INFN, Laboratori Nazionali di Frascati, 00044 Frascati, Italy

ZETA FUNCTIONS OF THE QUARTIC

(AND HOMOGENEOUS ANHARMONIC) OSCILLATORS

A. Voros*

We shall review some properties of the sequence of eigenvalues $\{\lambda_n\}_{n\in\mathbb{N}}$ for the differential operator called the *quartic oscillator*[1-3] :

$$\hat{H} = C^{4/3}\left(-\frac{d^2}{dq^2} + q^4\right) \quad , \quad C = \frac{1}{3}\,\Gamma(\tfrac{1}{4})^2\,(\tfrac{2}{\pi})^{1/2} \tag{1}$$

including some exact results about the *zeta function* of $\hat{H}$:

$$\zeta(s) = \mathrm{Tr}\,\hat{H}^{-s} = \sum_0^\infty \lambda_n^{-s} \quad . \tag{2}$$

The factor $C^{4/3}$ has been only put into (1) for our later convenience. Our methods will also apply to other *homogeneous anharmonic* oscillators :

$$\left(-\frac{d^2}{dq^2} + q^{2M}\right) \quad , \quad M \in \mathbb{N}^* \tag{3}$$

and our notations will allow easy conversion from M=2 to general M.

This article is meant as an ordered survey of existing results[4-5]. It is a slightly updated version of [5] in a more mathematical style, but with no claim to absolute rigor.

That $\hat{H}$ has a purely discrete spectrum $\lambda_0 < \lambda_1 < \lambda_2,\ldots,\ \lambda_n \to +\infty$ can be seen by abstract functional analysis ($\hat{H}$ is a positive operator on $L^2(\mathbb{R}, dq)$ with a compact resolvent). Besides, Stürm-Liouville theory shows [6] that $\lambda_0 > 0$, that all eigenvalues are simple and that they follow the asymptotic law :

$$\sigma_n = \sigma(\lambda_n) = 2\pi(n+\tfrac{1}{2}) + O(\tfrac{1}{n}) \quad , \quad n \to \infty \quad , \tag{4}$$

where $\sigma(\lambda)$ is the *classical action* for the energy λ :

$$\sigma(\lambda) = \int_{H(q,p)<\lambda} dq\,dp \quad , \quad H(q,p) = C^{4/3}(p^2+q^4) \quad . \tag{5}$$

With our normalization (1) :

$$\sigma(\lambda) = \lambda^{4/3} \text{ and } \sigma_n = \lambda_n^{4/3} \quad . \tag{6}$$

From the physical viewpoint, (4) is a semiclassical quantization condition (Bohr-Sommerfeld rule).

----

* Member of C.N.R.S.

We shall see that $\zeta(s)$ is a meromorphic function of $s \in \mathbb{C}$, and compute :

- in section 1 : its poles and their residues
- in section 2 : $\zeta(-n)$ $(n \in \mathbb{N})$
- in section 3 : $\zeta'(0)$
- in section 6 : $\zeta(n)$ $(n \in \mathbb{N}^*)$.

Section 4 lists the parallel results about the "modified zeta function" :

$$\zeta^P(s) = \sum_0^\infty (-1)^n \lambda_n^{-s} \quad . \tag{7}$$

Section 5 adapts all formulas to the general case (3) ; for M=1 (harmonic oscillator), classic results of arithmetic and analysis are recovered about Bernoulli numbers, the Stirling formula, the Euler gamma and Riemann zeta functions.

Finally section 7 gives (at present only for $M \leq 2$) the asymptotic expansions of $\zeta(s)$ and $\zeta^P(s)$ for $s \to -\infty$.

Among those results we can select two subsets (that share section 3 in common) :

- Sections 1 to 3 and 7 involve some form or another of semiclassical analysis, i.e. information about the *asymptotic* form of the sequence $\{\lambda_n\}$ for $n \to \infty$. They teach us about the deep nature of the semiclassical expansion, in a form hopefully generalizable to potentials other than q^{2M}.

- Sections 3 and 6 give *exact* results through a connection of the specific operators (3) with Bessel functions. While not generalizable, those results may have interesting *arithmetical* aspects.

Our whole study will be based on the interplay between several spectral functions of $\hat{H}$:

$$\left.\begin{aligned}
&\text{partition function : } \Theta(t) = \operatorname{Tr} \exp(-t\hat{H}) = \sum_0^\infty \exp(-t\lambda_n) \quad (\operatorname{Re} t > 0)\\
&\text{resolvent trace : } R(\lambda) = \operatorname{Tr}(\hat{H}-\lambda)^{-1} = \sum_0^\infty (\lambda_n-\lambda)^{-1} \quad (\lambda \in \mathbb{C} - \Lambda)\\
&\text{Fredholm determinant : } \Delta(\lambda) = \det(1-\lambda\hat{H}^{-1}) = \prod_0^\infty (1-\lambda/\lambda_n) \quad (\lambda \in \mathbb{C})\\
&\text{Zeta function : } \zeta(s) = \operatorname{Tr} \hat{H}^{-s} = \sum_0^\infty \lambda_n^{-s} \quad (\operatorname{Re} s > 3/4) \quad .
\end{aligned}\right\} \tag{8}$$

(and a few more exotic functions to be introduced as needed).

The series and infinite product in (8) converge to analytic functions in the indicated domains, thanks to Eq.(4). We furthermore have the obvious relations :

$$R(\lambda) = \int_0^\infty \Theta(t)\, e^{\lambda t}\, dt \quad (\operatorname{Re} \lambda < 0) \tag{9}$$

$$\Delta(\lambda) = \exp - \int_0^\lambda R(\lambda')\, d\lambda' \quad (\lambda \in \mathbb{C}) \tag{10}$$

$$\eta(s) = \Gamma(s)\zeta(s) = \int_0^\infty \Theta(t)\, t^{s-1} dt \quad (\operatorname{Re} s > 3/4) \quad . \tag{11}$$

1. POLES AND RESIDUES

This section and the next one are based on the possibility of expanding $\Theta(t)$ in fractional powers of t as $t \to 0$ ($|\mathrm{Arg}\, t| < \frac{\pi}{2} - \varepsilon$). This line of though has been pursued with success for several classes of partial differential operators[7-9], but not so much for (1) and similar potentials not vanishing at infinity ; our starting point will be Parisi's work[4] about the operators (3).

The expansion of $\Theta(t)$ for $t \to 0$ can be deduced from its asymptotic evaluation by symbol calculus. The Weyl formalism (i.e. the Wigner transformation) is the most efficient one for this purpose. The Weyl symbol of an operator $\hat{A}$ on $L^2(\mathbb{R}, dq)$ is a function $\hat{A}_W$ of $(q,p) \in \mathbb{R}^2$, linear in A, such that[10-11]:

- if $\hat{A}$ has the form $f(-id/dq)+g(q)$, then $\hat{A}_W(q,p)=f(p)+g(q)$; in particular $\hat{\mathbb{1}}_W=1$ and $\hat{H}_W = C^{4/3}(p^2+q^4) = H(q,p)$, the classical Hamiltonian of the quartic oscillator that appears in Eq.(5).

- when the trace is defined :

$$\mathrm{Tr}(\hat{A}\,\hat{B}) = (2\pi)^{-1} \int_{\mathbb{R}^2} \hat{A}_W\, \hat{B}_W \, dq\, dp \quad , \tag{12}$$

- the symbol of an exponential $\hat{A} = \exp(-t\hat{H})$ has an explicit combinatorial expansion[11]:

$$(\exp - t\hat{H})_W \;(q,p) \underset{t\to 0}{\sim} e^{-t\hat{H}_W(q,p)} \sum_{k=0}^{\infty} \frac{(-t)^k}{k!} \left[(\hat{H} - H_W(q,p).\, \hat{\mathbb{1}})^k \right] (q,p) \; . \tag{13}$$

Substituting (13) into (12) with $\hat{B} = \mathbb{1}$ and integrating term by term, we obtain a formal expansion : $\Theta(t) = \sum_0^\infty c_n(t)$. It would now be nice to identify some big class of (pseudodifferential) operators $\hat{H}$ for which this formal process can be carried out *and* $\sum c_n(t)$ is a true asymptotic series of some sort for $\Theta(t) = \mathrm{Tr}\,\exp(-t\hat{H})$. Here we just note that for positive polynomial symbols like p^2+q^{2M} all this holds, and furthermore homogeneity implies that $c_n(t) = c_n t^{i_n}$ where $i_n = \frac{M+1}{2M}(2n-1)$ and c_n are constants. More precisely, the leading singularity is the *classical* partition function , which for M=2 is :

$$\Theta_{cl}(t) = (2\pi)^{-1} \int_{\mathbb{R}^2} e^{-C^{4/3} t (p^2+q^4)} dq\, dp = \frac{3}{8\pi}\, \Gamma(\tfrac{3}{4})\, t^{-3/4} \tag{14}$$

and higher terms are the quantum corrections. To sum up, when M=2 :

$$\left.\begin{aligned} &\Theta(t) \underset{t\to 0}{\sim} \sum_{n=0}^{\infty} c_n t^{i_n} \;,\; i_n = \tfrac{3}{4}(2n-1) \\ \text{and} \qquad &c_o = 3\Gamma(\tfrac{3}{4}) \,/\, 8\pi \;,\; c_1 = -\Gamma(\tfrac{1}{4}) \,/\, 6 \;\ldots \end{aligned}\right\} \tag{15}$$

The numbers c_n and the various ways of obtaining them are the core of our study (for M=1, $\hat{H} = \pi(-d^2/dq^2+q^2)$, $\Theta(t) = (2\,\mathrm{sh}\,\pi t)^{-1}$, and $c_n = -\pi^{2n-1}(2^{2n-1}-1)B_{2n} \,/\, (2n)!$

where B_{2n} are the Bernoulli numbers). The expansion (15) is indeed basic to most of our results. It immediately implies, via Eqs.(9-11) :

- an expansion of the Fourier transform $R(\lambda)$ for $\lambda \to -\infty$:

$$R(\lambda) \sim \sum_0^\infty c_n \, \Gamma(1+i_n)(-\lambda)^{-1-i_n} \tag{16}$$

- the analytic continuation of the Mellin transform $\eta(s)$ to a *meromorphic* function by a standard technique[9] :

$$\eta(s) = \int_0^\infty \left[\Theta(t) - \sum_0^N c_n t^{i_n}\right] t^{s-1} dt \quad (-i_{N+1} < \mathrm{Re}\, s < -i_N) \tag{17}$$

and $\eta(s)$ has a simple pole at each point $(-i_n)$, with residues c_n . Since $\zeta(s)=\eta(s)/\Gamma(s)$ and i_n is never an integer, this implies that $\zeta(s)$ is meromorphic in $\mathbb{C}$, with a simple pole at every $(-i_n)$ $(n \in \mathbb{N})$, and :

$$\mathrm{Res}\, \zeta(s)_{s=-i_n} = c_n \,/\, \Gamma(-i_n) \, . \tag{18}$$

2. TRACE IDENTITIES

The last statement of the previous section also implies :

$$\zeta(-k) = 0 \qquad (k \in \mathbb{N}) \quad . \tag{19}$$

Formulas expressing the zeta function of an operator at the negative integers are known as *trace identities*. We shall now see that their nature is less trivial than it seems according to this derivation.

Formally $\zeta(-k) = \mathrm{Tr}\, \hat{H}^k = \sum_0^\infty \lambda_n^k$, but Eq.(2) for $\zeta(s)$ only holds for $\mathrm{Re}\, s > -i_o = \frac{3}{4}$, and $\mathrm{Tr}\, \hat{H}^k$ and $\sum_0^\infty \lambda_n^k$ are divergent expressions that need to be regularized. Formula (17) also shows that a new substraction term must appear in $\zeta(s)$ every time that $\mathrm{Re}\, s$ crosses a new pole $(-i_N)$ from right to left. These terms must refer to the asymptotic behaviour of the eigenvalues λ_n for $n \to \infty$: indeed, the first divergence of the series $\sum_0^\infty \lambda_n^{-s}$ is at $s=3/4$ because $\lambda_n = O(n^{4/3})$ by formula (4). We shall now express that relationship to all orders.

Formula (4) is only the beginning of a complete asymptotic expansion for the eigenvalues σ_n of the operator $\hat{H}^{3/4}$, in the implicit form (which we insist on keeping !) :

$$\overline{F}(\sigma_n) = 2\pi(n+\tfrac{1}{2}) \;\; (n \in \mathbb{N}) \;\; ; \;\; F(\sigma_n) \sim \sigma_n + \frac{b_1}{\sigma_n} + \frac{b_2}{\sigma_n^3} + \frac{b_3}{\sigma_n^5} + \ldots (\sigma_n \to +\infty) \; . \tag{20}$$

The coefficients b_n ($b_o = 1$, $b_1 = -\pi/3$ ...) can be computed recursively by the WKB (i.e. phase-integral) method[1-3,6] (Table 1).

Let us now attempt to compute $\zeta(s)$ explicitly from its defining series (2). In order to accelerate its convergence for $\mathrm{Re}\, s > 3/4$ and to obtain its explicit analytic continuation for $\mathrm{Re}\, s < 3/4$, we shall replace its late term by a *Euler-Mc Laurin summation formula* exploiting the fact that $\sigma_n = \lambda_n^{3/4}$ is given by Eq.(20), $\overline{F}$ being a smooth function such that $\overline{F}(\sigma) \sim \sum_0^\infty b_j\, \sigma^{1-2j} (\sigma \to +\infty)$. For any function $f(\sigma)$ integrable and C^∞ at $\sigma = +\infty$ the Euler-Mc Laurin formula for the function $f \circ \overline{F}^{-1}$ yields (B_{2m} are Bernoulli numbers) :

$$\sum_{k=0}^{\infty} f(\sigma_k) \sim \sum_{k=0}^{n-1} f(\sigma_k) + \frac{1}{2} f(\sigma_n) + \frac{1}{2\pi} \int_{\sigma_n}^{\infty} f(\sigma) \frac{d\overline{F}}{d\sigma} d\sigma - \sum_{m=1}^{\infty} \frac{B_{2m}}{(2m)!} \left[2\pi \left(\frac{d\overline{F}}{d\sigma}\right)^{-1} \frac{d}{d\sigma} \right]^{2m-1} f(\sigma)\Big|_{\sigma=\sigma_n}$$

The last sum is meant as an asymptotic expression in powers of σ_n^{-1} valid for $n \to +\infty$, in which $\frac{d\overline{F}}{d\sigma}$ itself only denotes the asymptotic representation $\sum_0^\infty (1-2j) b_j \sigma^{-2j}$. Letting $f(\sigma) = \sigma^{-4s/3}$ (Re s > 3/4), we find :

$$\zeta(s) = \sum_{k=0}^{\infty} \sigma_k^{-4s/3}$$

$$\sim \sum_{k=0}^{n-1} \sigma_k^{-4s/3} + \sigma_n^{-4s/3} \left[-\frac{1}{2\pi} \sum_{j\geq 0} \frac{(2j-1) b_j \sigma_n^{1-2j}}{(-1+2j+4s/3)} + \frac{1}{2} + \frac{4s}{3} \sum_{m=1}^{\infty} \frac{(2\pi)^{2m-1} B_{2m} C_{2m}(\sigma_n, s)}{(2m)!} \right] \tag{21}$$

where $C_{2m}(\sigma,s) = \sigma^{4s/3} \left[\left(\frac{d\overline{F}}{d\sigma}\right)^{-1} \frac{d}{d\sigma} \right]^{2m-2} \left[\left(\frac{d\overline{F}}{d\sigma}\right)^{-1} \sigma^{-4s/3-1} \right] = 0(\sigma^{-2m+1})$

The right-hand side of (21) is a power series in σ_n^{-1} with coefficients algebraic in the b_n and B_{2n} . Theoretically the asymptotic representation (21) remains valid throughout the whole complex s plane. Numerically however it becomes less and less precise as Re s decreases, as it yields $\zeta(s)$ as the difference of two numbers that increase much faster than $\zeta(s)$ itself. Table 2 lists some values of $\zeta(s)$ (and of $d\zeta/ds$) computed that way : we have chosen n=10, taken the values σ_0 to σ_{10} from the literature[1,12], and included into $\overline{F}(\sigma)$ a contribution of order $e^{-\sigma/2}$ that was computed in [2] and found to be numerically relevant.

We now interpret the trace identities (19) in the light of Eq.(21). The first trace identity $\zeta(0) = 0$ just restores the Bohr-Sommerfeld quantization rule $\overline{F}(\sigma_n) = 2\pi(n+\frac{1}{2})$; this is the analog for the confining potential q^4 of Levinson's theorem, which is the s=0 trace identity in the scattering case [7], and the famous term 1/2 in the Bohr-Sommerfeld rule is linked with the hidden role of $B_1=-1/2$ in the Euler-Mc Laurin formula. The next trace identity $\zeta(-1)=0$ teaches us that :

$$\sum_{k=0}^{n-1} \lambda_k + \frac{1}{2} \lambda_n = \frac{1}{2\pi} \frac{\sigma_n^{7/3}}{7/3} + \left(-\frac{1}{2\pi} \frac{b_1}{1/3} + 2\pi \frac{4B_2/3}{2} \right) \sigma_n^{1/3} + 0(n^{-5/3}) \tag{22}$$

and so on. Eq.(22) only explicits the nonvanishing terms of (21) for s=-1, but numerical convergence is greatly accelerated if more terms from (21) are included. If now n is not too small (even n=2 can do) the right-hand-side of (22) can be safely estimated from (20) and a *sum rule* for the smaller eigenvalues is found, by which these can be computed much more accurately than from Eq.(20) alone, which is unsatisfactory for very small n [4]. We note that trace identities constitute a form of *exact* information about the spectrum as a whole.

Up to this point we have essentially rephrased Parisi's results [4] , restricting them to the case M=2.

We now show that the two expansions (15) and (20) are in fact intimately related. This is heuristically obvious from the representation (21) for $\zeta(s)$, which has a pole at $s = \frac{3}{4}(1-2j) = -i_j$ for every $j \in \mathbb{N}$, with residue :

$$\operatorname{Res} \zeta(s)_{s=-i_j} = -\frac{i_j}{2\pi} b_j \quad . \tag{23}$$

Comparing with (18) we get the important formula :

$$b_n = 2\pi\, c_n / \Gamma(1-i_n) \quad . \tag{24}$$

Here follows a more rigorous derivation of (24). Starting from Eq.(15), let us proceed backwards to obtain an expansion for $t \to 0^+$ of the less familiar partition functions $\Theta_\alpha(t) = \operatorname{Tr} \exp(-t\hat{H}^\alpha)$, with $\alpha > 0$ arbitrary. We set :

$$\eta_\alpha(s) = \int_0^\infty \Theta_\alpha(t)\, t^{s-1} dt = \zeta(\alpha s)\Gamma(s) \quad , \tag{25}$$

since $\eta_\alpha(s)/\Gamma(s) = \operatorname{Tr}(\hat{H}^\alpha)^{-s} = \zeta(\alpha s)$. Hence $\eta_\alpha(s)$ two sets of poles : $(-\alpha^{-1} I)$ from $\zeta(\alpha s)$, where I is the set $\{i_n\}_{n\in\mathbb{N}}$, and $(-\mathbb{N})$ from $\Gamma(s)$; any common pole at $-k \in (-\mathbb{N}) \cap (-\alpha^{-1} I)$ is *double* and produces by inverse Mellin transformation a term $(t^k \log t)$ in the expansion of $\Theta_\alpha(t)$, which then reads :

$$\Theta_\alpha(t) \underset{t\to 0}{\sim} \sum_{i_n \in I \setminus \alpha N} \frac{\Gamma(-i_n/\alpha)}{\alpha\Gamma(-i_n)} c_n t^{i_n/\alpha} + \sum_{i_n \in I \cap \alpha N} \frac{(-1)^{1+i_n/\alpha}}{\alpha\Gamma(-i_n)\Gamma(1+i_n/\alpha)} c_n t^{i_n/\alpha} \log t$$

$$+ \sum_{k\in\mathbb{N}} \gamma_k(\alpha) t^k \quad , \tag{26}$$

where $\gamma_k(\alpha) = (-1)^k \zeta(-\alpha k)/k!$ if $\alpha k \notin I$ (and is even less explicit if $\alpha k \in I$) : we cannot compute it any better except for integer α (by the trace identities). By contrast, the coefficients of the terms *singular* around t=0 : $t^k \log t$ $(k \in \mathbb{N})$ or $t^{i_n/\alpha}$ $(i_n/\alpha \notin \mathbb{N})$, are in *one-to-one correspondence* with the terms of (15), and they depend on α in a perfectly known way.

The value $\alpha=3/4$ will be of special interest to us, as we shall deduce from the Bohr-Sommerfeld quantization rule (4) that $\Theta_{3/4}(t)$ has, among all $\Theta_\alpha(t)$, the closest functional resemblence to the *harmonic oscillator* partition function ; indeed $\hat{H}^{3/4}$ has an asymptotically harmonic spectrum : $\sigma_n \sim 2\pi(n+1/2)$.

For $\alpha=3/4$, Eq.(20) precisely reads :

$$\Theta_{3/4}(t) \sim \frac{c_o t^{-1}}{\frac{3}{4}\Gamma(\frac{3}{4})} + \sum_{n=1}^{\infty} \frac{c_n t^{2n-1} \log t}{\frac{3}{4}\Gamma(2n)\Gamma(-i_n)}$$

$$+ \sum_{k=1}^{\infty} \left[\frac{\zeta(-\frac{3}{4}(2k))}{\Gamma(2k+1)} t^{2k} - \left(\frac{c_k \Psi(2k)}{\frac{3}{4}\Gamma(-i_k)} + \zeta_o(-i_k) \right) \frac{t^{2k-1}}{\Gamma(2k)} \right] \tag{27}$$

with $\Psi = \Gamma'/\Gamma$, and

$$\zeta_o(-i_k) = \lim_{s\to -i_k} \left(\zeta(s) - \frac{c_k}{\Gamma(-i_k)(s+i_k)} \right).$$

Remark :

$$\zeta_o(3/4) = \lim_{n\to\infty} \left(\sum_0^{n-1} \sigma_k^{-1} - (\log \sigma_n)/2\pi \right) \tag{28}$$

(from Eq.(21)) reminds of Euler's constant γ in the case $\{\lambda_n\} = \mathbb{N}^*$. We estimated $\zeta_o(3/4) \simeq -0.026076728$, and similarly $\zeta_o(-3/4) \simeq 0.2716185$, $\zeta_o(-9/4) \simeq 5.995$.

Now the *singular part* (the top line) of the series (27) will be related to the asymptotic expansion of eigenvalues (20) by Laplace-Borel transformation. Considering an integrated density of levels for the operator $\hat{H}^{3/4}$:

$$F(\sigma) = 2\pi \sum_0^{\infty} \theta(\sigma - \sigma_k) = \int_C \Theta_{3/4}(t)\, e^{t\sigma} \frac{dt}{it} \tag{29}$$

(θ will denote the Heaviside step function ; C is the contour shown on Fig.1), we compare two Fourier-like decompositions of $F(\sigma)$. On the one hand, if we *extrapolate* to all real $\sigma \notin \{\sigma_n\}$ the function $\overline{F}$ of Eq.(20) so as to obtain a non-negative, monotonically increasing, C^∞ function $\overline{F}(\sigma)$ (Fig.2) still satisfying :

$$\overline{F}(\sigma) \sim \sum_0^{\infty} b_n \sigma^{1-2n} \quad , \quad \sigma \to +\infty \quad , \tag{30}$$

(none of our results will depend on the particular choice of $\overline{F}(\sigma)$), then $F(\sigma)$ and $\overline{F}(\sigma)$ are related by :

$$F(\sigma) = 2\pi \sum_{-\infty}^{+\infty} \theta(\overline{F}(\sigma) - 2\pi(k+\tfrac{1}{2})) = \overline{F}(\sigma) + i \sum_{m\neq 0} \frac{(-1)^m}{m} e^{-im\overline{F}(\sigma)} \tag{31}$$

where the last expression comes from the Poisson summation formula ; hence :

$$F(\sigma) \sim \sigma + \frac{b_1}{\sigma} + \frac{b_2}{\sigma^3} + \ldots + i \sum_{m\neq 0} e^{-im\sigma} \left[\frac{(-1)^m}{m} \exp\left\{-im\left(\frac{b_1}{\sigma} + \frac{b_2}{\sigma^3} + \ldots\right)\right\}\right] . \tag{32}$$

On the other hand, by shifting the contour C to the imaginary axis $i\mathbb{R}$ in (29), we recognize that each singularity $t_o \in i\mathbb{R}$ of $\Theta_{3/4}(t)$ contributes a term of order $e^{t_o\sigma}$ to $F(\sigma)$. Eq.(32) then means that the singularities of $\Theta_{3/4}(t)$ on $i\mathbb{R}$ are all the *integer* points. In particular $\left(\sigma + \frac{b_1}{\sigma} + \frac{b_2}{\sigma^3} + \ldots \right)$ is the Laplace transform of the singularity of $\Theta_{3/4}(t)$ at t=0, which singularity is *explicit in the first line* of Eq.(27). In this way we recover Eqs.(24) and (23).

Can we conclude sections 1 and 2 by saying that the two expansions (15) and (20) are equivalent ? Certainly not. Eq.(26) shows that the numbers b_n (or c_n) always govern the *singular part* of $\Theta_\alpha(t)$ at t=0, the regular part being of a more transcendental nature. Only for *integer* α can the *complete* expansion at t=0 be computed by the method of section 1 . Eq.(15) thus carries the additional information that the *regular part* of $\Theta(t)$ vanishes (i.e. has the form $\sum_{k\in\mathbb{N}} 0.t^k$) ; this piece of information is completely missing in a formula like (20), and is to be found in the set of trace identities (19) (see section 5 for the case of general M).

3. THE FREDHOLM DETERMINANT AND $\zeta'(0)$

We begin by writing the asymptotic behavior for $\lambda \to -\infty$ of the determinant $\Delta(\lambda)$ defined in (8). From the relation (10) : $\log \Delta(\lambda) = -\int_0^\lambda R(\lambda')d\lambda'$ and from the expansion (16) for $R(\lambda)$, we obtain, using $i_o = -3/4$:

$$\log \Delta(\lambda) \sim \sum_0^\infty c_n \Gamma(i_n)(-\lambda)^{-i_n} + L_o \qquad (\lambda \to -\infty) \tag{33}$$

where

$$L_o = \int_{-\infty}^0 [R(\lambda)-c_o\Gamma(\tfrac{1}{4})(-\lambda)^{-1/4}]\, d\lambda \quad . \tag{34}$$

For semi-classical purposes it is more natural to normalize the Fredholm determinant at $\lambda=-\infty$ by removing the constant L_o in (33), so we set :

$$D(\lambda) = e^{-L_o}\Delta(\lambda) = \exp\left\{-\int_{-\infty}^{\lambda} [R(\lambda')-c_o\Gamma(\tfrac{1}{4})(-\lambda')^{-1/4}]\, d\lambda' - c_o\Gamma(-\tfrac{3}{4})(-\lambda)^{3/4}\right\} \tag{35}$$

$$\Longrightarrow \log D(\lambda) \sim -\sum_0^\infty c_n\Gamma(i_n)(-\lambda)^{-i_n} \qquad (\lambda \to -\infty) \quad . \tag{36}$$

We first prove that :

$$-\log D(0) = L_o = \zeta'(0) \quad . \tag{37}$$

The first equality is obvious from $\Delta(0) = 1$. For the second one we note that since $\zeta(0)=0$, we have : $\zeta'(0) = \lim_{s\to 0} \frac{\zeta(s)}{s} = \eta(0)$, and by Eq.(17) :

$\eta(0) = \int_0^\infty [\Theta(t)-c_o t^{-3/4}]\, \frac{dt}{t}$, and finally by the relation (9) between $R(\lambda)$ and $\Theta(t)$:

$\eta(0) = \int_{-\infty}^0 [R(\lambda)-c_o\Gamma(\tfrac{1}{4})(-\lambda)^{-1/4}]d\lambda = L_o$. QED.

We shall next relate this new determinant $D(\lambda)$ to the *eigenfunctions* of the differential operator $\hat{H}$, via the *WKB approximation in the classically forbidden region*. For any $\lambda \leq 0$, the equation $\hat{H}\psi = \lambda\psi$ has two linearly independent solutions $\psi_\pm(\lambda,q)$ characterized by decrease conditions for $q \to \pm\infty$ precisely of the WKB type :

$$\psi_\pm(\lambda,q) \sim \pi(\lambda,q)^{-1/2} \exp \mp \int_0^q \pi(\lambda,q')dq' \quad (q \to \pm\infty) \tag{38}$$

where $\pi(\lambda,q) = (q^4-C^{-4/3}\lambda)^{1/2}$, and the integration lower bound is q=0 for pure convenience. We then define the analogs of the inverse transmission coefficients in scattering theory[13] :

$$\left.\begin{aligned}\psi_-(\lambda,q) &\sim a_-(\lambda)\pi(\lambda,q)^{-1/2}\exp\int_0^q \pi(\lambda,q')dq' \qquad (q\to+\infty)\\ \psi_+(\lambda,q) &\sim a_+(\lambda)\pi(\lambda,q)^{-1/2}\exp-\int_0^q \pi(\lambda,q')dq' \qquad (q\to-\infty)\ .\end{aligned}\right\} \tag{39}$$

We introduce here the notations $\frac{\partial}{\partial q} = {}'$, $\frac{\partial}{\partial\lambda} = \dot{}$. By computing the Wronskian $W(\lambda) = \psi_+\psi_-' - \psi_-\psi_+'$ at $q \to \pm\infty$, we find that :

$$\frac{W(\lambda)}{2} = a_+(\lambda) = a_-(\lambda) \text{ (henceforth written } a(\lambda)). \tag{40}$$

For $\lambda \leq 0$, the quantity $(a(\lambda)-1)$ is a measure of the departure of the eigenfunctions from semi-classical behavior ; hence $a(\lambda) \to 1$ for $\lambda \to -\infty$.

It is also well known that the *kernel of the resolvent operator* $(\hat{H}-\lambda)^{-1}$ admits the expression :

$$R(\lambda,q,q') = C^{-4/3}W(\lambda)^{-1}[\psi_-(q)\psi_+(q')\theta(q'-q)+\psi_+(q)\psi_-(q')\theta(q-q')] \tag{41}$$

(which satisfies both the differential equation and the boundary conditions for the Green's function). Therefore :

$$R(\lambda) = \int_{-\infty}^{\infty} R(\lambda,q,q)\,dq = C^{-4/3}W(\lambda)^{-1}\int_{-\infty}^{\infty} \psi_-(q)\psi_+(q)\,dq \ . \tag{42}$$

By combining the equations $\dot{\psi}_+[(\hat{H}-\lambda)\psi_-]=0$ and $\psi_-\dot{\overline{[(\hat{H}-\lambda)\psi_+]}} = 0$, and afterwards the same equations with ψ_+ and ψ_- permuted, we see that

$C^{-4/3}\psi_-(q)\psi_+(q) = f' = g'$ with $f(\lambda,q) = \psi_-'\dot{\psi}_+ - \psi_-\dot{\psi}_+'$, $g(\lambda,q) = \psi_+'\dot{\psi}_- - \psi_+\dot{\psi}_-'$,

and $f-g = \dot{W}(\lambda)$ (independent of q, as required). This allows to rewrite (42) as : $R(\lambda) = W(\lambda)^{-1}\,[f(\lambda,+\infty)-g(\lambda,-\infty)-\dot{W}(\lambda)]$, which can be reexpressed in terms of the asymptotic data (38-39) alone :

$$R(\lambda) = \frac{C^{-4/3}}{2}\int_{-\infty}^{\infty}\frac{dq}{\pi(\lambda,q)} - \frac{\dot{W}(\lambda)}{W} = c_o\Gamma(\tfrac{1}{4})(-\lambda)^{-1/4} - \frac{d}{d\lambda}\log a(\lambda) \ .$$

We now integrate with respect to λ to get our final result :

$$\log D(\lambda) + c_o\Gamma(-\tfrac{3}{4})\,(-\lambda)^{3/4} = \log\ a(\lambda) \ , \tag{43}$$

the integration constant being *zero* since both sides vanish for $\lambda \to -\infty$: the left-hand side because we substracted $\zeta'(0)$ from Eq.(33), and the right-hand side because semi-classical approximation amounts to $a(\lambda) \to 1$ for $\lambda \to -\infty$. The result (43) is similar to the equality of the Jost function with the Fredholm determinant of the Lippmann-Schwinger equation in the scattering case[13].

Eqs.(36) and (43) imply that $a(\lambda) \sim -\sum_1^\infty c_n\Gamma(i_n)(-\lambda)^{-i_n}$ for $\lambda \to -\infty$. But this expansion also follows from the definition of $a(\lambda)$ if we replace $\psi_\pm(\lambda,q)$ in (39) by its *complete WKB expansion* computed in the classicaly forbidden region. We thus have a third method (besides Eqs.(12-15) and (24)) of deriving the c_n .

Finally we remark that for the special value $\lambda=0$, the eigenfunctions $\psi_\pm$ of the quartic oscillator happen to be *exactly* expressible in terms of a Bessel function :

$$\psi_\pm(\lambda=0,q) = (\pm 2q/3\pi)^{1/2}\, K_{1/6}(\pm q^3/3) \, . \tag{44}$$

From the known asymptotic behaviour of $K_\nu(x)$ for $|x| \to \infty$[14], we draw :

$$a(0) = +2 \Longrightarrow \quad \zeta'(0) = -\log D(0) = -\log 2 \, . \tag{45}$$

The Euler-Mc Laurin formula (i.e. the derivative of Eq.(21) at s=0) then implies :

$$\lim_{n\to\infty} \left[\sum_{k=0}^{n-1} \log \lambda_k + \frac{1}{2} \log \lambda_n - (2\pi)^{-1} \lambda_n^{3/4} (\log\lambda_n - 4/3) \right] (= -\zeta'(0)) = \log 2$$

remembering that $\lambda_k = \sigma_k^{4/3}$ and $b_o = 1$. If we include one correction term and exponentiate, we find :

$$\prod_0^n \lambda_k \sim \left(e^{-4/3}\lambda_n\right)^{(2\pi)^{-1}\lambda_n^{3/4}} 2\sqrt{\lambda_n} \left[1 + \frac{b_1}{2\pi}\lambda_n^{-3/4}\log\lambda_n + \frac{4}{3}\left(\frac{b_1}{2\pi} + \frac{\pi}{6}\right)\lambda_n^{-3/4} + O(\lambda_n^{-9/4}\log\lambda_n)\right] \tag{46}$$

which generalizes Stirling's formula from the case $\{\lambda_n\} = \mathbb{N}^*$; higher order terms can also be computed from (21).

4. PARITY CONSERVATION AND ITS CONSEQUENCES

As the quartic potential is an even function, the operator $\hat{H}$ commutes with the space reflection (or parity) operator $\hat{P}$, and the eigenstates of $\hat{H}$ are even or odd according to their quantum number [6]. The even and odd parts of the spectrum can then be analyzed separately if we consider, in parallel with the spectral functions (8), the "alternating spectral functions" (for which relations (9-11) also hold) :

$$\left.\begin{aligned}
\Theta^P(t) &= \mathrm{Tr}\,(\hat{P}\exp(-t\hat{H})) = \sum_0^\infty (-1)^n \exp(-t\lambda_n) \\
R^P(\lambda) &= \mathrm{Tr}\,(\hat{P}(\hat{H}-\lambda)^{-1}) = \sum_0^\infty (-1)^n (\lambda_n - \lambda)^{-1} \\
\Delta^P(\lambda) &= \prod_0^\infty (1-\lambda/\lambda_{2n})(1-\lambda/\lambda_{2n+1})^{-1} \\
\zeta^P(s) &= \Gamma(s)^{-1}\,\eta^P(s) = \mathrm{Tr}(\hat{P}\,\hat{H}^{-s}) = \sum_0^\infty (-1)^n \lambda_n^{-s} \, .
\end{aligned}\right\} \tag{47}$$

The function $\Theta^P(t)$ was already analyzed in [8] for the radial Schrödinger equation in 3 dimensions ; in our 1-dimensional case, $\hat{P}$ plays the role of angular momentum.

To evaluate $\Theta^P(t)$ for $t \to 0$, we use the general formula (12) with $\hat{A} = \exp(-t\hat{H})$ and $\hat{B} = \hat{P}$.A crucial difference with the case of $\Theta(t)$ is that $\hat{P}$ has the Wigner function[15] :

$$P_W(q,p) = \pi\,\delta(q)\,\delta(p)$$

hence the integration in (12) is *suppressed*, and :

$$\Theta^P(t) = \frac{1}{2}\,(\exp -t\hat{H})_W\,(q=p=0) = \sum_0^\infty \frac{(-t)^j}{2\, j!}\,(\hat{H}^j)_W\,(q=p=0) \tag{48}$$

an expression containing only *integral* powers of t. For the quartic oscillator it has the specific form :

$$\Theta^P(t) \underset{t\to 0}{\sim} \sum_0^\infty d_n \frac{t^{3n}}{(3n)!} \quad (|\mathrm{Arg}\,t| < \frac{\pi}{2} - \varepsilon \,;\, d_o = \frac{1}{2}, d_1 = -\frac{9}{4} C^4, \ldots) \tag{49}$$

$$\Longrightarrow R^P(\lambda) \underset{\lambda\to\infty}{\sim} \sum_0^\infty d_n (-\lambda)^{-1-3n} \qquad (\varepsilon < \mathrm{Arg}\lambda < \pi - \varepsilon) \quad . \tag{50}$$

$\zeta^P(s)$ continues to an *entire function* (all poles of $\eta^P(s)$ are now killed by those of $\Gamma(s)$), that satisfies the *trace identities* :

$$\zeta^P(-3n) = (-1)^n d_n \text{ , otherwise } \zeta^P(-n) = 0 \qquad \forall n \in \mathbb{N} \; . \tag{51}$$

The first d_n are listed in Table 1.

The modified partition functions $\Theta^P_\alpha(t) = \mathrm{Tr}(\hat{P}\exp(-t\,\hat{H}^\alpha))$ have *no singular terms* in their expansion as $t \to 0^+$:

$$\Theta^P_\alpha(t) \sim \sum_0^\infty \frac{(-1)^k}{k!} \zeta^P(-\alpha k)\, t^k \; . \tag{52}$$

We treat in more detail the "alternating determinant" $\Delta^P(\lambda)$ to stress the differences with $\Delta(\lambda)$. As in § 3 , we integrate the expansion (50) to find :

$$\log \Delta^P(\lambda) = -\int_0^\lambda R^P(\lambda')d\lambda' \sim d_o \log(-\lambda) + L^P_o - \sum_1^\infty \frac{d_n}{3n} (-\lambda)^{-3n} \tag{53}$$

$$L^P_o = \int_{-\infty}^0 [R^P(\lambda) - d_o(-\lambda)^{-1}\theta(-1-\lambda)]d\lambda \; . \tag{54}$$

By an inverse Laplace transformation, followed by an integration by parts :

$$L^P_o = \int_0^\infty (\Theta^P(t) - d_o\, e^{-t}) \frac{dt}{t} \tag{55}$$

$$= -\int_0^\infty \left(\frac{d\Theta^P}{dt} + d_o e^{-t}\right) \log t \;\; dt \; . \tag{56}$$

Another integration by parts, upon the formula

$\zeta^P(s) = \frac{1}{\Gamma(s)} \int_0^\infty \Theta^P(t)\, t^{s-1} dt$, yields its analytic continuation down to $\mathrm{Re}\, s > -3$:

$$\zeta^P(s) = \frac{-1}{\Gamma(1+s)} \int_0^\infty \frac{d\Theta^P}{dt}\, t^s\, dt \tag{57}$$

$$\Longrightarrow (\zeta^P)'(0) = \Gamma'(1).\int_0^\infty \frac{d\Theta^P}{dt} dt - \int_0^\infty \frac{d\Theta^P}{dt} \log t \;\; dt \tag{58}$$

$$\Longrightarrow \qquad (\zeta^P)'(0) = L^P_o \; , \tag{59}$$

if we compare (58) with (56), remembering that $\Gamma'(1) = \int_0^\infty e^{-t} \log t\, dt \;(=-\gamma)$ and $\Theta^P(0) = d_o$.

As in § 3, we modify the normalization of our determinant :

$$D^P(\lambda) = e^{-L^P_o} \Delta^P(\lambda) \tag{60}$$

$$\Longrightarrow D^P(\lambda) \sim d_o \log(-\lambda) - \sum_1^\infty \frac{d_n}{3n} (-\lambda)^{-3n} \quad , \tag{61}$$

and we try to relate $D^P(\lambda)$ to the eigenfunctions $\psi_\pm(\lambda,q)$ of $\hat{H}$, subject to the asymptotic conditions (38) for $q \to \pm\infty$.

With the notations of § 3 , the kernel of $\hat{P}(\hat{H}-\lambda)^{-1}$ is :

$$R^P(\lambda,q,q') = C^{-4/3} W(\lambda)^{-1} [\psi_-(q)\psi_+(-q')\theta(-q-q') + \psi_-(-q')\psi_+(q)\theta(q+q')] \ . \tag{62}$$

Because $\psi_-(-q) = \psi_+(q)$, we get :

$$R^P(\lambda) = \int_{-\infty}^{\infty} R^P(\lambda,q,q)\, dq = 2W(\lambda)^{-1} \int_0^\infty \psi_+(q)^2\, dq \tag{63}$$

and for the Wronskian : $W(\lambda) = -2\psi_+(\lambda,0)\psi_+'(\lambda,0)$ (as computed at $q=0$).

Furthermore, by combining the equations $\dot{\psi}_+[(\hat{H}-\lambda)\psi_+] = 0$ and $\psi_+\dot{\overbrace{[(\hat{H}-\lambda)\psi_+]}} = 0$, we may express $C^{-4/3}\psi_+(q)^2$ as $[\dot{\psi}_+\psi_+' - \psi_+\dot{\psi}_+']'$, hence :

$$R^P(\lambda) = -\frac{d}{d\lambda} \log \left| \frac{\psi_+'(\lambda,0)}{\psi_+(\lambda,0)} \right| \ . \tag{64}$$

By integration we get :

$$\log D^P(\lambda) = \log|(\log\psi_+)'(\lambda,0)| + \frac{2}{3} \log C \ . \tag{65}$$

Precisely this integration constant arises, because for $\lambda \to -\infty$ the right-hand side behaves according to the *lowest order WKB approximation* :

$$|(\log\psi_+)'\ (\lambda,q=0)| \sim \pi(\lambda,q=0) = (-C^{-4/3}\lambda)^{1/2}$$

$$(+ \text{ terms of order } \leq \lambda^{-3/2})$$

$$\Longrightarrow \log|(\log\psi_+)'(\lambda,0)| - \frac{1}{2}\log(-\lambda) + \frac{2}{3}\log C \xrightarrow[\lambda\to-\infty]{} 0 \quad ,$$

which is to be compared with (61), remembering that $d_o = \frac{1}{2}$. Obviously, the complete WKB expansion of $(\log \psi_+)'$ for $\lambda \to -\infty$ now constitutes an alternate method to compute the coefficients d_n, thanks to Eqs. (61) and (65).

We have thus proved that

$$\left.\begin{aligned} D^P(\lambda) &= |C^{2/3}(\log\psi_+(\lambda,0))'| = \left|\frac{C^{2/3}\psi_+'(\lambda,0)}{\psi_+(\lambda,0)}\right| \\ \text{and} \qquad (\zeta^P)'(0) &= -\log D^P(0) \ . \end{aligned}\right\} \tag{66}$$

Relations similar to (66) have been proved in different contexts [16].

There is an interesting difference of *scaling behavior* between formulas (43) and (66). If we change $\hat{H}$ to $K\hat{H}$ ($K > 0$), $\zeta(s)$ becomes $K^{-s}\zeta(s)$, $\zeta'(0)$ becomes $\zeta'(0) - \log K.\zeta(0)$, hence $D(0)$ scales as $K^{\zeta(0)}D(0)$: since $\zeta(0) = 0$, it is scale invariant. Similarly $D^P(0)$ scales as $K^{\zeta^P(0)}\ D^P(0)$, but from (51) : $\zeta^P(0) = d_o = 1/2$ (a universal value according to the expansion (48) which is actually valid for *any* C^∞, confining, even potential). The condition for (66) to hold is only that the lowest order WKB approximation to $\psi_\pm(\lambda,q)$ should be valid whether $\lambda\to-\infty$ or $|q|\to\infty$.

The trace identities (51) provide a *new set of sum rules* for the eigenvalues. From a summation formula for alternating sums :

$$\sum_{k=0}^{\infty}(-1)^k f(\sigma_k) \sim \sum_{k=0}^{2n-1}(-1)^k f(k)+\frac{1}{2}f(2n) - \sum_{m=1}^{\infty}\frac{(2^{2m-1}-1)B_{2m}}{(2m)!}\left[2\pi(\frac{d\bar{F}}{d\sigma})^{-1}\frac{d}{d\sigma}\right]^{2m-1} f(\sigma)\Big|_{\sigma=\sigma_{2n}}$$

we deduce an asymptotic formula analogous to (21) : (67)

$$\zeta^P(s) \sim \sum_{k=0}^{2n-1}(-1)^k\sigma_k^{-4s/3}+\frac{1}{2}\sigma_{2n}^{-4s/3}+\frac{4s}{3}\sigma_{2n}^{-4s/3}\sum_{m=1}^{\infty}\frac{(2\pi)^{2m-1}(2^{2m-1}-1)B_{2m}}{(2m)!}C_{2m}(\sigma_{2n},s) \quad (68)$$

which allows to tabulate $\zeta^P(s)$ (Table 2) and to check the identities (51).

One consequence of (68) is a formula for $(\zeta^P)'(0)$ *in terms of the eigenvalues* :

$$(\zeta^P)'(0) = -\lim_{n\to\infty}\left[\sum_{0}^{2n-1}(-1)^k \log\lambda_k+\frac{1}{2}\log\lambda_{2n}\right] . \quad (69)$$

For the quartic oscillator, Eq.(44), together with (66) , implies [14] :

$$(\zeta^P)'(0) = -\log(4\pi^2(2/3)^{1/3}\Gamma(1/3)^{-4}C^{2/3}) = -\log(2/3(2\pi)^{5/3}\Gamma(1/4)^{4/3}\Gamma(1/3)^{-4}) . \quad (70)$$

5. THE CASE OF HOMOGENEOUS POTENTIALS

As stated in the introduction, all previous arguments extend to the following operators :

$$\hat{H}_n = C_M^{\frac{2M}{M+1}}\left(-\frac{d^2}{dq^2}+q^{2M}\right) , \quad C_M = \frac{\Gamma(1/2)\Gamma(1/2M)}{M\Gamma((3M+1)/2M)} . \quad (71)$$

As a rule, 3/4 should be replaced by (M+1)/2M everywhere in the text. Coefficients c_n, b_n, d_n depend on M, except : $b_o = 1$ by virtue of our normalization, and $d_o = 1/2$ by Eq.(48).

For odd M a fortuitous coincidence induces a qualitative change in the text of sections 1-2. Whenever (2n-1) is a (positive and odd) multiple of M, the exponent $i_n = \frac{M+1}{2M}(2n-1)$ is an *integer*, the pole of $\eta(s)$ is cancelled in Eq.(11) by a pole of $\Gamma(s)$, and for such n :

- $\zeta(s)$ is regular at $s=-i_n$ and $b_n = 0$;
- the trace identity at that point is now : $\zeta(-i_n)=(-1)^{i_n}(i_n)!c_n$. (72)

For M=1 this occurs for all $n\geq 1$, reminding us that the expansion (20) is trivial for the harmonic oscillator. The zeta function for M=1 is related to the Riemann zeta function $\zeta_R(s)$ by :

$$\zeta(s) = \pi^{-s}(1-2^{-s})\,\zeta_R(s) \quad (73)$$

and the trace identities (19) and (72) respectively restore the well-known values[17]: $\zeta_R(-2n)=0$ and $\zeta_R(1-2n)=-B_{2n}/2n$. As for the functions $R(\lambda)$ and $\Delta(\lambda)$, their definitions for M=1 invoke one more substraction :

$$R(\lambda) = \sum_0^\infty [(\lambda_n-\lambda)^{-1}-\lambda_n^{-1}] = \frac{1}{2\pi}[\Psi(\tfrac{1}{2}) - \Psi(\tfrac{1}{2}-\tfrac{\lambda}{2\pi})] \tag{74}$$

$$\Delta(\lambda) = \prod_0^\infty [(1-\lambda/\lambda_n)\exp(\lambda/\lambda_n)] = \frac{\sqrt{\pi}\, 2^{\lambda/\pi} e^{\gamma\lambda/2\pi}}{\Gamma(\frac{1}{2}-\frac{\lambda}{2\pi})}, \tag{75}$$

where Γ is the Euler gamma function, $\Psi = \Gamma'/\Gamma$, and γ is Euler's constant.

Further explicit changes in the text are, setting $\mu = \frac{1}{2M+2}$:

- Eq.(44) : $\psi_\pm(\lambda=0,q) = (4\mu q/\pi)^{1/2} K_\mu(2\mu q^{M+1})$ (76)
- Eq.(45) : $a(0)=(\sin\pi\mu)^{-1}$, $\zeta'(0)=\log\sin\pi\mu$. (77)

Replace accordingly $2\sqrt{\lambda_n}$ by $\frac{\sqrt{\lambda_n}}{\sin\mu\pi}$ (and always $\frac{3}{4}$ by $\frac{M+1}{2M}$) in Eq.(46).

- Eqs.(49-61) : replace 3n by (M+1)n.
- Eq.(70) :

$$(\zeta^P)'(0) = -\log\left[\frac{(\mu^{-1}C_M)^{2\mu M}\,\pi}{\sin\mu\pi.\Gamma(\mu)^2}\right]. \tag{78}$$

6. THE VALUES $\zeta(n)$, $n\in\mathbb{N}^*$.

Here we are in the domain of convergence of the representation (2) for $\zeta(s)$ and no semiclassical analysis is involved. We can readily take M arbitrary and treat the case of $\zeta^P(n)$ in parallel. We note that $\zeta(n)$ enters the expansions (which are obvious, and converge for $|\lambda| < \lambda_0$)

$$\log\Delta(\lambda) = -\sum_{n=1}^\infty \zeta(n)\lambda^n/n \;,\quad R(\lambda) = \sum_{n=1}^\infty \zeta(n)\lambda^{n-1} \quad (M>1)\,. \tag{79}$$

We start from the kernel (41) for the resolvent $(\hat{H}-\lambda)^{-1}$, which is explicit for $\lambda=0$ thanks to formula (44). We then iterate (41) n times to obtain an integral expression for $\zeta(n)=\mathrm{Tr}\,\hat{H}^{-n}$ in terms of Bessel functions (and likewise for $\zeta^P(n)=\mathrm{Tr}\,\hat{P}\,\hat{H}^{-n}$) :

$$\zeta(1) = C_M^{-4\mu M}\frac{4W(0)^{-1}}{\pi(M+1)^2}\int_0^\infty K_\mu(2\mu q^{M+1})K_\mu(2\mu(e^{i\pi}q)^{M+1})q\,dq \tag{80}$$

$$\zeta^P(1) = C_M^{-4\mu M}\frac{4W(0)^{-1}}{\pi(M+1)^2}\int_0^\infty K_\mu(2\mu q^{M+1})^2 q\,dq \tag{81}$$

etc ... (we recall that $\mu=(2M+2)^{-1}$ and $W(0)=2D(0)=2/\sin\pi\mu$). The problem of interest here is to reduce such expressions to simpler arithmetic forms. For n=1 this is possible thanks to the Weber-Schafheitlin formulas [14] :

$$\zeta^P(1) = \sum_0^\infty (-1)^k\lambda_k^{-1} = \frac{\sin\pi\mu}{2\sqrt{\pi}}(C_M/2\mu)^{-4\mu M}\,\Gamma(\mu)\Gamma(2\mu)\Gamma(3\mu)/\Gamma(\tfrac{1}{2}+2\mu) \tag{82}$$

and, remarkably :

$$\zeta(1) = \sum_0^\infty \lambda_k^{-1} = \frac{\mathrm{tg}\ 2\pi\mu}{\mathrm{tg}\ \pi\mu}\ \zeta^P(1) \quad . \tag{83}$$

For M=1, $\zeta^P(1)=1/4$ and $\zeta(1)=\infty$ as expected. For M=2 :

$$\zeta^P(1) = C^{-4/3}(2/3)^{1/3}\Gamma(1/3)^5/16\pi^2 = \frac{3}{8}(2\pi)^{-4/3}\Gamma(1/4)^{-8/3}\Gamma(1/3)^5 \tag{84}$$

and

$$\zeta(1) = 3\zeta^P(1) \quad . \tag{85}$$

The latter relation means that the sum of the inverse *even* eigenvalues of the quartic oscillator equals exactly *twice* the sum of the inverse *odd* ones.

In collaboration with D. and G. Chudnovsky we have also somewhat reduced the integrals for $\zeta(2)$ and $\zeta^P(2)$ (work in preparation).

7. THE LIMIT $s \to -\infty$.

We shall now derive asymptotic expansions for $\zeta(s)$ and $\zeta^P(s)$ as $s \to -\infty$ in the case M=2, and explain their importance in the structure of the spectrum of $\hat{H}$. In the harmonic case M=1 the corresponding result is a (weak) consequence of the Riemann functional equation for $\zeta_R(s)$; with the normalization of Eq.(73) :

$$\zeta(s) \sim -\Gamma(1-s)\pi^{-1} \sin\frac{\pi s}{2}\ (1+O(2^{-s})) \quad (s \to -\infty) \ . \tag{86}$$

Our reasoning for M=2 will actually mimic one proof of the Riemann result [17], but it will stop short from yielding an exact functional equation and we shall content ourselves with the asymptotic expression :

$$\zeta(-3s/4) \sim \Gamma(1+s)\ \frac{2^{s/2}}{\pi}\ \frac{\sin 3\pi s/4}{\cos \pi s/2}\left(1+\frac{\alpha_1}{s-1}+\frac{\alpha_2}{(s-1)(s-2)}+\ldots\right)\ (s \to +\infty) \tag{87}$$

or equivalently :

$$\frac{\cos 2\pi s/3}{\sin \pi s}\ \zeta(-s) \sim \frac{s}{3\pi}\ 2^{2(1+s/3)} \sum_{j=0}^\infty \alpha_j \Gamma\left(\frac{4s}{3}-j\right) \qquad (s \to +\infty) \tag{88}$$

where the coefficients α_j are curiously given by the generating function :

$$\sum_0^\infty \alpha_j\ \sigma^{-j} \equiv \exp\left(\frac{b_1}{2\sigma}+\frac{b_2}{2^2\sigma^3}-\left(\frac{b_3}{2^3\sigma^5}+\frac{b_4}{2^4\sigma^7}\right)+\left(\frac{b_5}{2^5\sigma^9}+\frac{b_6}{2^6\sigma^{11}}\right)-\ldots\right) \ . \tag{89}$$

The relative error in (87), i.e. the relative discrepancy from the exact result, in the sense of [2], should be of the order of $2^{-s/2}$. Similarly :

$$\zeta^P(-s) \sim \frac{4s}{3\pi}\ 2^{4s/3}\ \left(1+2\cos\frac{2\pi s}{3}\right) \sum_{j=0}^\infty \beta_j\ \Gamma\left(\frac{4s}{3}-j\right) \tag{90}$$

with the β_j generated by the relation :

$$\sum_0^\infty \beta_j \sigma^{-j} \equiv \exp\left(\frac{1}{2}\sum_{n=1}^\infty (-1)^{n+1}\ 4^{-n}\ b_n\ \sigma^{1-2n}\right) . \tag{91}$$

We shall see that the blow-up of $\zeta(s)$ for $s \to -\infty$ has to do with a subdominant semiclassical feature of the spectrum $\{\lambda_n\}$: as $n \to \infty$, the eigenvalues *deviate* by exponentially small (yet numerically meaningful) amounts from their approximate values taken from the expansion (20) [2-3]. In other words, $\zeta(s)$ for $s \to -\infty$ serves as a microscope to probe the fine structure of the spectrum. The computation of those effects amounts to the evaluation of certain *Stokes multipliers* for a barrier penetration (i.e. quantum tunneling) problem with four turning points.

The key to the expansion (87) is a Mellin representation of $\zeta(s)$ in terms of *an analytic function with well controlled singularities*. We explain this point in general terms. Consider the integral :

$$\xi(s) = \frac{1}{\Gamma(s)} \int_0^\infty f(t)\, t^{s-1}\, dt \quad , \tag{92}$$

where the function f(t) is *analytic around t=0*, precisely in a "keyhole domain" $\{|t|<R_o\} \cup \{|\mathrm{Arg}t| < \delta\}$, except for an isolated branch point at $t_\varepsilon (0<|t_\varepsilon|=R<R_o$, $\mathrm{Arg}\, t \neq 0)$; we place a cut for f(t) on $t_\varepsilon\times(1,\infty)$ (Fig.3) and we assume an expansion :

$$i^{-1}[f(t_\varepsilon(x+i0))-f(t_\varepsilon(x-i0))] = \sum_{j\in J} f_j^{(\varepsilon)}(x-1)^j/\Gamma(j+1) \qquad (x \to 1^+) \quad . \tag{93}$$

Here J is a sequence of reals increasing to $+\infty$, the cut for the function z^j is on $\{z<0\}$, and the expression $(x-1)^j/\Gamma(j+1)$ is to be understood as the generalized function $\delta^{(-1-j)}(x-1)$ whenever $(-1-j) \in \mathbb{N}$ [18].

Then the expansion (93) controls the behavior of $\xi(s)$ for $s \to -\infty$: to see this, we write $\xi(s)$ as the contour integral (Fig.3) :

$$\xi(s) = -i(2\pi)^{-1}\Gamma(1-s) \int_C f(t)\, (-t)^{s-1}\, dt$$

and for s<0 large enough we may shift the contour to obtain :

$$\xi(s) = -i(2\pi)^{-1}\Gamma(1-s)\Big(\int_{C'} f(t)(-t)^{s-1}dt + \xi_r(s)\Big) \tag{94}$$

where the remainder $\xi_r(s)$ is $O(R'^s)$ for any $R<R'<R_o$. If we substitute the discontinuity formula (93) into (94) and integrate term by term, we obtain the large s expansion of $\xi(s)$:

$$\xi(s) \underset{s\to-\infty}{\sim} (2\pi)^{-1}(-t_\varepsilon)^s \sum_{j\in J} f_j^{(\varepsilon)}\Gamma(-s-j)[+\Gamma(-s)\times O(R'^s)]. \tag{95}$$

If there are several isolated branch points, their contributions of the form (95) add up, but the factor $(-t_\varepsilon)^s$ establishes the dominant influence of the nearest branch point(s) from t=0.

Concerning our function $\zeta(s)$, the Mellin representation (11) cannot be used as we hardly control the singularities of $\Theta(t)$ (we expect it to have the imaginary axis as a natural boundary). There is however the alternate representation (25) for $\alpha=3/4$:

$$\zeta(3s/4) = \Gamma(s)^{-1} \int_0^\infty \Theta_{3/4}(t)\, t^{s-1}\, dt \tag{96}$$

and we saw in section 2 that $\Theta_{3/4}$ has isolated singularities on the imaginary axis. Now the expansion of $\zeta(s)$ for $s\to-\infty$ will be dominated by the *nearest* singularities of $\Theta_{3/4}(t)$ anywhere in the complex t plane. Besides that, $\Theta_{3/4}(t)$ also has a singularity at t=0, which must be resolved before we can proceed along Eqs.(93-95). An answer to those problems is the final result of [3]. Because of our different conventions (and of a few misprints in [3], listed in reference), we repeat here the main steps of the argument :

- a Feynman path integral representation suggests that the singularities of $\Theta_{3/4}(t)$ lie at the values taken by the classical action on *complex* periodic orbits of the Hamiltonian (5), in accordance with the Balian-Bloch analysis of the Schrödinger equation [19] . Those orbits are elliptic "cn" functions, so the singular points form a *lattice* : $L = (\frac{1+i}{2})\mathbb{Z} + (\frac{1-i}{2})\mathbb{Z}$. Computation shows them [3] to be *branch points* : all this means that $\Theta_{3/4}(t)$ could be continued from {Ret>0}, first to an analytic function in a cut complex plane, with cuts parallel to the semi-axis $(-\infty,0]$ (this region will be called "first sheet"), and from there to an analytic function *on a Riemann surface ramified over* $\mathbb{C}$, all branch points of which lie over the lattice L (Fig.4) ; we intend to focus upon the four points of L nearest from t=0 : $t=(\pm 1\pm i)/2$.

- in the integral (29) defining the integrated density of levels $F(\sigma)$, we may now shift the contour until it wraps around all cuts on the first sheet. By Laplace transformation, the discontinuities of $\Theta_{3/4}(t)$ on that sheet generate a double expansion of the regularized density $\overline{F}(\sigma)$ in powers of σ^{-1} and of $e^{-\sigma/2}$: terms of order $e^{-m\sigma/2}$ come from the branch points on $\{\mathrm{Re}\, t=-m/2\}$. For m=0, we recover the series (30) and its relation to the discontinuity of $\Theta_{3/4}(t)$ at t=0, as given by Eqs.(27) (first line) and (24). Moreover (27) can now be understood as a decomposition :

$$\Theta_{3/4}(t) = (2\pi)^{-1}[\frac{b_o}{t} + \Theta_{disc}(t)\,\log t] + \Theta_{reg}(t) \tag{97}$$

where

$$\Theta_{disc}(t) = -\sum_1^\infty b_j t^{2j-1}/\Gamma(2j-1) \tag{98}$$

and $\Theta_{reg}(t)$ are *both analytic in the disk* $|t|<1/\sqrt{2}$, whose radius equals the distance of the four nearest branch points $(\pm 1\pm i)/2$. By the same Laplace transform argument, the discontinuities of $\Theta_{3/4}(t)$ at $t_\varepsilon=2^{-1/2}e^{3i\pi/4}$ and at t_ε^* (angles are measured from the positive real axis on the first sheet, in which t_ε and t_ε^* lie : Fig.5), correspond to corrections of order $e^{-\sigma/2}$ to the series (30) for $\overline{F}(\sigma)$. Such corrections are computable by the *complex WKB method* through the evaluation, to dominant order in $e^{-\sigma/2}$, of a *Stokes constant* associated with a certain barrier penetration problem. We found [2,3] :

$$\overline{F}(\sigma) \sim \sum_0^\infty b_j\sigma^{1-2j} + e^{-\sigma/2}\,(\exp -\frac{1}{2}\sum_1^\infty(-1)^j b_j\sigma^{1-2j})(-2\sin\frac{1}{2}\sum_0^\infty b_j\sigma^{1-2j}) + O(e^{-\sigma}) \tag{99}$$

which immediately translates to :

$$\left.\begin{array}{l} i^{-1}[\Theta_{3/4}(t_\varepsilon(x+i0))-\Theta_{3/4}(t_\varepsilon(x-i0))] \\ -i^{-1}[\Theta_{3/4}(t_\varepsilon^*(x+i0))-\Theta_{3/4}(t_\varepsilon^*(x-i0))] \end{array}\right\} = -2ix \sum_{j=0}^{\infty} \alpha_j (x-1)^{j-1}/\Gamma(j) \qquad (100)$$

with the α_j given by Eq.(89).

We also know that $\Theta_{3/4}(t)$ is analytic for $\mathrm{Re}\, t > 0$ in the first sheet ($\Theta_{3/4}(t) = \sum_0^\infty e^{-t\sigma_k}$ there) ; t_ε and t_ε^* are thus the *only* first sheet singularities of $\Theta_{3/4}(t)$ at distance < 1.

- an additional difficulty is the singularity at t=0 itself, which will force us to look at other sheets of $\Theta_{3/4}(t)$ around t=0. In order to postpone that part of the analysis, we make here a digression to the parallel treatment of $\zeta^P(s)$, which can rely directly on Eqs.(92-95) since by Eq.(52) $\Theta^P_{3/4}(t)$ is analytic at t=0. The Feynman path integral suggests that the singularities of $\Theta^P_{3/4}(t)$ lie at the centers of the squares of L, thus forming the lattice L' *dual* to L). Formula (99), analyzed separately for even and odd levels, does give the discontinuities of $\Theta^P_{3/4}(t)$ at the points $\{(m+\frac{1}{2})i\}$ and $\{-\frac{1}{2}+mi\}$ $(m \in \mathbf{Z})$. At the three points of L' nearest to t=0, namely $-1/2$ and $\pm i/2$ ($+\,1/2$ is excluded as $\Theta^P_{3/4}(t)$ is analytic for $\mathrm{Re}\, t > 0$ in the first sheet), we get :

$$\left.\begin{array}{l} i^{-1}[\Theta^P_{3/4}(-\frac{1}{2}(x+i0)) - \Theta^P_{3/4}(-\frac{1}{2}(x-i0))] = i^{-1}[\Theta^P_{3/4}(\frac{i}{2}(x+i0)) - \Theta^P_{3/4}(\frac{i}{2}(x-i0))] \\ \quad = i^{-1}[\Theta^P_{3/4}(-\frac{i}{2}(x+i0)) - \Theta^P_{3/4}(-\frac{i}{2}(x-i0))] = -2x \sum_0^\infty \beta_j \dfrac{(x-1)^{j-1}}{\Gamma(j)} \end{array}\right\} \qquad (101)$$

where the coefficients β_j admit the generating function (91).

The asymptotic behavior of $\zeta^P(-s)(s \to +\infty)$ is now the sum of the three contributions from the discontinuities (101), which amounts precisely to formula (90).

Remark : Eq.(90) predicts the *correct* locations of the zeros of $\zeta^P(-s)$, and for $d_n = (-1)^n \zeta^P(-3n)$ it predicts the large order behavior :

$$d_n \sim (-1)^n \frac{12}{\pi} n\, 2^{4n} \sum_0^\infty \beta_j\, \Gamma(4n-j) \quad (n \to +\infty) \qquad (102)$$

which is very accurately satisfied as early as $n \simeq 10$.

- returning to Eq.(96), we treat the t=0 singularity of $\Theta_{3/4}(t)$ by substituting the decomposition (97) into the integral and cutting off the latter at some $R' < 1$ (because Θ_{disc} and Θ_{reg} certainly have unknown singularities at $t = +1$) :

$$\zeta(3s/4) = \frac{1}{\Gamma(s)}\left[\int_0^{R'}(2\pi)^{-1}\,\Theta_{disc}(t)\log t.t^{s-1}dt + \int_0^{R'}\Theta_{reg}(t)t^{s-1}dt\right] + \Gamma(-s)\times O(R'^s)$$

$$= (2\pi)^{-1}[\Psi(s)\phi(s)+d\phi/ds] + \frac{1}{\Gamma(s)}\int_0^{R'}\Theta_{reg}(t)t^{s-1}dt + \Gamma(-s)\times O(R'^s) \tag{103}$$

where $\phi(s) = \frac{1}{\Gamma(s)}\int_0^{R'}\Theta_{disc}(t)\;t^{s-1}dt$. The integral (103) will again be dominated by a sum of contributions from the nearest branch points $(\pm 1\pm i)/2$, but now evaluated *separately* for Θ_{disc} and Θ_{reg}. Eq.(97) uniquely determines the discontinuities of Θ_{disc} and Θ_{reg} at those four points in terms of the discontinuities of $\Theta_{3/4}$ taken at the same points but in *both sheets* of the sector $-2\pi < \operatorname{Arg} t < 2\pi$ (those will be the only singularities of Θ_{disc} and Θ_{reg} in the disk $|t|<1$).

- there only remains to investigate $\Theta_{3/4}(t)$ in the neighboring sheets across the cut from t=0, precisely for $\pi < |\operatorname{Arg} t| < 2\pi$. The main idea here is that there is no singularity above t_ε and t_ε^*, i.e. that $\Theta_{3/4}(t)$ is *regular* at $t=2^{-1/2}e^{\pm 5i\pi/4}$: by reapplying the complex WKB method to the equation $\hat{H}\psi = \lambda\psi$ with $\operatorname{Arg}\lambda \neq 0$, we find that the exponentially small term in (99) *disappears* (*Stokes phenomenon*) when $|\operatorname{Arg}\lambda| = \pi/3$, values at which the cuts in Fig.4 get rotated by $\pm\pi/4$: this means that the branch points $(-1\pm i)/2$ are absent on the neighboring sheets [20], which fact was implicitly assumed in [3]. Then, because $\Theta_{disc}(t)$ is an *odd* function of t, the discontinuity of $\Theta_{3/4}(t)$ at $t'_\varepsilon = 2^{-1/2}e^{7i\pi/4}$ has to be the *reflection* of the one at $t_\varepsilon = 2^{-1/2}e^{3i\pi/4}$, and similarly for t'^*_ε (Fig.5) :

$$\left.\begin{array}{l} i^{-1}[\Theta_{3/4}(t'_\varepsilon(x+i0)) - \Theta_{3/4}(t'_\varepsilon(x-i0))] \\ -i^{-1}[\Theta_{3/4}(t'^*_\varepsilon(x+i0)) - \Theta_{3/4}(t'^*_\varepsilon(x-i0))] \end{array}\right\} = 2ix\sum_{j=0}^{\infty}\alpha_j(x-1)^{j-1}/\Gamma(j) \tag{104}$$

- a tedious computation finally establishes that the contributions of the discontinuities of Θ_{disc} and Θ_{reg} at $t=(\pm 1\pm i)/2$ to formula (103) precisely add up to (88). Corrections to (88) would involve the next nearest discontinuities (with $|t|=1$): they should be smaller than (88) by a factor $2^{-|s|/2}$.

As a check, we note that Eq.(88) predicts the correct location of the zeros and poles of $\zeta(s)$, and for the coefficients b_n (related to the residues of $\zeta(s)$ by (23)), it implies the expansion :

$$b_n \sim \frac{2^{n+3/2}}{\pi}\cos\left(\frac{3\pi}{4}+\frac{n\pi}{2}\right).\sum_0^\infty \alpha_j\Gamma(2n-1-j) \qquad (n\to+\infty) \tag{105}$$

already derived in [2,3] ; formula (105) is obeyed very accurately when tested up to b_{60} (there is actually a very small discrepancy, but it just has the form expected from the contribution of the t=-1 singularity !).

In principle this treatment should extend to general $M>2$ with $\Theta_{3/4}$ replaced by $\Theta_{(M+1)/2M}$, but we do not yet understand the analytic structure of this function away from the half-plane $\mathrm{Re}\, t \geq 0$.

We end this section by indicating two directions towards which the preceding discussion requires improvement :

- it should be made rigorous, by a proper extension (to functions analytic in the large) of the works dealing with Fourier integral operators and/or microfunctions [21]

- we need to analyze the more distant singularities of $\Theta_{3/4}$ and the global topology of its Riemann surface, in order to resum the divergent eigenvalue expansion (20) correctly, and also to understand the more complicated situation that arises for $M > 2$. What is required here is to fully exploit the information given by connection formulas in relation with the Stokes phenomenon ; limited progress in that direction will be reported elsewhere [20].

8. CONCLUSION

The study of $\zeta(s)$ for the quartic oscillator has uncovered several features of the semiclassical expansions for that system. In particular there exists a unique set of coefficients $\{b_n\}$ (table 1), related to the sequence of residues of $\zeta(s)$, that controls simultaneously the small-t expansion of $\Theta(t)$ by Eqs.(15) and (24), the large-n expansion of the eigenvalues λ_n by Eq.(20), the large $-|q|$ behavior of the eigenfunctions in the "deep forbidden region" $\lambda \to -\infty$ by Eqs.(36) and (43), and the $s \to -\infty$ behavior of $\zeta(s)$ by Eqs.(87-89). Furthermore the first b_j control the $n \to \infty$ behavior of b_n itself through formulas (105) and (89).

We have benefited from helpful discussions with many colleagues, among whom we are especially grateful to R. Balian, E. Brézin, D. Chudnovsky and G. Chudnovsky, Y. Colin de Verdière, H. Cornille, H. De Vega, A. Douady, L.D. Faddeev, S. Graffi and B. Malgrange. We also thank Mrs. C. Bourgois for her assistance in computer work.

REFERENCES

[1] C. BENDER, K. OLAUSSEN, P. WANG, *Phys. Rev.* D16, 1740 (1977)
[2] R. BALIAN, G. PARISI, A. VOROS, *Phys. Rev. Lett.* 41, 1141 (1978)
[3] R. BALIAN, G. PARISI, A. VOROS, in : "*Feynman Path Integrals*", Springer Lecture Notes in Physics, vol. 106 (1979). That article contains the following misprints :
- Eqs.(4.7) and (4.10) : a factor i is missing in front of $\sum_{m\neq 0}(\ldots)$
- in the two lines above Eq.(4.13) : replace $\sum_{k=0}^{\infty}$ by $\sum_{k=-2}^{\infty}$
- in Eq.(4.13) : replace $\sum_{\ell=0}^{\infty}$ by $\sum_{\ell=-1}^{\infty}$
- in the second line after Eq.(4.13) : replace $\rho_{2j-1}^{(o)}$ by $\rho_{2j-2}^{(o)}$

- in the seventh line after Eq.(4.13) : replace $\rho_{2j}^{(o)}$ by $\rho_{2j-1}^{(o)}$.

[4] - G. PARISI, *Trace Identities for the Schrödinger Operator and the WKB Method*, Preprint LPTENS 78/9 (Ecole Normale Supérieure, Paris, March 1978)
[5] - A. VOROS, Nucl. Phys. B165, 209 (1980)
[6] - E.C. TITCHMARSCH, *Eigenfunctions Expansions*, Part I (Oxford U.P. 1961)
[7] - S. MINAKSHISUNDARAM, A. PLEIJEL, *Canad. J. Math.* 4, 26 (1952)
L. HÖRMANDER, *Acta Math.* 121, 93 (1968)
[8] - I.C. PERCIVAL, *Proc. Phys. Soc.* 80, 1290 (1962)
[9] - I.M. GELFAND, *Usp. Mat. Nauk* 11, 191 (1956) ; L.A. DIKII, *Izv. Akad. Nauk SSSR, Ser. Mat.* 19, 187 (1955) and *Usp. Mat. Nauk* 13, 111 (1958) (translated in *Transl. Amer. Math. Soc.* (2) 18, 81) ;
R.T. SEELEY, AMS *Proc. Symp. Pure Math.* 10, 288 (1967)
[10]-E.P. WIGNER, *Phys. Rev.* 40, 749 (1932)
[11]-A. VOROS, Thesis (Orsay 1977, unpublished) ; B. GRAMMATICOS and A. VOROS, *Ann. Phys.* 123, 359 (1979)
[12]-C.E. REID, *J. Molec. Spectrosc.* 36, 183 (1970)
[13]-R.G. NEWTON, *Scattering Theory of Waves and Particles, Mc Graw Hill 1966* ;
V.E. ZAKHAROV, L.D. FADDEEV, *Funkt. Analiz i Ego Pril. (Russian)* 5, 18 (1971) ;
K. CHADAN, P.C. SABATIER, *Inverse Problems in Quantum Scattering Theory, Springer 1977*
[14]-A. ERDELYI et al., "*Higher Transcendental Functions*", *vol.2* (Bateman Manuscript Project, Mc Graw Hill, New York) ;
W. MAGNUS, F. OBERHETTINGER, R.P. SONI, *Formulas and Theorems for the Special Functions of Mathematical Physics, Springer Verlag (1966)*
[15]-A. GROSSMANN, *Comm. Math. Phys.* 48, 191 (1976)
[16]-F.R. GANTMAKHER, M.G. KREIN, "*Oscillating Matrices and Kernels : Small Oscillations of Mechanical Systems*",Akademie Verlag (Berlin, 1960) ;
J. SCHONFELD et al., *Ann. Phys.* (1980 to appear)
[17]-E.C. TITCHMARSCH, *The Riemann Zeta Function*, Clarendon Press (Oxford 1951)
[18]-I.M. GELFAND, G.E. SHILOV, *Generalized Functions I* (Academic Press 1968)
[19]-R. BALIAN, C. BLOCH, *Ann. Phys.* 85, 514 (1974)
[20]-A. VOROS, in preparation.
[21]-J.J. DUISTERMAAT and V.V. GUILLEMIN, *Invent. Math.* 29, 39 (1975)
H. KOMATSU (ed.), *Hyperfunctions and Pseudo-Differential Equations*, Springer Lecture Notes in Mathematics vol. 287 (1973)
[22]-A.C. HEARN, REDUCE *User's Manual* (University of Utah, 1973).

Service de Physique Théorique
CEN Saclay BP n°2
91190 Gif-sur-Yvette, France.

TABLE 1

The first coefficients b_n and d_n .

(REDUCE [22] output)

$$B(0) := 1$$

$$B(1) := (-\ \pi)/3$$

$$B(2) := (11\ \Gamma(1/4)^{8})/(10368\ \pi^{2})$$

$$B(3) := (4697\ \Gamma(1/4)^{8})/(466560\ \pi)$$

$$B(4) := (-\ 390065\ \Gamma(1/4)^{16})/(501645312\ \pi^{4})$$

$$B(5) := (-\ 53352893\ \Gamma(1/4)^{16})/(1934917632\ \pi^{3})$$

$$B(6) := (122528437805\ \Gamma(1/4)^{24})/(24519276232704\ \pi^{6})$$

$$B(7) := (457890455939\ \Gamma(1/4)^{24})/(1073234313216\ \pi^{5})$$

$$B(8) := (-\ 15168742752828973\ \Gamma(1/4)^{32})/(9244212944751 8208\ \pi^{8})$$

$$B(9) := (-\ 1185170778607084 37515\ \Gamma(1/4)^{32})/(4714548601823428608\ \pi^{7})$$

$$B(10) := (29115708968576993 1700559\ \Gamma(1/4)^{40})/ (182103599641255068 30336\ \pi^{10})$$

$$D(0) := 1/2$$

$$D(1) := (-\ 9\ C^{4})/4$$

$$D(2) := (1701\ C^{8})/4$$

$$D(3) := (-\ 1097307\ C^{12})/2$$

$$D(4) := (10333100421\ C^{16})/4$$

$$D(5) := (-\ 258327414331263\ C^{20})/8$$

$$D(6) := (1768210026160034583\ C^{24})/2$$

$$D(7) := (-\ 9321807810423365 3681937\ C^{28})/2$$

$$D(8) := (1725448516471637935 68351189061\ C^{32})/4$$

$$D(9) := (-\ 5230118845170110681 9010145067664 33\ C^{36})/8$$

$$D(10) := (1229564553606524469 27230976728749 2145587\ C^{40})/8$$

s	$\zeta(s)$	$\zeta'(s)$	$\zeta^P(s)$	$(\zeta^P)'(s)$
-6.0	-.0	-634115.	9490563.	-35928404.
-5.9	-45851.8	-312862.	6411470.	-26027490.
-5.8	-67247.6	-133189.	4212089.	-18307784.
-5.7	-75375.2	-40896.	2685370.	-12521508.
-5.6	-77371.0	-7673.	1654743.	-8326564.
-5.5	-78734.5	-30985.	978653.	-5375102.
-5.4	-87739.4	-193407.	548676.	-3356899.
-5.3	-159992.7	-2309745.	284766.	-2015790.
-5.2	84397.2	-2316352.	129688.	-1151233.
-5.1	10910.9	-210549.	43703.	-612543.
-5.0	-.000	-49900.93	.00	-290317.0
-4.9	-2673.086	-11188.49	-18981.27	-107523.7
-4.8	-3068.769	1049.76	-24376.83	-11477.1
-4.7	-2733.062	4868.87	-22988.02	32819.6
-4.6	-2195.397	5581.95	-18777.39	47902.3
-4.5	-1654.207	5135.15	-13908.18	47780.8
-4.4	-1180.014	4325.13	-9445.91	40799.9
-4.3	-789.695	3493.83	-5817.85	31656.0
-4.2	-476.800	2794.05	-3104.41	22777.5
-4.1	-223.429	2319.73	-1216.74	15242.8
-4.0	--.000	2245.215	-.000	9371.19
-3.9	264.475	3468.966	710.728	5091.24
-3.8	1103.232	23473.005	1063.232	2159.40
-3.7	-1129.287	23254.142	1177.778	283.28
-3.6	-334.912	2794.109	1145.767	-814.96
-3.5	-162.830	1060.332	1032.808	-1370.80
-3.4	-86.883	538.131	883.455	-1569.63
-3.3	-46.077	304.542	726.142	-1549.35
-3.2	-22.446	179.910	577.602	-1407.50
-3.1	-8.385	107.509	446.482	-1209.67

s	$\zeta(s)$	$\zeta'(s)$	$\zeta^P(s)$	$(\zeta^P)'(s)$
-3.0	-.00000000	63.7867658	336.1292802	-997.466975
-2.9	4.93609705	37.1479617	246.6396654	-795.354310
-2.8	7.78637962	21.3018393	176.2899527	-616.045587
-2.7	9.43911389	12.7995466	122.4959734	-464.622036
-2.6	10.54303783	10.2889505	82.4195173	-341.494985
-2.5	11.75123572	15.7223257	53.3258018	-244.463601
-2.4	14.44318459	46.5061407	32.7714718	-170.080482
-2.3	29.25971086	454.2842312	18.6827185	-114.511868
-2.2	-17.27583077	454.0301596	9.3662199	-74.043040
-2.1	-2.51880707	45.4816279	3.4823349	-45.344404
-2.0	-.000000000	13.08985438	.00000000	-25.5832999
-1.9	.802699732	4.47424248	-1.85444489	-12.4418875
-1.8	1.062545873	1.18691027	-2.64730639	-4.0823606
-1.7	1.098707020	-.27990239	-2.78265252	.9134198
-1.6	1.031461908	-.98179856	-2.54085176	3.6129146
-1.5	.913975724	-1.32839029	-2.11021849	4.8005823
-1.4	.771195002	-1.51015263	-1.61219030	5.0389834
-1.3	.613801084	-1.63562452	-1.12082682	4.7210474
-1.2	.442987029	-1.79485671	-.67754793	4.1130482
-1.1	.249302359	-2.12621840	-.30201818	3.3888503
-1.0	.000000000	-3.02193299	-.00000000	2.6564860
-.9	-.431162606	-6.49013542	.23111849	1.9782961
-.8	-2.186973245	-50.84887994	.39853930	1.3858560
-.7	2.732234696	-50.76794435	.51154892	.8907928
-.6	.992546255	-6.24929374	.57992925	.4924450
-.5	.592217121	-2.62762286	.61300996	.1831478
-.4	.390523758	-1.59114226	.61915739	-.0482283
-.3	.256499903	-1.14158948	.60555089	-.2139975
-.2	.155388222	-.90326157	.57814015	-.3262642
-.1	.072528059	-.76651556	.54170964	-.3961364

TABLE 2

s	$\zeta(s)$	$\zeta'(s)$	$\zeta^P(s)$	$(\zeta^P)'(s)$
0.0	-.00000000000000	-.6931471805599	.500000000000	-.4333441145
.1	-.06784867141886	-.6727661898828	.455853654203	-.4461174358
.2	-.13657047965339	-.7137026863965	.411363981827	-.4412218325
.3	-.21374632236383	-.8512671102779	.368016126546	-.4240818430
.4	-.31329130969004	-1.1892550135090	.326813011938	-.3989466822
.5	-.46918717525590	-2.0857573118444	.288383976691	-.3690685248
.6	-.80341300384056	-5.4485482759026	.253075725381	-.3368757100
.7	-2.40789793378155	-47.8626032270793	.221026647392	-.3041310392
.8	2.35631676196944	-47.8396973092926	.192226259915	-.2720705506
.9	.75643642198897	-5.3791265574583	.166561814104	-.2415214624
1.0	.43156237827051	-1.9676530763547	.143854126090	-.2129999574
1.1	.29006214682228	-1.0186597195285	.123884566486	-.1867905789
1.2	.21043337229281	-.6223766566509	.106414936312	-.1630095054
1.3	.15939946848792	-.4180785345630	.091201721057	-.1416541111
1.4	.12406586817392	-.2981805457922	.078005977126	-.1226411298
1.5	.09834787941793	-.2214122433868	.066599882666	-.1058355340
1.6	.07897645572923	-.1691318916240	.056770786517	-.0910719792
1.7	.06402485422711	-.1318819516086	.048323417742	-.0781703873
1.8	.05227586119348	-.1044252686579	.041080774173	-.0669469807
1.9	.04291813457772	-.0836554644691	.034884089584	-.0572218353
2.0	.03538750884214	-.0676249358029	.029592182906	-.0488238143
2.1	.02927821335407	-.0550555697173	.025080416138	-.0415935604
2.2	.02429034495007	-.0450760773557	.021239427237	-.0353850818
2.3	.02019729503601	-.0370732034538	.017973757428	-.0300663372
2.4	.01682472434523	-.0306034984114	.015200456518	-.0255191353
2.5	.01403651351832	-.0253388904180	.012847722690	-.0216385795
2.6	.01172508808264	-.0210319235204	.010853613119	-.0183322309
2.7	.00980457824390	-.0174928629256	.009164846979	-.0155191149
2.8	.00820587024750	-.0145741967847	.007735711832	-.0131286592
2.9	.00687295334649	-.0121598834671	.006527076955	-.0110996254
3.0	.00576017525548	-.0101577242616	.005505512132	-.0093790738

TABLE 2 : Some values of $\zeta(s)$, $\zeta^P(s)$ and their derivatives. The last digits are *estimated* to be accurate, but *not guaranteed* to be so (especially for $s < -5$).

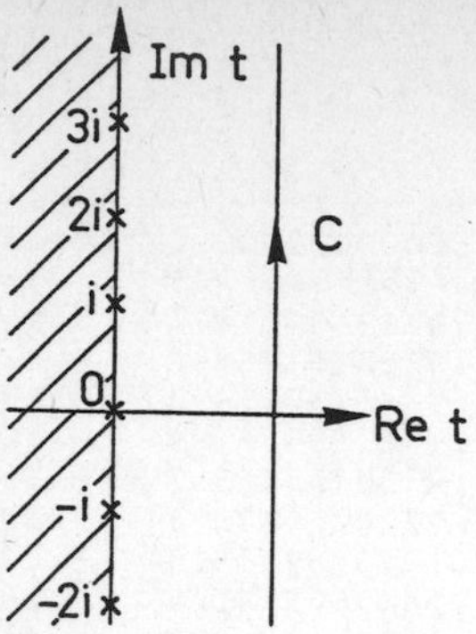

Fig.1

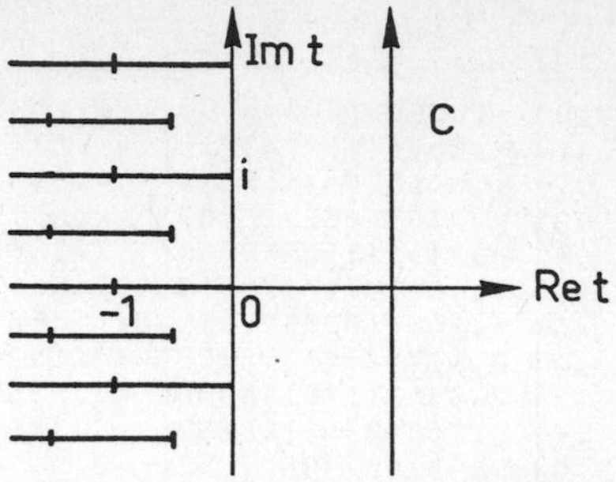

Fig.4

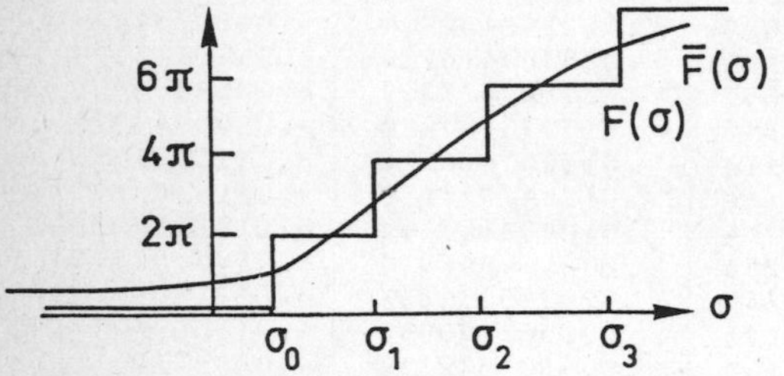

Fig.2

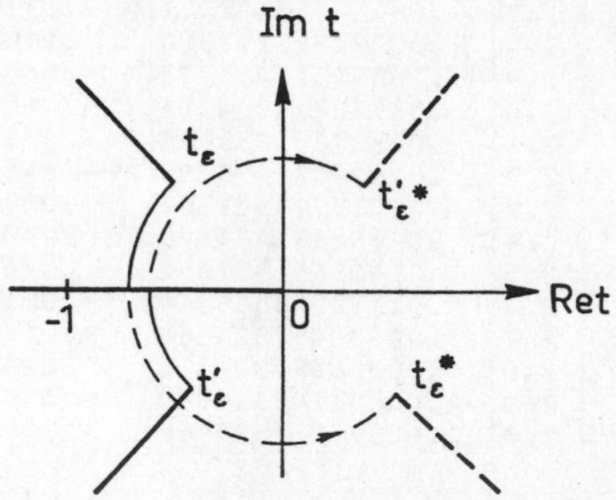

Fig.5

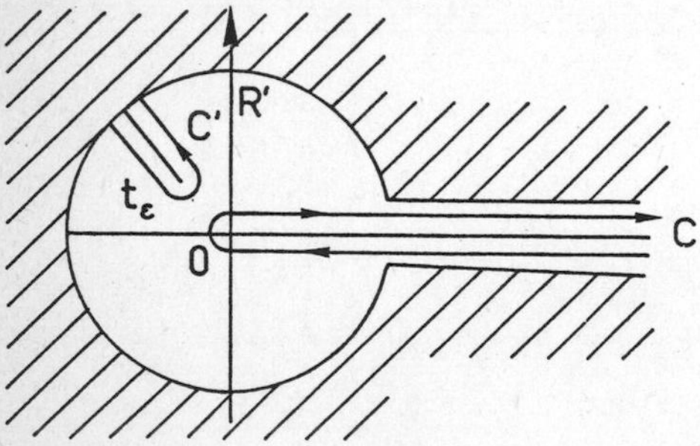

Fig.3

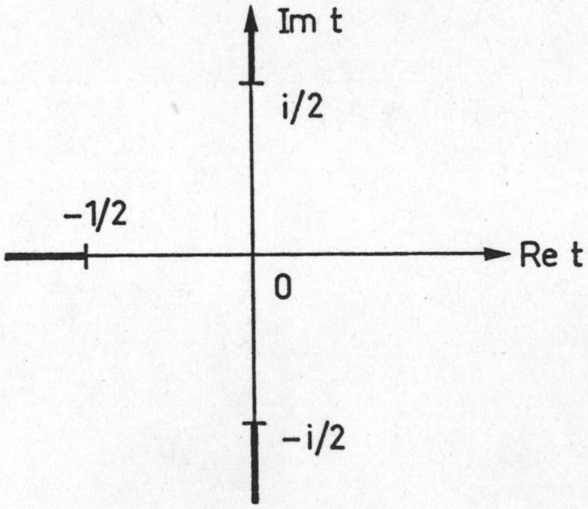

Fig.6

On Trace Formula*

by

Lipman Bers

Here is an elementary proof of the identity you presented in the Chudnovsky seminar. After noticing this proof I found that the idea (using Hadamard's theorem) and the result itself, for the case $n = 1$, are in an old paper by Levinson (on the inverse Sturm-Liouville problem, Mat. Tidsskr B, 1949, pp. 25-30).

We consider the Sturm-Liouville system

$$\Phi''(t) + \lambda R(t)\Phi(t) = 0 \tag{1}$$

$$\Phi(0) = \Phi(\tau) = 0. \tag{2}$$

Here Φ is an unknown n-vector valued function, R a given, say continuous, $n \times n$-matrix valued function, and $\tau > 0$ a given number. The matrix $R(t)$ is supposed to be Hermitean positive semi-definite for all t, with $R(t_0)$ positive definite for some t_0, $0 \leq t_0 < \tau$. We want to compute the number

$$D(1) = \prod_{j=1}^{\infty} (1 - 1/\lambda_j)$$

or, more generally, the entire function

$$D(\lambda) = \prod_{j=1}^{\infty} (1 - \lambda/\lambda_j) \tag{3}$$

where $\lambda_1, \lambda_2, \ldots$ are the nonvanishing eigenvalues of the problem, each written down as often as its multiplicity indicates. (The multiplicity is the dimension of the space of eigenfunctions.)

*This is an extract from a letter of Professor Lipman Bers to Professor Marc Kac from December 5, 1979 reproduced with kind permission of Professor Bers.

Instead of (2) one can consider other boundary conditions, for instance

(4) $$\Phi(0) = \Phi'(\tau) = 0$$

(5) $$\Phi'(0) = \Phi'(\tau) = 0$$

or, more generally,

(6) $$\Phi'(0) - A\Phi(0) = \Phi'(\tau) + B\Phi(\tau) = 0.$$

In the latter case A and B are positive definite Hermitean matrices. We also treat the periodicity problem

(7) $$\Phi(0) = \Phi(\tau), \ \Phi'(0) = \Phi'(\tau).$$

In each case all eigenvalues are positive except that in case (5) and (7) $\lambda = 0$ is an n-fold eigenvalue.

In all cases the system can be written as a symmetric Fredholm integral equation, and the function $D(\lambda)$ is the Fredholm determinant.

(Indeed, the Fredholm determinant is characterized by the relations

(8) $$D(0) = 1, \quad D'(\lambda)/D(\lambda) = - \Sigma_{i=0}^{\infty} T_{i+1} \lambda^{i}$$

where T_i is the trace of the i-th iterated kernel; cf., for instance Tricomi's text on integral equations, p. 27. Now (3) clearly implies the first relation above, the second follows from (3) by noting that

$$- \log \prod_{j=1}^{\infty} (1 - \lambda/\lambda_j) = - \Sigma_{j=1}^{\infty} \log(1 - \lambda/\lambda_j)$$

$$= \Sigma_{j=1}^{\infty} \Sigma_{k=1}^{\infty} \lambda^k/k\lambda_j^k = \Sigma_{k=1}^{\infty} (\Sigma_{j=1}^{\infty} \lambda_j^{-k}) \lambda^k/k,$$

so that

$$D'(\lambda)/D(\lambda) = \Sigma_{i=0}^{\infty} (\Sigma_{j=1}^{\infty} \lambda_j^{-(i+1)}) \lambda^{i}.$$

But for a symmetric kernel

$$T_i = \Sigma_{j=1}^{\infty} \lambda_j^{-1},$$

by the so-called bilinear relation.)

We proceed to compute (3). Define the matrix valued functions $X_j(t)$, $j = 0,1,\ldots$ by the relations

$$X_0(t) = I,\ X_1(t) = tI,\ X_{j+2}(t) = \int_0^t (t - s)(R(s)X_j(s)ds,$$

where I is the $n \times n$ unit matrix. It follows easily that there is a function $m(\tau) > 0$ such that,

$$\|X_j(t)\| + \|X'(t)\| = O(m(t)^j/j'), \quad j \to \infty$$

uniformly for $|t| \leq \tau$.

Now set

(9) $$C(z,t) = \Sigma_{j=0}^{\infty}(-1)^j z^{2j} X_{2j}(t), \quad S(z,t) = \Sigma_{j=1}^{\infty}(-1)^{j-1} z^{2j-1} X_{2j-1}(t).$$

For a fixed z, $C(z,t)$ and $S(z,t)$ as functions of t satisfy the differential equation

(10) $$\Psi''(t) + z^2 R(t) = 0.$$

More precisely, the columns of C form a basis of (vector valued) solutions of (10) which have derivative 0 at the origin, while the columns of S form a basis of (vector valued) solutions of (10) which vanish at the origin.

For a fixed t, the functions $C(z,t)$ and $S(z,t)$ are entire functions of z of order at most 1.

A number $\lambda = z^2 > 0$ is an eigenvalue of multiplicity k of (1), for the boundary conditions (2), if and only if the rank of the matrix $S(z,\tau)$ is precisely $n - k$. If so, there exists an invertible constant matrix Q such that k columns of $S(z,\tau)Q$ vanish. We conclude that $\lambda > 0$ is an eigenvalue of multiplicity k if and only if $x = \sqrt{\lambda}$ is a zero of det $S(z,\tau)$ of order k' with

(11) $$k' \geq k.$$

We proceed to show that equality holds in (11), for all eigenvalues. To do this consider, for any number ε, $0 \leq \varepsilon \leq 1$, the problem (1), (2) with $R(t)$ replaced by

(12) $$R_\varepsilon(t) = (1 - \varepsilon)I + \varepsilon R(t).$$

For each ε, the positive eigenvalues $\lambda_j^{(\varepsilon)}$ and the positive roots $x_j^{(\varepsilon)}$ of det $S(z,\tau)$ can be arranged in ascending sequences

$$\lambda_1^{(\varepsilon)} \leq \lambda_2^{(\varepsilon)} \leq \ldots, \quad x_1^{(\varepsilon)} \leq x_2^{(\varepsilon)} \leq \ldots$$

taking into account their multiplicities. It is known that each $\lambda_j^{(\varepsilon)}$ depends continuously on ε, and it is easily seen that each $x_j^{(\varepsilon)}$ does. Also, each $x_i^{(\varepsilon)}$ equals a $\lambda_j^{(\varepsilon)}$, each $\lambda_j^{(\varepsilon)}$ equals an $x_i^{(\varepsilon)}$, and if k of the $\lambda_j^{(\varepsilon)}$ are equal, so are at least k of the $x_i^{(\varepsilon)}$. For $\varepsilon = 0$,

$$C(z,t) = (\cos zt)I, \quad S(z,t) = (\sin zt)I$$

and det $S(z,t) = (\sin z\tau)^n$. The positive roots of det $S(z,\tau)$ are $\nu\pi/\tau$, $\nu = 1,2,\ldots$, each of order n; the positive eigenvalues are $\nu\pi/\tau$, $\nu = 1,2,\ldots$, each of order n. Thus $x_j^{(0)} = \lambda_j^{(0)}$ for all j. Hence $x_j^{(\varepsilon)} = \lambda_j^{(\varepsilon)}$ for all j and all ε, in particular for $\varepsilon = 1$. This proves our assertion.

Since $S(z,\tau) = z\tau I + O(z^3)$, $z \to 0$, and S is an odd function of z, of order at most 1, $S(\sqrt{\lambda},\tau)/(\sqrt{\lambda}\tau)$ is an entire function of λ of order at most 1/2, and so is its determinant

$$\Delta(\lambda) = [\det S(\sqrt{\lambda},\tau)]/(\sqrt{\lambda}\tau)^n.$$

Since $\Delta(0) = 1$, Hadamard factorization yields

$$\Delta(\lambda) = \prod (1 - \lambda/\lambda_j)$$

where λ_j are all roots of $\Delta(\lambda)$, with multiplicities. But we just proved that those are the eigenvalues of (1), (2). Hence

$$(\sqrt{\lambda}\tau)^n D(\lambda) = \det S(\sqrt{\lambda},\tau).$$

For $\lambda = 1$, this becomes

$$t^n \prod (1 - 1/\lambda_j) = \det S(1,\tau).$$

Other boundary conditions can be treated similarly. For (4) we obtain

$$(\sqrt{\lambda})^n D(\lambda) = \det S'(\sqrt{\lambda},\tau)$$

where the prime denotes differentiation with respect to τ, and for (5)

$$D(\lambda) = \det C(\sqrt{\lambda},\tau).$$

In case (6) we observe that the columns of $C(z,t) + AS(z,t)$ are a basis for solutions of (10) satisfying the first boundary condition (6). Set

$$M(z,t) = C'(z,t) + AS'(z,t) + BC(z,t) + BAS'(z,t).$$

We reason as before, accompanying the deformation (12) of the differential equation by the deformation

$$A_\varepsilon = (1 - \varepsilon)I + A, B_\varepsilon = (1 - \varepsilon)I + B$$

of the boundary conditions, and conclude that

$$(\det B)D(\lambda) = \det M(\sqrt{\lambda},\tau).$$

Finally, in the case of the periodicity conditions (7) and for $\lambda > 0$, we are looking for constant vectors σ and γ, not both 0, such that

$$C(\sqrt{\lambda},\tau)\sigma + S(\sqrt{\lambda},\tau)\gamma = \sigma$$

$$C'(\sqrt{\lambda},\tau)\sigma + S'(\sqrt{\lambda},\tau)\gamma = \gamma.$$

Defining the $2n \times 2n$ matrix

$$Q(\sqrt{\lambda},\tau) = \begin{pmatrix} C(\sqrt{\lambda},\tau) - I & S(\sqrt{\lambda},\tau) \\ C'(\sqrt{\lambda},\tau) & S'(\sqrt{\lambda},\tau) - I \end{pmatrix}$$

we obtain as before that

$$\lambda^{2n}D(\lambda) = \det Q(\sqrt{\lambda},\tau).$$

In all cases, the Freholm determinant turns out to be an actual determinant, obtained by solving an initial value problem.

Department of Mathematics
Columbia University
New York, NY
USA

Resolvent and Trace Identities in the One Dimensional Case

by

D. Chudnovsky and G. Chudnovsky

One of the basic tools for investigations of completely integrable systems associated with inverse scattering method is the method of trace identities or resolvent expansion. It is especially nice in the one-dimensional case and for operators of the second order.

For the Sturm-Liouville equation

(1) $$-\varphi'' + u(x)\varphi = \lambda\varphi$$

we have an asymptotic expansion of the diagonal of the resolvent $R(x,\lambda)$ of (1) when $\lambda \to -\infty$

(2) $$R(x,\lambda) = \Sigma_{L=0}^{\infty} \frac{R_L[u]}{\lambda^{L+1/2}}$$

Here $R_K[u]$ are differential polynomials in u, i.e. polynomials in $u, u', u'', \ldots$ or weight $2K$, where $u^{(j)}$ is of weight $j + 2$:

$$R_0 = \frac{1}{2}, \quad R_1 = \frac{1}{4}u, \quad R_2 = \frac{1}{16}(3u^2 - u''), \ldots.$$

The coefficients $R_K[u]$ are computed using equations satisfied by the resolvent:

$$-2RR'' + (R')^2 + 4(u-\lambda)R^2 = 1$$

or

$$-R''' + 4(u-\lambda)R' + 2u'R = 0$$

(the so-called "Lenard" formula).

The quantities $R_K[u]$ are known as Hamiltonian densities for the $K-1^{th}$ Korteweg-de-Fries equation. If an integral

$$\tilde{\mathcal{R}}_K = \int R_K[u]\,dx$$

is defined (e.g. for periodic in x functions $u(x)$), then

$$u_t = \frac{\partial}{\partial x} \frac{\delta \tilde{\mathcal{R}}_K}{\delta u}$$

is called an $(K-1)^{th}$ KdV. The solutions of the stationary higher KdV equations are called finite band potentials.

Of course, the resolvent helps us to compute the ζ-function of Sturm-Liouville problem the and trace identities. We have two types of coprresponding "θ-functions" investigated in details by Gelfand and Dikij.

Theorem A: *For the elementary solution* $e(t,x,y)$ *of* $\partial e/\partial t = -Qe$; $Q = -d^2/dx^2 + u$, *corresponding to the periodic boundary conditions we have an asymptotic expansion for* $t \to 0+$:

$$e(t,x,x) \underset{\sim}{=} \frac{1}{\sqrt{4\pi t}} \Sigma_{m=0}^{\infty} \frac{(-t)^m}{(2m-3)\ldots 3.1} I_{m-1}, \tag{3}$$

where $I_{m-1} = I_{m-1}[u]$ *is a differential polynomial in* u.

Here $e(t,x,x)$ have the eigenfunction expansion

$$e(t,x,x) = \Sigma_{i=0}^{\infty} e^{-\lambda_i t} f_i^2(x).$$

Hence for a natural "θ-function"

$$\theta(t) = \Sigma_{i=0}^{\infty} e^{-\lambda_i t}$$

we can get a nice representation:

Theorem B: *The trace*

$$\theta(t) = \Sigma_{i=0}^{\infty} e^{-\lambda_i t}$$

of e^{-tQ} *can be asymptotically expanded near* 0: $t \to 0+$ *as*

(4) $$\theta(t) \underset{\sim}{=} \frac{1}{\sqrt{\pi t}} \Sigma_{m=0}^{\infty} \frac{(-t)^m}{(2m-3)\ldots 3.1} H_{m-1}$$

$H_{-1} = 1$ and

(5) $$H_n = \int I_n[u,u',\ldots]dx.$$

In (5) the integral is defined properly, e.g. for functions periodic in x (say, $u(x)$ with period 1).

Identification of symbols: Here in theorem A I_m and R_{m+1} are related:

$$I_{m-1}(2m+1) \quad \text{and} \quad 2^{m+1} R_m$$

are different only by a complete derivative in x. Hence, for $u(x)$ periodic with the period 1,

$$H_{n-1} = \frac{2^{n+1}}{2n-1} \int_0^1 R_n[u]dx = \frac{2^{n+1}}{2n-1} : \tilde{\mathcal{R}}_n.$$

Remark: If $u \equiv 0$ and, of course,

$$\theta(t) = \Sigma_{\ell=1}^{\infty} e^{-\ell^2\pi^2 t}$$

then (4) gives nothing more than does the Jacobi identity for theta function

$$\theta(t) \sim \frac{1}{\sqrt{4\pi t}} - \frac{1}{2}$$

but the exponential terms in $e^{-1/t}$ disappear. So the right side of (4) is not true expression for $\theta(t)$!

How do we compute $\theta(t)$? This is unclear for $u \neq$ const. However the asymptotics of $\theta(t)$ in (4) can be found for all, say, periodic potentials (or rapidly decreasing ones) using the inverse scattering method. For a finite band potential these expressions are only in terms of hyperelliptic integrals, Drach [4].

The nth stationary Korteweg-de-Vries equation can be written as

$$\Sigma_{i=0}^{n+1} d_i \, R_i[u] = 0,$$

or after normalization as

$$(\mathrm{KdV}_n) \qquad \Sigma_{i=0}^{n+1} 2^{i+1} c_i \, R_i[u] = 0, \quad c_{n+1} = 1.$$

Now for u(x) being periodic function and the solution of KdV_n (e.g. $u(x) = n(n + 1)\wp(x)$ - Lamé potentials) the spectrum of $Qy = \lambda y$ consists of n finite intervals and one half-finite (n forbidden zonae or n bands):

$$\lambda_0 < \underbrace{\lambda_1 \leq \lambda_2}_{1 \text{ band}} < \ldots < \underbrace{\lambda_{2n-1} \leq \lambda_{2n}}_{n\text{-th band}} < +\infty$$

Here λ_j: $j = 0,\ldots,2n$ are the first integrals of the dynamical system KdV_n and e.g. c_i are symmetric polynomials in λ_j.

If we define now n numbers λ_i': $i = 1,\ldots,n$ by

$$2^n(\lambda - \lambda_1')\ldots(\lambda - \lambda_n') = \Sigma_{i=1}^{n+1} \Sigma_{j=1}^{i} c_i \tilde{\mathcal{R}}_{j-1} 2^{-j-1} \lambda^{i-j},$$

now we are ready to write the closed expression in the right side of (4):

$$\underset{\sim}{\theta}(t) = \frac{\sqrt{-1}}{\pi}\left(\int_{\lambda_0}^{\lambda_1} + \int_{\lambda_2}^{\lambda_3} + \ldots + \int_{\lambda_{2n}}^{\infty}\right) e^{-\lambda t} \, \Pi_{i=1}^{n} (\lambda - \lambda_i') \frac{d\lambda}{r(\lambda)};$$

$r(\lambda) = \sqrt{-1} \sqrt{\Pi_{j=0}^{2n}(\lambda - \lambda_k)}$ (the integration over n bands).

This is only an asymptotical formula!

Trace identities: Of course $\zeta(s)$ corresponds to $Q = -\dfrac{d^2}{dx^2} + u$ and a certain boundary problem. Usually it is $y(0) = y(t) = 0$. If μ_n are the corresponding eigenvalues, then we can build the ζ-function

$$\zeta_Q(s) = \Sigma_n \mu_n^{-s}$$

For $u \equiv 0$ and $T = \pi$, $\zeta_{Q_0}(s) = \zeta(2s)$ for usual Riemann ζ-function. Then the explicit expressions for $\zeta_Q(-k): k \geq 1$ are called trace formulae.

Remark: For $\zeta(2s)$ there is no such thing as trace formulae.

The numbers $\zeta_Q(-k)$ can be used to evaluate the "regularized" value of

$$\Sigma^* \mu_n^k.$$

For this (such a form belongs essentially to Gelfand and Levitan) we exclude in μ_n^k the terms in the asymptotics of $\mu_n \sim n^2 + C_0 + \frac{C_1}{n^2} + \ldots$ that spoil the convergence, e.g. according to Dikij (1952) if we write $\zeta_Q(s)$ as

$$\zeta_Q(s) = \zeta(2s) + d_2(s)\zeta(2s+2) + \ldots + d_{2k}(s)\zeta(2s+2k) + \Sigma\, O(n^{-2s-2k})$$

then

$$\Sigma_n\{\mu_n^k - n^{2k} - d_2(-k)n^{2k-2} - \ldots - d_{2k}(-k)\} = \zeta_Q(-k) + \frac{1}{2} d_{2k}(-k)$$

(here for determinateness $T = \pi$ for a moment).

In fact, all these quantities are determined explicitely in terms of $\tilde{\ }_k$ only. The best way to do this is to restrict us to $C^\infty u(x)$, say, periodic with the period 1 and $T = 1$, so that $\mu_n \sim \pi^2 n^2 + \ldots$.

1. The asymptotic for μ_n is

$$\mu_n \sim \pi^2 n^2 + k_0 + k_1(\pi n)^{-2} + k_2(\pi n)^{-4} + \ldots$$

and all the coefficients k_j can be obtained from such an asymptotics:

$$\mu_n^p = \Sigma_{i+j=p} \frac{p!\,(n^2\pi^2)^j 2^{j+1}\tilde{\mathcal{R}}_j}{i!\,(2j-1)\ldots 3.1} + O(1): n \to \infty,$$

e.g. $k_0 = 4\tilde{\mathcal{R}}_1 = \int_0^1 u\,dx$, $k_1 = \frac{8}{3}\tilde{\mathcal{R}}_2 - 8\tilde{\mathcal{R}}_1^2, \ldots$.

In particular we know the expressions for all the coefficients $d_{2j}(-k)$ and we can write such a complicated but rigorous statement (Dikij-Mc Kean-van-Moerbeke).

<u>Corollary</u>: <u>For any</u> $k \geq 1$

$$\Sigma_{n=1}\{\mu_n^k - \Sigma_{i+j=k} \frac{k!\,(n\pi)^{2i} 2^{j+1} \tilde{\mathcal{R}}_j}{i!\,(2j-1)\ldots 3.1}\} = \frac{1}{2} \frac{k!\,2^{k+1} \tilde{\mathcal{R}}_k}{(2k-1)\ldots 3.1} + \zeta_u(-k).$$

<u>Here we have</u>

$$\Sigma_{k=1}^{\infty} \frac{2^k \epsilon^k}{k} \zeta_u(-k) = -\log\{\Sigma_{j=0}^{\infty} 2^{j+1} \epsilon^j \mathcal{R}_j(0)\}$$

Here $\mathcal{R}_j(0) = R_j[u(0), u'(0), u''(0), \ldots]$.

References

[1] D.V. Chudnovsky, G.V. Chudnovsky, Spectral interpretation of classical completely integrable systems, Preprint IHES/M/78/236, July 1978, pp. 1-12.

[2] H. Mc Kean, P. van Moerbeke, Inventiones Math. 30, 217 (1978).

[3] L. Dikij, Zeta-function of the ordinary differential equation on the finite interval. Izvestija Akad. Sci., Ser. math., 19 (1955), 187-200.

[4] J. Drach, C.R. Acad. Sci. Paris, 167, 744-746 (1918), 168, 47-50 and 337-340 (1919).

Department of Mathematics
Columbia University
New York, NY
USA

THE DEVIL'S STAIR CASE TRANSFORMATION IN INCOMMENSURATE LATTICES

Serge AUBRY

Abstract

Free energy models describing defects in crystal structures contain necessarily strong anharmonic terms allowing metastable configurations (defectible models). If such a model involves conflicting forces (frustrated model), its response to the relative variation of these forces is qualitatively different from a linear or quasi-linear response, and the structure evolves by discontinuous processes of defects creation or annihilation. These transformations turn out to be described by pathological functions at the macroscopic scale.

A model used for epitaxy and incommensurate structures illustrates these concepts. An elastic chain of atoms is submitted to a periodic modulating potential with a period different from the atomic spacing. This model allows to study the many defects structures (epitaxy dislocations) at fixed concentration or at fixed pressure (or chemical potential) which corresponds to different physical situations.

An incommensurate structure with a given wave-vector is represented by a fixed defect concentration. When the amplitude of a modulating potential increases, the defects structure exhibits a transition from a "fluid" regime (called analytic regime) with a zero frequency phason mode to a locked regime (called non-analytic regime) with a finite gap in the phonon excitation spectrum. Frustration variation in the model corresponds to pressure variation. The wave-vector of the modulated structure (i.e. the defect concentration) varies continuously with piece wise constant parts at each commensurate value (the resulting curve is called a devil's stair case).

When the modulating potential is small enough, the devil's stair case is rather smooth. The locking forces which oppose to any structure change are either null or small. True incommensurate structures (with phasons) are possible with finite probability (this variation curve is called an incomplete devil's stair case). When the modulating potential increases, the devil's stair case becomes steeper. Locking forces appear and oppose to any structure change which results physically into hysteresis associated with this continuous transformation. Commensurate configurations only are obtained (but possibly with high order which appears experimentally as incommensurate). Phasons at zero frequency do not exist (the devil's stair case is then called complete).

Recent experiments on thiourea ($S{=}C{-}(NH_2)_2$) have been found in satisfactory qualitative agreement with the predictions of this model.[2]

1. Introduction

We present results on a phenomenological model for the incommensurate structures which are actually observed in many crystals [2]. An incommensurate structure is a crystal structure in which the atoms are displaced from their crystal site by a periodic modulation, the period of which is generally incommensurate with the three periods of the crystal. Such structures appear spontaneously in certain range of temperature and pressure in a large number of crystals. Despite the crystal looses in fact its periodicity in the direction of the wave-vector of the modulation, it is not a disordered system. It is quasi-periodic because its reciprocal lattice is only composed with Bragg spots and superimposed surstructure spots which have this wave-vector. Each of these spots has, in principle, a zero width.

Although it is in practice very difficult to know the details of the atomic interactions, it is rather obvious that these structures must be the consequence of conflicting forces. Such systems are called frustrated. (This concept was first introduced for spin glasses [1]) The effect of frustration becomes interesting only when the system is defectible : A structure is defectible when it can accept localized defects*. Such a property, cannot be found for example in the structure of harmonic models or of those where the anharmonic terms are not sufficient to stabilize defects (quasi-harmonic models). When the frustration is varying in a defectible structure, this structure may evolve discontinuously by defects creation or annihilation from which results a new behaviour in the physical transformation. This microscopic behaviour will require, to be described precisely at the macroscopic scale, the use of highly pathological mathematic functions, as for example in the model studied in this paper, the devil's stair case functions [12,20] . Conversely, if the system is undefectible, the structure transformation is smooth and can be described with ordinary functions. This paper shows, on a crude model, an example of the effect of the defectibility and of the frustration and the physical meaning of the mathematical pathologies which can be useful to shed some lights on unexplained experiments [2]. The model that we study,has a ground-state which can be either an incommensurate structure or a commensurate one but which can exhibit also many types of stable defects due to the existence of an underlying discrete lattice. Distribution of these defects allows to interpret the ground-state structure and to find its properties.

Section 2 describes the model. In section 3, we recall the continuous approximation which is often used in phenomenological approaches but which, neglects the lattice effect or eventually considers it as a small perturbation [18]. We find the limit of validity of the continuous approximation. Section 4 studies the defects properties in the lattice when their concentration is fixed by boundary conditions. It refers to rigorous results published elsewhere [8]. Although this situation can be useful for understanding certain metal insulator transitions, its study is not developed further, but is however necessary, to find in Section 5, the behaviour of the defect structure in the lattice at fixed chemical potential . We find the uncomplete or the complete devil's stair case transformation. The conclusion discusses the possible extensions of this work to other models for incommensurate structures.

2. Incommensurate structures. A model.

Incommensurate structures and their phase transitions are studied on the simplest phenomenological model which exhibits them and on which many exact properties can be obtained and interpreted. This model is represented by a linear chain of atoms with coordinates u_i for the ith atom with potential energy

$$\Phi(\{u_i\}) = \sum_i \lambda V(u_i) + W(u_{i+1}-u_i) - \mu(u_{i+1}-u_i) \quad . \tag{1}$$

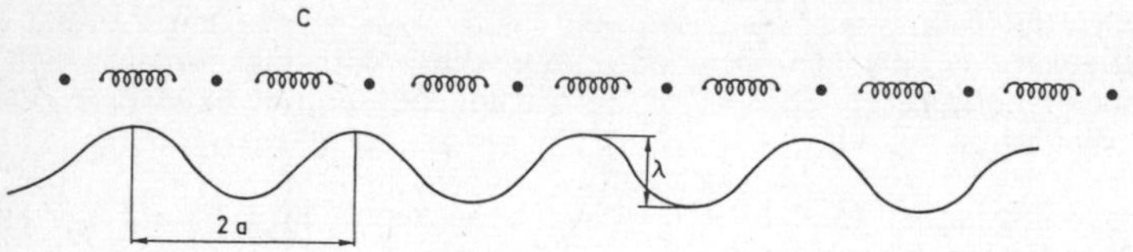

Fig.1-a

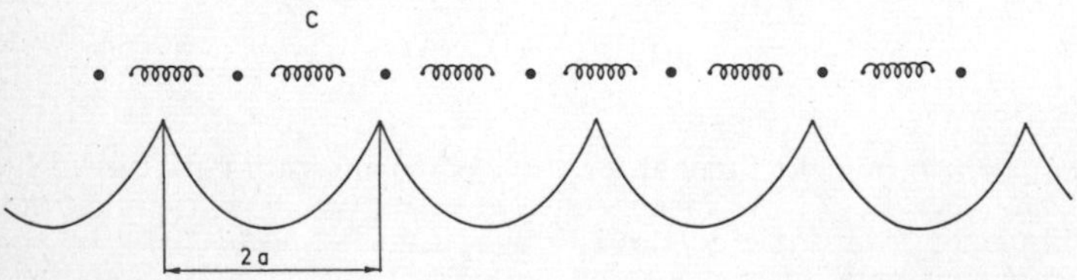

Fig.1-b

Fig.1 - Scheme of the chain of atoms of model (1) with a) a sinus potential, b) the potential considered in appendix A.3.

----

* Note that frustrated spin models are defectible by walls, vortices etc ...

Neighbouring atoms are coupled by the convex potential W and the chain is submitted to a periodic and symmetric potential V with period 2a and amplitude λ . μ is a chemical potential (or a pressure). V and W are both minimum for u=0 in order that, in the absence of chemical potential (μ=0), the ground-state is obtained for $u_i \equiv 0$ and has a zero energy. We study the ground state of this model versus the two parameters λ and μ .

This model has been used for many applications. A particular version with

$$V(u) = \frac{1}{2}(1 - \cos\pi u) \qquad W(u) = \frac{1}{2}u^2 \tag{2}$$

(in reduced units) has been originally proposed by Dehlinger [3] and next studied with some details by Frenkel and Kontorova [3] (1939) and Franck and Van der Merwe [4] (1949) as a model for crystal dislocations. We argue that when quantum and thermal fluctuations become sufficiently small below some critical temperature (roughening transition) the defects which make the superimposed modulated structure can be flat and well defined *.

In this range of temperature, the ground state of the classical model (1) gives some qualitative ideas of 3 dimensional incommensurate structures with only one direction of modulation. The parameters λ and μ are then considered as temperature, pressure dependent (of course no informations are obtained by this way on critical behaviour). Thus u_i must be interpreted as some collective variable which does not fluctuate, which could be for example the position of the ith wall, while V represents the potential produced by the lattice.

In the absence of periodic potential $V(\lambda=0)$, the ground-state of this model is

$$u_i = i\ell + \alpha \tag{3}$$

where α is the position of the first atom and ℓ the distance between neighbouring atoms which is given by the equation $W'(\ell)=\mu$. Since ℓ is generally different of a multiple of 2a, a conflict (frustration) arises between the potential V and the elastic term (for $\lambda\neq0$). The ground-state satisfies the equation :

$$\frac{\partial\Phi}{\partial u_i} = \lambda V'(u_i) - W'(u_{i+1}-u_i) + W'(u_i-u_{i-1}) = 0 \quad . \tag{4}$$

The solutions of this equation can be represented in the frame of a general formalism described in [8] by trajectories $\{u_i\}$ with respect to a "discrete time" i in the phase space $\{u_i,p_i\}$ of a dynamical system with action Φ (p_i is the conjugate variable of u_i) . It is allowed to expect stochastic trajectories in most dynamical systems as (4). A consequence is that the

----

* To be more precise, let us consider for example a 3 dimensional Ising model with first and second neighbour coupling constant in the x direction J_1 and J_2 and first neighbour coupling constant J in the perpendicular directions (ANNNI model) [30] . The critical temperature T_c corresponding to the occurrence of the ferromagnetic state in mean field approximation is

$$k_B T_c = 2(J_1+J_2)+4J$$

while the roughening transition T_R of the walls perpendicular to the x direction and separating two regions with opposit ferromagnetic order, depends essentially of J . In usual non-frustrated Ising models, T_R is smaller than T_c but in this example when J_2 becomes negative with an increasing modulus, T_c decreases and reaches zero while T_R does not practically vary. It results that for $-J_2$ large enough, T_R becomes larger than T_c which indicates the freezing of the walls which are perpendicular to Ox in the paramagnetic phase for $T_c<T<T_R$. Thus, a static wall distribution should result, yielding an incommensurate or commensurate structure for $T_c < T < T_R$. A closely similar argument for the spin-glass transitions has been developed by G. Toulouse and al. [18]. In two dimensions, T_R is zero so that there is no incommensurate structure with long range order. R. Bidaux [20] and L. de Seze solved exactly a particular model with conflicting interactions between first and second neighbours and found that there is no ordered incommensurate structures. M. Selke and M. Fisher [20] found by Monte Carlo calculations, the possible existence of "quasi ordered" phases without any long range order as in the XY model in two dimensions.

general solution of (4) for $\lambda \neq 0$ cannot be explicited analytically . However for small enough λ there still exist solutions which keep a non-stochastic but smooth and analytical behaviour. The problem is to find among the solutions of (4) which extremalize the energy (1), the solution which yields its absolute minimum and thus will be the classical ground-state. The theory of this paper deals essentially with the stochastic character of this solution (i.e. its analyticity).

3. Continuous approximation

This model has been studied with standard approximations which force integrability* and therefore loose a part of the physics far from the adequate limits [3,4,5]. A well-known approximation strictly valid for λ small and $\ell \ll 2a$, is to assume that u_i which varies slowly with i, can be replaced by a smooth function u(x) of its index x=i such that $u_{i+1}-u_i \# \frac{du}{dx}$ in (1). (The standard equivalent approximation on dynamical systems is the adiabatic approximation) Equation (4) becomes a differential equation which can be integrated by quadrature

$$\lambda V'(u(x)) - W''(0) \frac{d^2u}{dx^2} = 0 \quad . \tag{5}$$

With potential (2) all the calculations are explicitely tractable [3,4] (using the properties of a sine-Gordon equation) and yield a second order transition at a critical μ_c (proportional to $\sqrt{\lambda}$) such that for $|\mu| < \mu_c$, the ground-state of (1) is $u_i \equiv 0$ and such that for $|\mu| > \mu_c$ it can be written as

$$u_i = i\ell + \alpha + g(i\ell+\alpha) = f(i\ell + \alpha) \tag{6}$$

f is a monotonous increasing analytic function, g is periodic with period 2a, α is an arbitrary phase and ℓ expands proportionally to $\sqrt{\lambda}$ / $\mathrm{Log}|\mu - \mu_c|$. It is a modulated structure, the period ℓ of which is generally incommensurate with the lattice spacing, since the discreteness of the lattice has been neglected in (5).

This continuous approximation implies the existence of a phason which is the zero frequency mode corresponding to the phase translation in (6), the energy of (1) being independent of α. This solution is found with the linearized equation for the small motions ε_i of the atoms i, of mass m, around their ground-state position u_i^o :

$$m\ddot{\varepsilon}_i - W''(u_{i+1}^o - u_i^o)(\varepsilon_{i+1}-\varepsilon_i) + W''(u_i^o - u_{i-1}^o)(\varepsilon_i-\varepsilon_{i-1}) + \lambda V''(u_i^o)\varepsilon_i = 0 \tag{7}$$

for which $\varepsilon_i = f'(i\ell + \alpha)$ is the time independent solution (as can be checked by differentiating (4) with repect to α with u_i given by (6)).

The kink solution of (4) (epitaxy dislocation) defined as the minimum energy configuration with the limiting condition :

$$\lim_{N\to+\infty, N'\to-\infty} (u_N - u_{N'}) = 2a \tag{8}$$

(-2a for an antikink) is explicitely obtained with potential (2) as a solution of the static sine-Gordon equation. In fact only exceptional equations exhibit true solitons.* Close to the continuous limit (small λ), a method due to R E Peierls allows to estimate the barrier energy E_B which is necessary to jump, to move a kink by one lattice spacing. (See appendix A1). With analytic potential

$$E_B \# K_1 \sqrt{\lambda} \exp - \frac{K_2}{\sqrt{\lambda}} \quad , \tag{9}$$

where K_1 and K_2 are some constants related to the convergence radius of the Taylor series of analytic functions. Note that $1/\sqrt{\lambda}$ is proportional to the half-size of the kink in lattice spacing unities. Thus, when the size of the kink is much larger than one, formula (9) confirms that the lattice locking is neglegible which justifies the use of the continuous model (5). But

---

* Systems for which the solutions can be explicited analytically, are "integrable" and do not exhibit any stochastic properties. They are exceptional and thus might be non-representative for certain physical applications.

when the size of a kink is of the order or smaller than unity, the lattice locking cannot be neglected. This is correlated with the occurrence of strong stochasticity in the associated dynamical system. Of course, it is just in this regime that the discrete model (1) exhibits important qualitative deviations with the continuous model.

4. Many defects structure in a lattice

Equation (4) has infinitely many solutions which are determined recursively from the knowledge of two consecutive atomic positions [7,8] and are represented by the trajectories of the associated and fictitious dynamical system. Each of these trajectories corresponds to a certain random distribution of kinks (8) which can be either stable or unstable. The Lyapounov exponent γ which determines the stability of the trajectory in the fictitious system, (but does not correspond to the stability of the associated configuration) is interpreted as the inverse of a coherence length $1/\xi$.

4.a. Lyapounov exponent and coherence length

Having a solution of (4) $\{ u_i^o \}$, the neighbouring solutions $u_i = u_i^o + \varepsilon_i$ with ε_i small are given by the linear expansion of (4) and yields the time independent equation (7) (with $\ddot{\varepsilon}_i = 0$). The solution of (7) is determined recursively by the linear relation :

$$\varepsilon_{i+1} = (\lambda V''(u_i^o)/W''(u_{i+1}^o - u_i^o) + 1)\, \varepsilon_i + \frac{W''(u_i^o - u_{i-1}^o)}{W''(u_{i+1}^o - u_i^o)} (\varepsilon_i - \varepsilon_{i-1}) . \qquad (10)$$

When there is no zero frequency mode in equation (7), ε_i must diverge for i going either to $+\infty$ or to $-\infty$, for any initial choice of $(\varepsilon_o, \varepsilon_1)$, in order that the assumption ε_i small, becomes unconsistent. In fact, it diverges exponentially (as proved by the Oseledec theorem [21]) as

$$\varepsilon_i \propto \exp(\gamma |i|) \qquad (11)$$

which defines the characteristic exponent $\gamma \geq 0$ of the solution $\{u_i\}$. Thus $\{u_i\}$ cannot be close to $\{u_i^o\}$ for any i . This exponent $\gamma = 1/\xi$ determines the distance ξ over which the solution $\{ u_i \}$ is leaving (or approaching) the solution $\{u_i^o\}$ for $i \to \infty$ ..

When there is a zero frequency mode, called phason, in equation (7), ε_i does not diverge and $\gamma = 0$.

If the configuration $\{u_i\}$ is locally perturbed, for example when the atom 0 is displaced by δu_o and maintained at this position, the far atoms of the perturbed configuration (which satisfy equation (4) except for i=0) are displaced proportionally to

$$|\delta u_n| \propto |\delta u_o| \exp - |n| / \xi . \qquad (12)$$

When $\gamma = \frac{1}{\xi} = 0$, the atoms are moved at infinity without any restoring force and this is the phason mode. We found that u_i depends <u>continuously</u> on u_o and thus can be written with a continuous hull function f such that

$$u_i = f(i\ell + \alpha) \qquad (13)$$

with an arbitrary phase α. This continuous function f is such that equation (4) is satisfied for any phase α. In fact when the hull function f is continuous it is generally analytic, see [8]. (Except at $\lambda = \lambda_c(\ell)$ see the next).

When $\gamma = 1/\xi \neq 0$, there is no continuous hull function f as in (13), because a small change $\delta\alpha$ of the phase α would provide a neighbouring solution $u_i^1 = f(i\ell + \alpha + \delta\alpha)$ of the solution $u_i^0 = f(i\ell + \alpha)$ for any i. But this is just impossible since equation (10) would yield $\gamma = 0$.

Then, the configuration $\{u_i^0\}$ is represented by a trajectory of the associated dynamical system imbedded in the stochastic region and, following previous studies of stochasticity [8], is defectible. It means that there exist other configurations u_i^1 satisfying equation (4) such that

$$\lim_{|i| \to \infty} |u_i^1 - u_i^0| = 0 \quad , \tag{14}$$

but which are different in some finite region of the space i. Some of these configurations u_i^1 are interpreted as elementary defects, while the others can be viewed as built from these elementary defects. The knowledge of the characteristic exponent γ determines the size of the elementary defects since the behaviour of $|u_i^1 - u_i^0|$ is $\exp - |i| / \xi$ for large i. (Clearly when $\gamma = 0$ the size of the defect would be infinite but the configuration is then generally undefectible).

4.b. Ground-state at fixed volume

The ground-state of model (1) is found among the solutions of (4) [8] and depends on the boundary conditions at infinity (μ is then a void parameter)

$$\lim_{|N-N'| \to \infty} \frac{u_N - u_{N'}}{N-N'} = \ell = 2ac \quad , \tag{15}$$

which determines in fact the concentration c of kinks (8). The properties of the kink structure turn out to be strongly dependent on the rationality of $\ell/2a$ but obviously this mathematical result must be physically interpreted.

1) Rational kink concentration

The ground-state is proved to be indeed a commensurate configuration [8]. Setting $\ell_{com}/2a = \frac{r}{s}$ with r and s two irreducible integers, we find

$$u_{i+s} = u_i + 2ra \quad . \tag{16}$$

The unit cell contains s atoms and its size is 2ra.

The elementary phase defects of a commensurate configuration are defined and their existence is proved [8]. Particularly, we find again the kinks (8) for the commensurate configuration with s=1 in (16). They physically correspond to a mean shortening of the infinite chain or a phase shift of

$$\delta\alpha = 2a / s \quad , \tag{17}$$

for a delayed phase defect (and to the same lengthening for an advanced phase defect) and are also the minimum energy configurations for this boundary condition. See Fig.2

The half-size ξ of the defect is the inverse of the Lyapounov exponent γ which is non zero for any commensurate configuration. Phase defects as well as kinks are locked on the lattice by the Peierls force (Appendix A1). This locking becomes important when the phase defect size is of the order of the lattice spacing.

For large enough λ, it is very easy to prove that γ has non-zero lower bound (see ref [8] theorem 5) and that ξ has a finite upper bound. Thus the energy barrier of the lattice locking remains always finite whatever is the commensurability s. (See the scheme of figure 4). (Note that $\gamma(\ell)$, the Lyapounov exponent of the ground-state with the boundary condition (15), is likely a continuous but non differentiable function. Let us point out here

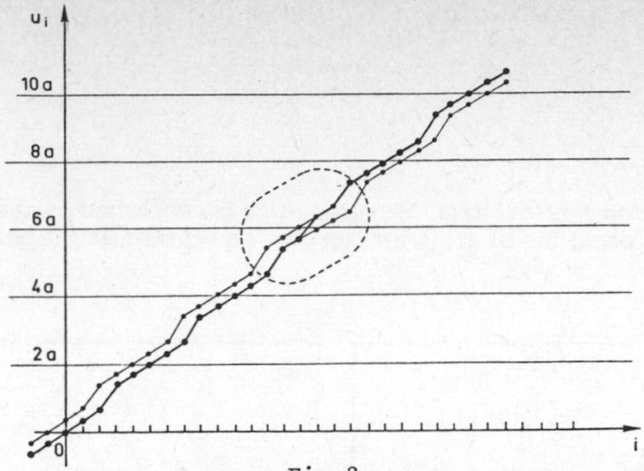

Fig.2

Fig.2 - Scheme of an elementary phase defect (full dots); the two shifted commensurate configurations are represented by light dots.

that this result cannot be obtained in phenomenological theories of incommensurate systems which consider the lattice effects by perturbative expansion terms [17] and assume that these series are convergent.

Such an approach would imply that the lattice locking necessarily disappears for high order commensurability s. This is indeed true for small enough λ (typically $\lambda \leq 0.2$ in model (2)) but becomes qualitatively wrong for larger λ. But it is just in this regime that our theory deviates significantly (complete devil's stair case) from the results obtained by these theories which are only convenient in the uncomplete devil's stair case regime.

2) Irrational defect concentrations. The transition by breaking of analyticity.

A long study [8] which cannot be described in this limited size paper shows that the ground state with condition (15) is indeed an incommensurate one. This result is expressed by the fact that u_i possesses a hull function f i.e. such that

$$u_i = f(i\ell+\alpha) = i\ell+\alpha + g(i\ell+\alpha) \tag{18}$$

where f is a monotonous increasing function and g is periodic with the period 2a of V. These functions are dependent on ℓ and on the model parameters, but the phase α can be chosen arbitrarily. However f is not necessarily a continuous function.

This function is analytic* for most irrational $\ell/2a$ only when λ is smaller than a critical value $\lambda_c(\ell)$. f becomes discrete for $\lambda > \lambda_c(\ell)$ and is then the sum of an infinite number of Heaviside functions

$$f(x) = \sum_{i=0}^{\infty} f_i \; Y(x-x_i) \tag{19}$$

----

*This transition by breaking of analyticity can exist only if V and W are analytic functions but an equivalent one, with the same physical properties, exists if V and W are more than 4 times differentiable. It likely does not exist if V and W are less differentiable.

where Y(x) can be chosen either with the determination

$$Y(x) = 0 \text{ for } x < 0 \quad \text{and} \quad Y(x) = 1 \text{ for } x \geq 0$$

or

$$Y(x) = 0 \text{ for } x \leq 0 \quad \text{and} \quad Y(x) = 1 \text{ for } x > 0 \ .$$

The x_i form a dense set on the real axis. This transition is confirmed by the numerical check of Fig.3 showing the transformation of the trajectory which represents the solution of (4).

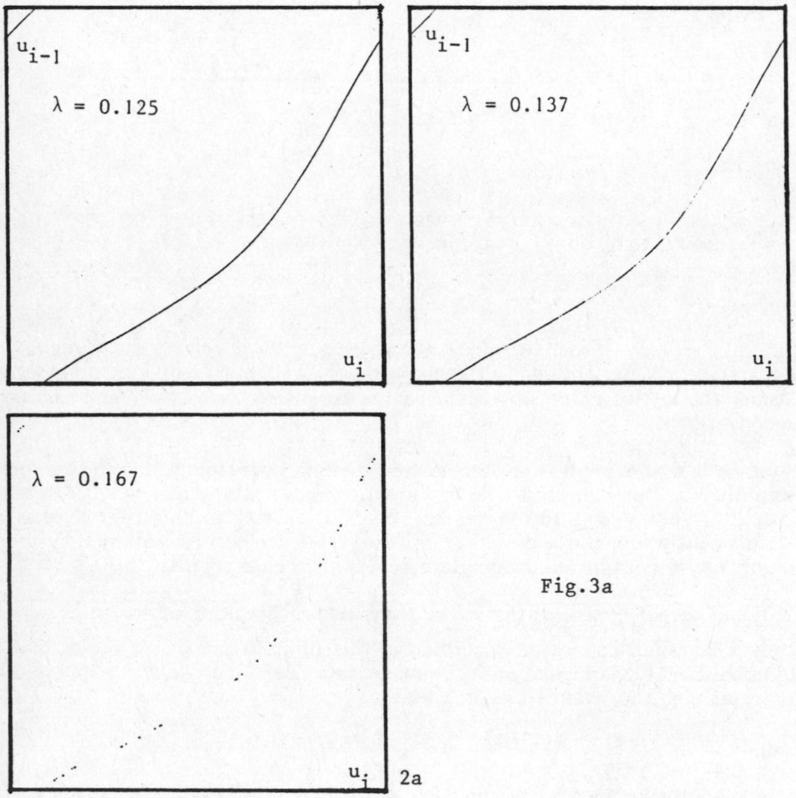

Fig.3a

Fig.3a - From G. André[11].

Figure plotted by the sequence of points (u_i, u_{i-1}) *mod 2a from* $i=0$ *to* $i=1003$, *for the ground-state of model (1) with potential (2) for* $\ell/2a=158/1003$ *which is practically an irrational number. For* $\lambda > \lambda_c$, *the figure is a smooth curve. For* $\lambda \neq \lambda_c(\ell) = 0.136$, *critical fluctuations in the point density are distinguishable. For* $\lambda > \lambda_c$, *the figure is highly disconnected : it is a Cantor set with zero measure.*

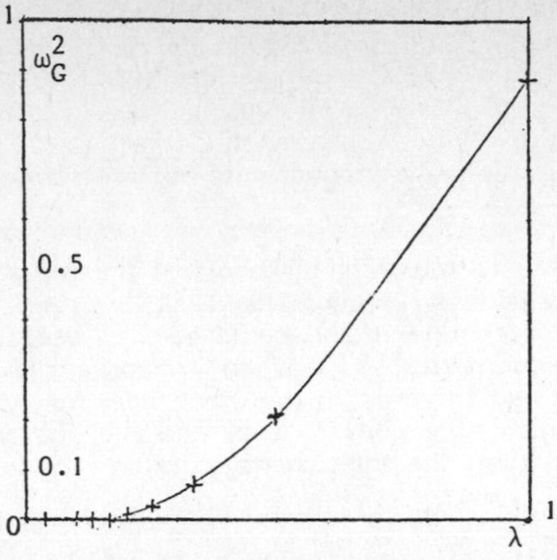

Fig.3b

Fig.3b - Variation of the square of the lowest frequency ω_G^2 of the phonon (Eq.(7)) corresponding to figure (3a). For $\lambda \neq \lambda_c(\ell)$, note that ω_G^2 takes off from zero.

The scheme of Fig.4 shows that when $\lambda < \lambda_c(\ell)$ the phase variation allows the atoms to occupy any position on the barriers of the potential $\lambda\, V(u)$. While for $\lambda > \lambda_c(\ell)$ the barrier energy of $\lambda V(u)$ becomes high enough to confine the atoms close to the minimum. Clearly the function f is then discontinuous at some points. A detailed investigation shows that in fact by a "mirror effect", the function f has infinitely many discontinuities and takes the form (19) see ref. [8].

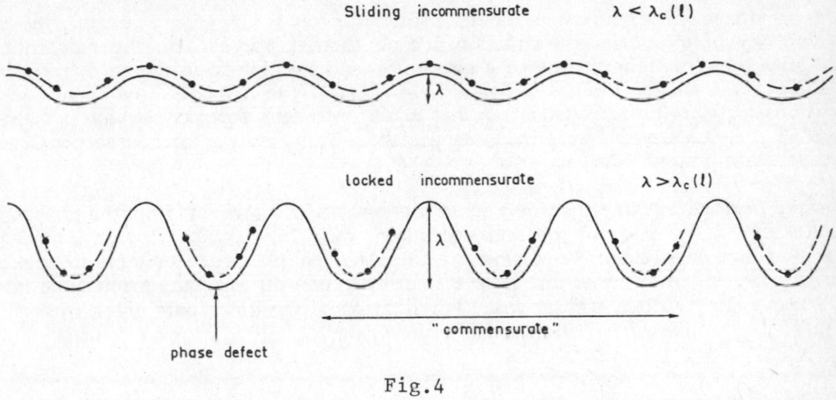

Fig.4

Fig.4 - Scheme showing the atomic positions before and after the transition by breaking of analyticity.

Thus for $\lambda < \lambda_c(\ell)$, the incommensurate configuration has a phason mode. The analytic modulation g contains harmonics which decrease exponentially when their order goes to infinity. g is close to a sinus. The coherence length ξ is infinite : this configuration does not accept any defect. The kink concentration c, determined by (17), makes a"fluid phase", in which the kinks are physically undistinguishable.

Thus for $\lambda > \lambda_c(\ell)$, the coherence length is finite and the configuration does not have any phason mode and is locked. There is a finite gap in the phonon spectrum given by equation (4).

Since function f is discontinuous, the harmonics of g decrease slowly at large order. Moreover the configuration is defectible : Fig. 4.b shows that it can be better considered as an array of phase defects. Clearly the existence of energy barriers allows to put a few disorder in the choice of the wells for the atoms. This disorder must be weak if $\gamma=1/\xi$ is small but can be much more important for large γ (or equivalently large λ). Of course these disordered configurations have more energy than the ground state but are stable, so that they can have a physical existence..

$\lambda_c(\ell)$ is nothing else than the stochasticity threshold of the torus on which the trajectory, representative of the configuration with condition (15), is lying. Renormalization group attempts have been done to determine it and could give us useful informations about the universal behaviour of critical quantities [23]. When λ approaches $\lambda_c(\ell)$ by upper values, the phonon gap and $\gamma(\ell)=1/\xi$ go to zero. On the other side, for λ smaller than $\lambda_c(\ell)$, the rate of decay* of the harmonics of g goes to zero and the phason velocity diverges ! At $\lambda=\lambda_c(\ell)$, as for usual transition, the solution of equation (10) $\{\varepsilon_i\}$ would behave with some power law**.

Numerical evaluation [9,11] of $\lambda_c(\ell)$ shows that $\lambda_c(\ell)$ is rather small. For example, with potential (2), $\lambda_c(\ell)$ does not exceed 0.2 (in reduced units) for any ℓ. In fact, $\lambda_c(\ell)$ is zero at each rational $\ell/2a$, but this result requires to be physically interpreted. When $\ell_{com}/2a=r/s$ is rational, but of large order s, the transition occurs at $\lambda_c(\ell_{com}) = 0$, but there is a sharp cross-over at finite λ for which the very low frequency of a phonon (which is "almost the phason") increases suddenly [8]. In fact, the transition of Fig.(3) is calculated for a commensurate configuration but of order 1003. The cross-over which is shown appears as a transition. Practically the physical determination of $\lambda_c(\ell)$ would be very dependent on the accuracy of the measure and $\lambda_c(\ell)$ could be seen as vanishing only close to low order rational $\ell/2a$ such as 1, 1/2, 1/3, ...

It was believed by many that a phason must exist in any incommensurate structure as a consequence of the Goldstone theorem applied to the continuous group which shifts the phase of the ground state. This theorem is indeed applicable for $\lambda < \lambda_c(\ell)$, because the phase α has the topology of a continuous rotation group. Then it proves also the existence in the vicinity of this zero frequency mode of a phonon branch starting from the zero frequency with a finite slope : the phason velocity. For $\lambda > \lambda_c(\ell)$, the discontinuity of f and its two determinations show that the phase group has a discontinuous topology (which is Cantor-like) and thus that the Goldstone theorem is unapplicable. This result is of course consistent with the fact that there is no phason mode for $\lambda > \lambda_c(\ell)$.

Study of wave propagation in incommensurate lattices [8,12] suggests that the phason mode is well defined for $\lambda < \lambda_c(\ell)$ in the analytic phase while for $\lambda > \lambda_c(\ell)$ the modes do not propagate and are localized. We found on this basis a physical connection between the transition at $\lambda_c(\ell)$ and metal insulator transition in one-dimensional deformable lattices in which it is associated with a transition by localization of the electronic eigen wave functions [12].

5. The devil's stair case transformation [9]

Model (1) is useful to understand the transformation of the incommensurate structures when the chemical potential is varying without boundary conditions (15). For that we will use the results of the previous section on the transition by breaking of analyticity. We minimize the energy per atom (1) $\psi(\ell) - \mu\ell$ with $\psi(\ell)$ defined as

$$\psi(\ell) = \lim_{N-N'\to\infty} \frac{1}{N-N'} \sum_{N<n\leq N} [\lambda V(u_n) + W(u_{n+1}-u_n)] \quad . \tag{20}$$

----

*This rate of decay is proportional to the width of the band of the complex domain parallel to the real axis in which function g is analytic (i.e. the distance from the real axis of the closest pole of g). For $\lambda>\lambda_c(\ell)$, the harmonics g_n of g decay as 1/n, which can be readily proven from the fact that x+g(x) is monotonous increasing.

**This exponent could be compared with the intermittency exponent as studied by Y. Pomeau and P. Manneville [24].

in order to determine $\ell(\mu)$. We have proved that the atomic mean distance $(u_N-u_{N'})/(N-N')$ for the system in one if its ground-states, tends uniformly to the same limit ℓ if N' or N' or both go to infinity. This result proves that the system cannot share into two phases with different atomic mean distance ℓ whatever are the boundary conditions (15). As a consequence $\ell(\mu)$ must have no discontinuities when μ varies. If there was a discontinuity in μ, for example between ℓ_1 and ℓ_2, it would be possible, by choosing boundary conditions (15) with $\ell_1 < \ell < \ell_2$, to get the coexistence of two phases with different atomic mean distance ℓ_1 and ℓ_2. But this is impossible.

Let us note however that this result is the consequence of the convexity of the potential W and is not maintained when this condition is not satisfied.

$\psi(\ell)$ is thus a convex function of ℓ and consequently $\psi(\ell)$ has monotonous increasing left and right derivatives $\psi'_-(\ell)$ and $\psi'_+(\ell)$ which are equal almost everywhere (see ref.[26]). $\psi'_-(\ell)$ is left continuous while $\psi'_+(\ell)$ is right continuous. $\ell(\mu)$ is then determined by the implicit inequation :

$$\psi'_-(\ell) \leq \mu \leq \psi'_+(\ell) \tag{21}$$

which becomes an equation when $\psi'_-(\ell) = \psi'_+(\ell)$.

For the commensurate case ℓ_{com}=2a r/s, the two derivatives in (21) are unequal [12] as a consequence of the existence of phase defects. An arbitrary ground-state with atomic mean distance $\ell > \ell_{com}$ can be considered as a periodic array of $(\ell - \ell_{com})$ s/2a advanced elementary phase defects (17) per atom. When ℓ goes to ℓ_{com}, the interaction energy between far equidistant defects vanishes exponentially which yields the creation energy of an advanced elementary phase defect for μ=0 :

$$e_+(\ell_{com}) = \lim_{\ell\to\ell^+_{com}} (\psi(\ell) - \psi(\ell_{com})) / (\ell-\ell_{com})s/2a) = \frac{2a}{s}\psi'_+(\ell_{com}) . \tag{22}$$

Identically, the energy required to annihilate a delayed elementary phase defect for μ=0 is

$$e_-(\ell_{com}) = \frac{2a}{s}\psi'_-(\ell_{com}) . \tag{23}$$

The two energies (22) and (23) are different. This is particularly obvious when ℓ_{com}=0 (or is a multiple of 2a) because $e_+(0)=-e_-(0)$ is the non zero energy of a kink (8) or of an antikink and then $\psi'_+(0)\neq\psi'_-(0)$.

As a result, the curve $\ell(\mu)$, which is continuous monotonous increasing, has a constant step at each commensurate value ℓ_{com} for $\psi'_-(\ell_{com}) \leq \mu \leq \psi'_+(\ell_{com})$. According to B. Mandelbrot [13] such a curve with infinitely many steps is called a devil's stair case. Similar curves are found in [14] about the rotation number of diffeomorphismes of the circle onto the circle. We examine now the width of the steps in order to know the possible consequences about the experimental observations. There are essentially two qualitatively different regimes :

1) the uncomplete devil's stair case for which the sum of the step widths for all commensurate ℓ in the interval $\ell_1 \leq \ell_{com} = 2a\frac{r}{s} \leq \ell_2$

$$I(\ell_1,\ell_2) = \sum_{\ell_1\leq\ell_{com}\leq\ell_2} (\psi'_+(\ell_{com}) - \psi'_-(\ell_{com})) \tag{24}$$

is smaller than $(\ell_2 - \ell_1)$.

By the decomposition Lebesgue theorem [26], the continuous monotonous increasing function $\ell(\mu)$ can be decomposed into a sum

$$\ell(\mu) = \ell_{ac}(\mu) + \ell_{s.c}(\mu) \tag{25}$$

where $\ell_{ac}(\mu)$ is a monotonous increasing absolutely continuous function (a smooth function) and $\ell_{s.c}(\mu)$ a monotonous increasing singular continuous function (which has a zero derivative almost everywhere). The absolutely continuous part $\ell_{ac}(\mu)$ is non-zero in the case of an uncomplete devil's stair case.

2) The complete devil's stair case for which the sum of the steps $I(\ell_1, \ell_2)$ is equal to $(\ell_2 - \ell_1)$. The absolutely continuous part $\ell_{ac}(\mu)$ in (25) is zero.

5.1. Completeness of the devil's stair case for large λ

We show first that the devil's stair case $\ell(\mu)$ is complete in any variation interval $[\ell_1 \ell_2]$ such that for $\ell_1 < \ell < \ell_2$, the Lyapounov exponent $\gamma(\ell)$ has a non-zero lower bound γ_o:

Let us consider the expansion of $\psi(\ell)$ close to a commensurate $\ell_{com} = 2a\ r/s$. The interacting energy $e_{int,\ell}$ between two advanced (or delayed) elementary phase defects varies exponentially with their distance n. Since the half size of a phase defect is $\xi = 1/\gamma(\ell_{com})$ the overlapping at half distance between the two defects (see Fig.5) corresponds to an atomic displacement δu proportional to $\exp -\gamma(\ell_{com})n/2$. The additional energy due to this overlapping is proportional to $(\delta u)^2$ which yields the following behaviour for the interacting energy :

$$e_{int,\ell}(n) \# 2A(\ell_{com})[\exp -\gamma(\ell_{com}) \frac{n}{2}]^2 = 2A(\ell_{com})\exp -\gamma(\ell_{com})n \qquad (26)$$

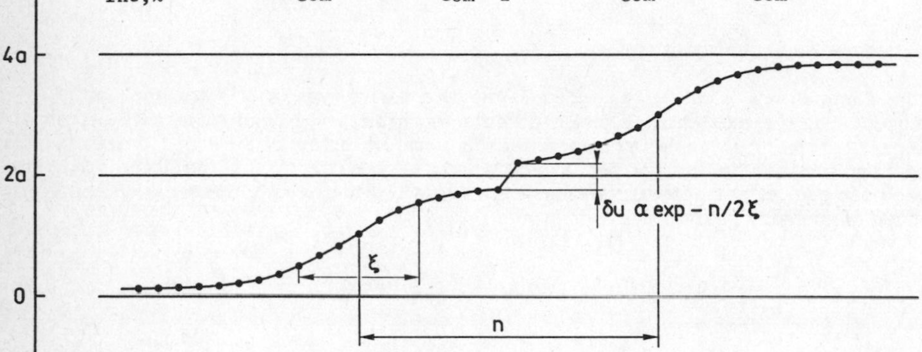

Fig.5

Fig.5 - Scheme of two overlapping defects showing their interaction.

In the case of an advanced and of a delayed phase defect the sign of $A(\ell_{com})$ is reversed. $A(\ell_{com})$ can be estimated for n=0 in this situation. The interacting energy $-2A(\ell_{com})$ cancels the defect energies $e_+(\ell_{com})$-$e_-(\ell_{com})$ and thus $A(\ell_{com})$ is of the order of $(e_+(\ell_{com}) - e_-(\ell_{com}))/2$. Obviously, it has a finite upper bound. A configuration with atomic mean distance $\ell = \ell_{com} + \delta\ell$ ($\delta\ell > 0$) can be considered as a commensurate configuration with $(\delta\ell/2a)s$ equidistant advanced phase defects per atom which are then at distance $2a/(s\delta\ell)$. Its energy per atom can be estimated for small $\delta\ell$ as the sum of the energy of the commensurate configuration plus the energy of the phase defects plus the interacting energy between the phase defects (we neglect the interactions between next nearest neighbour defects).

$$\psi(\ell_{com}+\delta\ell) = \psi(\ell_{com}) + \frac{s\delta\ell}{2a} e^+(\ell_{com}) + \frac{s\delta\ell}{2a} A(\ell_{com}) \exp - \frac{2a}{s\delta\ell} \gamma(\ell_{com}) \qquad (27)$$

$e^+(\ell_{com})$ is determined by (23). So, $\ell(\mu)$ has a logarithmic behavior for $\mu \to \psi'_+(\ell_{com})$ by upper values (or to $\psi'_-(\ell_{com})$ by lower values)

$$\ell(\mu) - \ell_{com} \# - \frac{2a\gamma(\ell_{com})}{s \operatorname{Log}(\mu - \psi'_+(\ell_{com}))} \qquad (28)$$

(see Fig.6). This result can be obtained in the continuous approximation but only for $\ell_{com}=0$. Thus larger is the Lyapounov exponent $\gamma(\ell_{com})$, steeper is the edge of the devil's stair case.

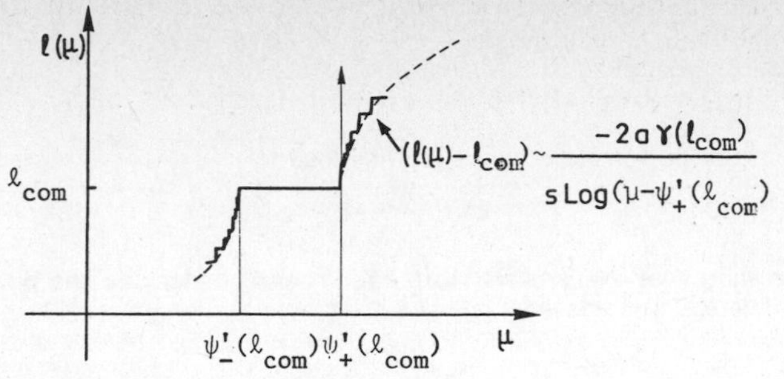

Fig.6

Fig.6 - Behaviour of the devil's stair case at the edge of each step . Note that it is complete at least in a small interval.

If $\gamma(\ell_{com})$ has a strictly positive lower bound in some interval $[\ell_1, \ell_2]$ there exists constants A_b and γ_b

$$A_b > \sup_{\ell_1 \leq \ell_{com} \leq \ell_2} A(\ell_{com}) \tag{29a}$$

$$0 < \gamma_b < \inf_{\ell_1 \leq \ell_{com} \leq \ell_2} \gamma(\ell_{com}) \tag{29b}$$

such that for $\ell_1 < \ell_{com} < \ell < \ell_2$

$$0 \leq \psi(\ell) - (\ell-\ell_{com})\psi'_+(\ell_{com}) \leq s \frac{(\ell-\ell_{com})}{2a} A_b \exp - \frac{2a\,\gamma_b}{s(\ell-\ell_{com})} . \tag{30}$$

Now we use appendix 2 which proves that if $\psi(\ell)$ satisfies (30), then $\psi'_+(\ell)$ (or $\psi'_-(\ell)$) is a discrete function which varies only by discontinuity jumps at the commensurate ℓ_{com}. As a result, $\ell(\mu)$ determined by (21) is a complete devil's stair case.

For large enough λ ($\lambda > \operatorname{Sup} \lambda_c(\ell)$), $\gamma(\ell)$ has a non-zero lower bound in any interval. This can be easily proven when the atoms are localized in the convex part of the potential V[8] (see Fig.4.b). (With potential (2) $\lambda > 0.2$ is sufficient). For any λ, assuming that $\gamma(\ell)$ is a continuous function of ℓ, we find that the devil's stair is complete in a finite interval in ℓ around each step at $\ell_{com} = 2ar/s$, but these intervals become very small when λ goes to zero. Despite each rational determines such an interval, the sum of their measures (which is given by a series) can be smaller than the measure of the whole interval of variation of ℓ. We are going to show now that the devil's stair case becomes necessarily uncomplete for small λ.

5.2. Uncompleteness of the devil's stair case for small λ

For small enough λ, $\lambda < \lambda_c(\ell)$, there exists incommensurate configurations with phason so that $\gamma(\ell)$ is zero. Inequality (30) does not hold with a non-zero γ_b and the proof of completeness of appendix A.2 does not work.

In fact, the width of the commensurate phases can be estimated and show that their sum goes to zero when λ goes to zero. Incommensurate phases are obtained for a set of values of μ, the measure of which tends to be the measure of the whole interval of variation of μ. Moreover, these incommensurate phases must have a phason mode ($\gamma(\ell)=0$) because the previous subsection proves(under the assumption of continuity of $\gamma(\ell)$) that if $\gamma(\ell)$ is non-zero at ℓ, the devil's stair case is complete in some neighbourhood of ℓ.

When λ is small, the ground-state of model (1) is close to

$$u_i = i\ell + \alpha \quad . \tag{31}$$

If $\ell_{com}=2a\,r/s$, the energy of the ground-state becomes dependent on the phase α. The locking potential per atom versus this phase α is

$$\nu_s(\alpha) = \lim_{N-N'\to\infty} \frac{1}{N-N'} \sum_{i=N'}^{N} V(i\ell_{com}+\alpha) \quad . \tag{32}$$

Expanding $V(x)$ as a Fourier series

$$V(x) = \sum_n V_n \exp(i \frac{2\pi}{2a} nx) \tag{33}$$

yields

$$\nu_s(\alpha) = \sum_n V_{ns} \exp(i \frac{2\pi}{2a} ns\, \alpha) \quad . \tag{34}$$

A neighbouring solution of (31) can be described with a slowly varying phase α_i as

$$u_i = i\ell_{com} + \alpha_i \, .$$

Since α_i is close to α_{i+1}, the index i can be replaced by a continuous variable x and expanding in $\alpha(x)$ the energy (1) yields

$$\phi(\{\alpha_i\}) = (N-N')\,(W(\ell_{com}) - \mu\ell_{com}) +$$

$$\int_{N'}^{N} \left[\frac{1}{2} W''(\ell_{com}) \frac{\partial^2\alpha}{\partial x^2} + (W'(\ell_{com})-\mu)\frac{\partial\alpha}{\partial x} + \lambda\, \nu_s(\alpha)\right] dx \tag{35}$$

and $\alpha(x)$ satisfies the equation

$$\lambda\, \nu_s'(\alpha) - W''(\ell_{com}) \frac{\partial^2\alpha}{\partial x^2} = 0 \tag{36}$$

$\nu_s(\alpha)$ has the period $2a/s$ and the elementary phase defects of the commensurate configuration $\ell_{com}=2a\,r/s$ are obtained as the kink solutions of (34) determined by

$$\lim_{x\to-\infty} \alpha(x) = 0 \quad \text{and} \quad \lim_{x\to+\infty} \alpha(x) = \pm \frac{2a}{s} \quad . \tag{37}$$

When $V(x)$ is an analytic function $\nu_s(\alpha)$ is close to a sinus and equation (37) is close to a static sine-Gordon equation. By homogeneity relations the energy of an elementary phase defect is of the order of magnitude of

$$e_+(\ell_{com}) \;\#\; K \sqrt{W''(\ell_{com})\lambda A_s \left(\frac{2a}{s}\right)^2} + (W'(\ell_{com})-\mu)\frac{2a}{s} \quad , \tag{38}$$

and

$$e_-(\ell_{com}) \# -K\sqrt{W''(\ell_{com})\lambda A_s(\frac{2a}{s})^2} + (W'(\ell_{com})-\mu)\,2a/s \qquad (39)$$

where A_s is the amplitude of variation of $v_s(\alpha)$ and K is a constant factor or order unity (which can be shown to be smaller than $\sqrt{2}$). By formula (22) and (23), we get

$$\psi'_+(\ell_{com}) - \psi'_-(\ell_{com}) < K'\sqrt{\lambda}\quad\sqrt{A_s} \qquad (40)$$

where K' is some finite constant. Then if the series $\sum_s s\sqrt{A_s}$ converges,*

$$I(\ell_1,\ell_2) < \sqrt{\lambda}\,(\ell_2-\ell_1)\,K'\sum_s (s+1)\sqrt{A_s} \qquad (41)$$

goes to zero when λ goes to zero which proves that for λ small enough the devil's stair case becomes uncomplete.

If the series $\sum_s s\sqrt{A_s}$ diverges, no conclusion can be obtained. This situation occurs when the harmonics V_n of V decrease slower than $1/n^5$, i.e. when V has less than four derivatives. In fact, when V is only continuous, example of appendix A3, proves that the devil's stair case remains always complete.

As a result of these considerations, when the ends of the chain are let free, the devil's stair case varies continuously in order that the non-analytic incommensurate configurations without phason are not obtained (except with zero probability). An other way to understand physically this result is to note that if f has harmonics a small change of ℓ into a close commensurate ℓ_{com} allows to gain a locking energy by the choice of the phase α. This energy gain can be shown generally to become larger than the elastic energy lost for a certain ℓ_{com} close to ℓ, when the nth harmonics of the modulation decay slow enough i.e. as $1/n$**. This implies again that the hull function of the modulation is non-analytic.

6. Physical consequences. Concluding remarks

The results of this model can be considered as generic for physical situations where the wave-vector of an incommensurate modulation varies continuously when some physical parameters vary. There are several physical situations.

1) When λ is small, the incommensurate configurations evolve smoothly because the widths of the steps are very small (see Figure 6). Despite there are tiny complete parts of the devil's stair case at the edge of each step, they are undistinguishable. The incommensurate configurations have a phason mode at zero frequency while the commensurate ones have a very low frequency phonon mode (quasi-phason). As a result, this sliding mode makes reversible the transformation when the parameters are varying.

This situation is what is predicted by the continuous models which represent the lattice by perturbative terms of an assumed convergent series [17] (see Figure 7).

----

* Note that if by accident V has no harmonics (example.(2)) the perturbation calculation of $I(\ell_1, \ell_2)$ should be expanded at higher order in λ but still show the uncompleteness of the devil's stair case at small λ.

** This result is still obtained using the theory of Diophantine approximations of numbers as shown in appendix A2.

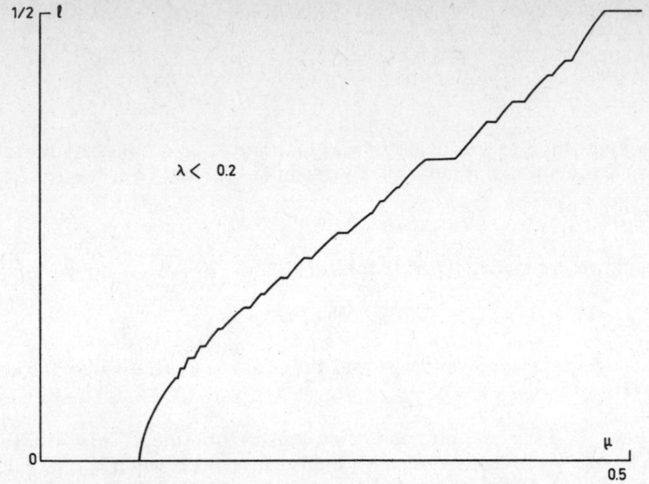

Fig. 7 - Uncomplete devil's stair case. The width of the steps are small and becomes negligible for small λ.

2) When λ increases the complete part of the devil's stair case close to the rational ℓ /2a grows so that the whole devil's stair case becomes complete everywhere. (λ > 0.2 with model 2).

Strictly speaking, our theory predicts that there is a zero probability of having a true incommensurate configuration but there still exists high order commensurate configurations which can be physically considered as incommensurate within the experimental accuracy (see Fig.8).

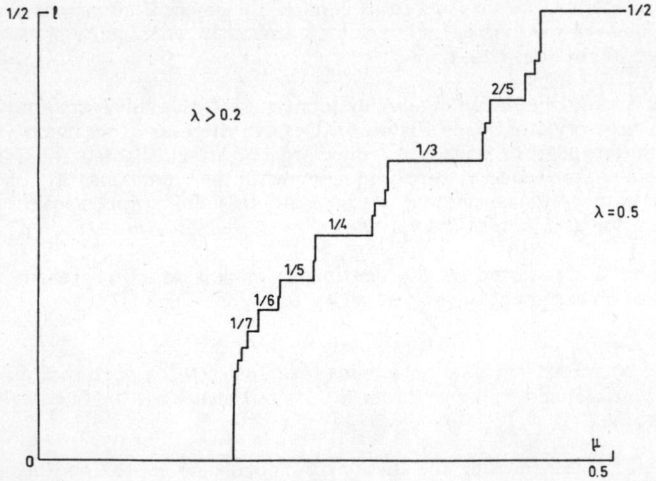

Fig. 8 - Complete devil's stair case ℓ(μ) (in the example of appendix A3).

The phason mode disappears despite there still exists low frequency modes of the kinks or of the phase defects locked on the lattice. Simultaneously, the transformation becomes irreversible and we get a global hysteresis. Model of appendix A3 which is exactly calculable, is a qualitatively good approximation. (For small λ , it overestimates the width of the incommensurate phases). It shows that there exists a finite energy barrier given by formula A.3.18, whatever is the commensurability to which the delay of the hysteris could be empirically related. An essential feature is that the Lyapounov exponent γ is not zero which allows the existence of metastable locked structures without any long range order. These structures are out of equilibrium but could have physically a very long or even infinite life time. When λ is not too large, these structures must be composed of large micro-domains in which the structure is incommensurate ; these structures are the consequence of the existence of stochastic trajectories in the fictitious dynamical system which, close to the stochasticity threshold, exhibit intermittency [23].

The analysis of such structures remains to be done and is clearly determined by the stochastic properties of the corresponding trajectories. In any case, careful experimental examination of the long range order of incommensurate structures in the complete devil's stair case region should be done.

3) For larger λ, the Lyapounov exponent increases so that the devil's stair case becomes cliff-like. The transformation looks like a sequence of few first order transitions at the simplest commensurabilities with an usual important hysteresis (see the model of appendix A3 for large λ). See Fig.9.

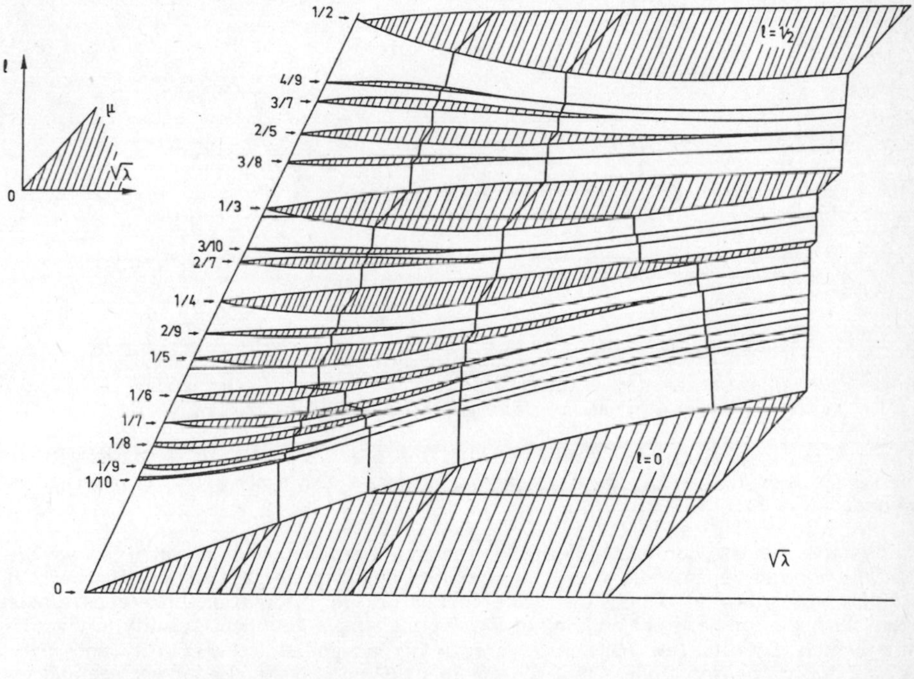

Fig. 9 - Three dimensional representation of the "devil's hill" $\ell(\mu, \sqrt{\lambda})$ with respect to the two parameters μ and $\sqrt{\lambda}$, for the exactly soluble model of appendix A3.

4) Devil's stair case versus temperature

To simulate the situation where the wave-vector of an incommensurate modulation varies with the temperature it is not appropriate to represent the devil's stair case versus the parameter μ. The temperature acts essentially on the fluctuations and so determines in some sense, the height of the energy barriers produced by the lattice. Temperature is thus better represented by the parameter λ. Figure (10) shows a section of the surface represented by figure (9) at a constant λ corresponding at λ=0 to $\ell/2a = 1/7$. It shows that the devil's stair case progressively transforms from an incomplete devil's stair for small λ ($T \lesssim T_i$) into a complete one ($T \gtrsim T_c$), while the hysteresis gradually appears. The end of the transformations looks first order.

This picture appears qualitatively very close to the recent experimental observations of the wave-vector versus temperature in thio-urea ($S{=}C{-}(NH_2)_2$)[2],[27]

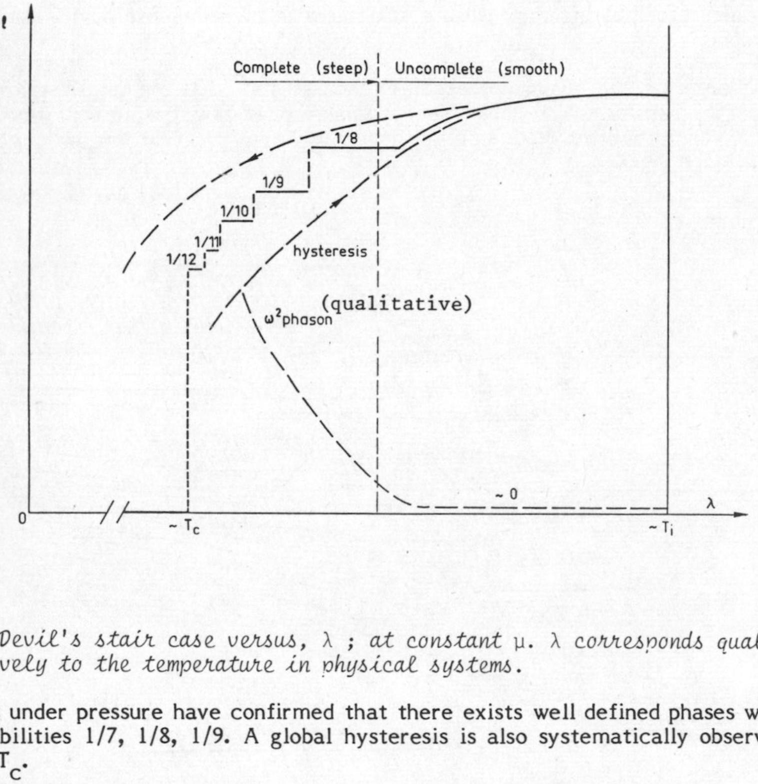

Fig.10 - Devil's stair case versus, λ ; at constant μ. λ corresponds qualitatively to the temperature in physical systems.

Experiments under pressure have confirmed that there exists well defined phases with simple commensurabilities 1/7, 1/8, 1/9. A global hysteresis is also systematically observed in the region near T_c.

Global hysteresis is also commonly observed in many other compounds in which the wavevector varies, which could be an indication of a complete devil's stair case region. However, it has been found insulators in which the wave-vector of the modulation undergoes almost no variation with the temperature and do not lock at any simple commensurability. Biphenyl is an example which exhibits few harmonics, practically no global irreversibility and moreover seems to exhibit a phason mode [28]. Clearly this system stays in the incomplete and smooth devil's stair case regime.

Finally, it appears that the more important observable result of the complete devil's stair case transformation is the occurrence of irreversibility. This result cannot be obtained by standard "integrable" theories since the transition by "breaking of analyticity" is just the underlying feature of this behaviour. Let us note that, the complete devil's stair case transformation can be experimentally distinguished from a true first order transition, despite the fact that both transformation are irreversible. Indeed, a first order transition exhibits during the transformation, superimposed spots in the X-ray or neutron spectrum, corresponding to the coexistence of two phases, while the devil's stair case exhibits only one kind of spots corresponding to a homogeneous phase which transforms continuously.

Discussion of the model

As we already mentioned this model can have application at finite temperature in systems where the fluctuations have not a crucial role. For example, this excludes applications for two dimensional adsorbed layers of atoms. But we can also wonder if the devil's stair case properties are maintained in other possible models for incommensurate structures. J. Villain and M. Gordon [24] have studied models at 0K in which they found that the devil's stair case becomes "harmless", i.e. composed only of a finite number of first order transitions. Their arguments are only valid in the region where the lattice effects are very important (The incommensurate structure are represented by defects).In this region, our model predicts a very steep complete devil's stair case.

This is indeed a serious peculiarity of our model to have no first order transitions. The assertion of these authors could be true for T close to T_c and can be also considered as in agreement with certain experiments. However farther from this limit, it becomes possible to show on many models, using similar methods as in 6.2 that there exists necessarily an infinite number of constant steps, despite at the edge of each step there is a small discontinuity (but as these authors, we assumed without any proof that the ground-state must be only incommensurably or commensurably modulated which is not obvious).Thus, the physical behaviour of all these models should be quite similar (except close to T_c) ; there is a region with either a complete or a harmless devil's stair case and a region with an incomplete devi'ls stair case which can be distinguished as in figure 10.

In fact, there is little hope that more complicated models give rise to a simpler behaviour such as the harmless devil's stair case except close to certain limits. The underlying mathematics related to stochasticity problems become much more complicated so that important new questions arise : is it possible that the ground-state be something else that the commensurate or incommensurate modulated structure ? Indeed there is a counter example in two dimensions for which the ground state is proved to have no periodicity [25]However, we proved on a large family of models that some "weak properties of periodicity" are maintained because the ground state is necessarily represented in the fictitious dynamical system as a minimal invariant closed set [8].But, if there exists sliding modes, we shew that the ground state is necessarily an incommensurate structure with eventually many modulations with different periods [25].

It appears as probable that new kinds of structure exist in rather simple models but with sufficiently "strong frustration". However, proving their existence appears actually as a "tour de force". Already, mysterious structures are found experimentally [29] which have no periodicity property and their real understanding should involve stochastic theory.

Acknowledgment

We thank F. Denoyer, M. Lambert and A.H. Moudden for useful discussions and communications of their experimental results prior to publication.

APPENDIX A1 - LATTICE LOCKING OF A KINK : PEIERLS FORCE

We present here the Peierls calculation of the energy barrier which locks a defect in a discrete lattice.

Let us consider for example a discrete model with energy

$$\Phi(\{u_i\}) = \sum_i \frac{1}{2}(u_{i+1} - u_i)^2 + \lambda V(u_i) \qquad (A1.1)$$

where $V(u_i)$ can be either a symetric double-well or a periodic potential. The stationary solutions satisfy the equation :

$$\frac{\partial \Phi}{\partial u_i} = 2u_i - u_{i+1} - u_{i-1} + \lambda V'(u_i) = 0 \quad . \qquad (A1.2)$$

For λ small $u_{i+1} \# u_i$, for any i, can be described as a continuous function u(x) of x=i and equation

$$\lambda V'(u) - u'' = 0 \quad , \qquad (A1.3)$$

yields a kink solution

$$u(x) = f(\sqrt{\lambda}(x+\alpha)) \qquad (A1.4)$$

with α an arbitrary phase. $1/\sqrt{\lambda}$ is thus proportional to the size of the kink.

In the discrete lattice, (A1.1) the energy of the kink depends on the phase α and using (A1.4) is

$$\phi_{kink} = \sum_i \lambda[V(u_i) - \frac{1}{2} u_i V'(u_i)] = \lambda \sum_i F(\sqrt{\lambda}(i+\alpha)) \quad , \qquad (A1.5)$$

with

$$F(x) = V(f(x)) - \frac{1}{2} f(x) V'(f(x)) = V(f) - \frac{1}{2} f f'' \quad . \qquad (A1.6)$$

(A1.5) is readily written as

$$\phi_{kink} = \lambda \int_{-\infty}^{+\infty} F(\sqrt{\lambda}(x+\alpha)) \sum_i \delta(x-i)\, dx \qquad (A1.7)$$

and using the identity

$$\sum_i \delta(x-i) = \sum_n \exp(i\, 2\pi\, nx) \quad , \qquad (A1.8)$$

becomes

$$\phi_{kink} = \sqrt{\lambda} \sum_n G\left(\frac{2\pi n}{\sqrt{\lambda}}\right) \exp(-\, 2\pi n\alpha) \quad , \qquad (A1.9)$$

with the Fourier transform of F(x)

$$G(Q) = \int_{-\infty}^{+\infty} \exp(i\, Q\, X)\, F(X)\, dX \quad . \qquad (A1.10)$$

For small λ, ϕ_{kink} is expanded as

$$\phi_{kink} \# \sqrt{\lambda}\, G(0) + \sqrt{\lambda}\left|G\left(\frac{2\pi}{\sqrt{\lambda}}\right)\right| \cos(2\pi(\alpha-\beta)) \quad , \qquad (A1.11)$$

where β is the phase of $G(\frac{2\pi}{\sqrt{\lambda}})$. The locking barrier E_B, to move the kink of one lattice-spacing, is the double of amplitude of the energy variation with respect to the phase α :

$$E_B \;\#\; 2\sqrt{\lambda}\,\left|G(\frac{2\pi}{\sqrt{\lambda}})\right| \;. \qquad (A1.12)$$

If F(X) is an analytic function and if R is the distance of its closest pole to the real axis

$|G(Q)|$ behaves as exp-R $|Q|$ for large real Q. If F(X) is only ν times differentiable G(Q) behaves as $|Q|^{-(\nu+1)}$. These results are easily obtained with the inverse formula of (A1.10)

$$F(X) = \int_{-\infty}^{+\infty} G(Q)\exp(-iQX)dQ. \qquad (A1.13)$$

Thus we find that the locking barrier behaves, for small λ, either as

$$E_B \propto \sqrt{\lambda}\;\exp-(\frac{2\pi R}{\sqrt{\lambda}}) \;,$$

if the initial potential V is analytic or as

$$E_B \propto (\sqrt{\lambda})^{\nu+1} \;,$$

if V is only ν times differentiable.

APPENDIX A2 - PROOF OF COMPLETENESS OF THE DEVIL'S STAIR CASE

We prove that if inequality (30) holds, then $\psi'_+(\ell)$ is a discrete function. The same result can be obtained for weaker conditions ; when, for $\ell_1 \leq \ell_{com} < \ell \leq \ell_2$, there exists A_b and $\varepsilon > 0$ such that the convex function ψ satisfies

$$0 \leq \psi(\ell) - \psi(\ell_{com}) - (\ell-\ell_{com})\psi'_+(\ell_{com}) \leq A_b\,(s(\ell-\ell_{com}))^{4+\varepsilon} \;. \qquad (A2.1)$$

This condition is a fortiori satisfied when (30) is satisfied.

We prove the result by showing that the variation of $\psi'_+(\ell)$ on the irrational numbers I_∞ in $[\ell_1, \ell_2]$ is zero. (For simplicity we drop out 2a : ℓ_{com} = r/s with r and s two irreducible integers). We consider the interval union

$$I_S^{\nu}(x) = \bigcup_{s>S} [\frac{r}{s}, \frac{r}{s} + \frac{x}{s^{1+\nu}}[\qquad (A2.2)$$

where x is a positive number, and $0 < \nu < 1$; the r/s are the irreducible rational numbers in $[\ell_1\; \ell_2]$ and S a given integer.

It is known that any irrational θ can be approximated by infinitely many rationals h/k such that [15]

$$0 < \theta - \frac{h}{k} < \frac{1}{k^2} \qquad (A2.3)$$

It is a variation on the Hurwitz theorem . A fortiori the weaker inequality

$$0 < \theta - \frac{h}{k} < \frac{x}{k^{1+\nu}} \qquad (A2.4)$$

which implies

$$\theta \in [\frac{h}{k}, \frac{h}{k} + \frac{x}{k^{1+\nu}}[\qquad (A2.5)$$

can be satisfied by an infinite number of rationals h/k. Thus, all the irrationals of $[\ell_1 \ell_2]$ are in the union I_s^ν (x) and for any $x > 0$, $s > 0$ and $\nu < 1$, the set I_∞ is included in I_s^ν (x). The variation of $\psi'_+(\ell)$ on the irrational I_∞ written as Var(ψ'_+ , ℓ_∞) is positive and bounded as

$$0 < \mathrm{Var}(\psi'_+, I\ ) \le \mathrm{Var}(\psi'_+, I_s^\nu(x)) \le \sum_{\substack{s>S \\ r/s\in[\ell_1 \ell_2]}} \mathrm{Var}(\psi'_+, [\frac{r}{s}, \frac{r}{s} + \frac{x}{s^{1+\nu}}[\ ) \qquad (A2.6)$$

ψ'_+ is monotonous increasing so that the right member of (A2.6) can be calculated and

$$0 < \mathrm{Var}(\psi'_+, I_\infty) \le \sum_{\substack{s>S \\ r/s}} [\psi'_+(\frac{r}{s} + \frac{x}{s^{1+\nu}}) - \psi'_+(\frac{r}{s})] \ . \qquad (A2.7)$$

Since (A2.7) is true for any x, we can replace the right number of (A2.7) by its average over $0 < x < 1$.

Integrating over x, the positive monotonous increasing function ψ'_+ , we get

$$0 < \mathrm{Var}(\psi'_+, I_\infty) \le \sum_{\substack{s>S \\ r/s\in[\ell_1, \ell_2]}} s^{1+\nu} [\psi(\frac{r}{s} + \frac{1}{s^{1+\nu}}) - \psi(\frac{r}{s}) - \psi'_+(\frac{r}{s}) \frac{1}{s^{1+\nu}}] \ . \qquad (A2.8)$$

Using the hypothesis (A2.1),(A2.8) becomes

$$\begin{aligned} 0 \le \mathrm{Var}(\psi'_+, I_\infty) &\le \sum_{\substack{s>S \\ r/s\in[\ell_1, \ell_2]}} s^{1+\nu} A_b (s. \frac{1}{s^{1+\nu}})^{4+\varepsilon} \\ &\le A_b \sum_{s>S} ((\ell_2 - \ell_1)s + 1)\ 1/s^{(3+\varepsilon)\nu - 1} \end{aligned} \qquad (A2.9)$$

ν is arbitrary close to 1 so that it can be chosen in order that the series in s in (A2.9) converges. But S is arbitrary. The right member of (A2.9) is the remainder at order S of a convergent series. For S going to infinity, it vanishes so that

$$\mathrm{Var}\ (\psi'_+, I_\infty) = 0 \ . \qquad (A2.10)$$

APPENDIX A3 - AN EXACTLY CALCULABLE MODEL*

The ground state of model (1) with

$$W(u) = \frac{1}{2} u^2 \qquad (A3.1)$$

---

*Note that there exists some minor mistakes in the initial calculation in ref[9] .

and

$$V(u_i) = \frac{1}{2}(u_i - 2m_i a)^2 \qquad (A3.2)$$

with $m_i = \text{Int}(u_i/2a + 1/2)$ can be exactly calculated (see Fig.1-b). The singularities of $V(u)$ for $u_i=2ka$ (k an integer) do not allow the existence of an incomplete devil's stair case for small λ. Moreover, all the properties which determine a complete devil's stair case (see section 5) can be exactly checked so that the arguments of subsection 5.1 become a rigorous proof. The completeness is proven in this appendix by an explicit calculation of this devil's stair case. Equation (4) becomes

$$(2+\lambda)u_i - u_{i+1} - u_{i-1} = 2m_i a\lambda \quad . \qquad (A3.3)$$

We know from (ref.[8] theorem 1) that there exists a phase α and ℓ for any ground state such that

$$m_i = \text{Int}\left(\frac{i\ell+\alpha}{2a}\right) \quad . \qquad (A3.4)$$

The linear equation in u_i, (A3.3), can be easily solved and yields

$$u_i = A\sum_n \eta^{|n|} m_{n+i} \qquad (A3.5)$$

with

$$\eta = 1 + \frac{\lambda}{2} - \frac{1}{2}\sqrt{4\lambda+\lambda^2} \qquad (A3.6)$$

and

$$A = \frac{2a\lambda}{\sqrt{4\lambda+\lambda^2}} \quad . \qquad (A3.7)$$

Let us note that the Lyapounov exponent of the configuration is

$$\gamma(\ell) = -\text{Log}\,\eta \quad . \qquad (A3.8)$$

It is independent of ℓ and depends only on λ. Clearly it is never zero (unless $\lambda=0$). The devil's stair case must be complete from section 5.

The energy Φ of model (1) becomes, with u_i given by (A3.5),

$$\Phi = \frac{1}{2}Aa\lambda \sum_{ij} \eta^{|i-j|}(m_i - m_j)^2 \qquad (A3.9)$$

which yields the mean energy per atom

$$\psi(\ell) = \frac{2a^2\lambda^2}{\sqrt{4\lambda+\lambda^2}} \sum_{n>0} \eta^{|n|}\,\psi_n(\ell) \qquad (A3.10)$$

where

$$\psi_n(\ell) = \lim_{N\to\infty} \frac{1}{N}\sum_{i=1}^{N}(m_{i+n} - m_i)^2 \quad . \qquad (A3.11)$$

This mean value is easily calculated with (A3.4). $(m_{i+n}-m_i)$ can take only two possible values

$$m_{i+n} - m_i = \text{Int}\left(\frac{n\ell}{2a}\right) = s_n$$

or

$$m_{i+n} - m_i = s_n + 1$$

It is s_n with probability

$$p_n = (s_n+1) - \frac{n\ell}{2a} \qquad (A3.12)$$

in order that

$$p_n s_n + (1-p_n)(s_n+1) = \langle m_{i+n} - m_i \rangle = \frac{n\ell}{2a} . \qquad (A3.13)$$

Thus

$$\psi_n(\ell) = p_n s_n^2 + (1-p_n)(s_n+1)^2 = \frac{n\ell}{2a}\,(2\mathrm{Int}(\frac{n\ell}{2a})+1) - (\mathrm{Int}\,\frac{n\ell}{2a})^2 - \mathrm{Int}(\frac{n\ell}{2a}) . \qquad (A3.14)$$

The derivative of (A3.10)

$$\psi'_+(\ell) = \frac{a\lambda^2}{\sqrt{4\lambda+\lambda^2}} \sum_{n>0} n(1+2\mathrm{Int}\,\frac{n\ell}{2a})\eta^n \qquad (A3.15)$$

depends only on ℓ by the step function Int(n ℓ/2a) so that the derivative $\psi''_+(\ell)$ is zero almost everywhere. $\psi'_+(\ell)$ has a discontinuity for each rational $\ell_{com}/2a=r/s$, which is produced by the terms of order ns in the series (A3.15). The inverse function $\ell(\mu)$ of $\psi'_+(\ell)$, in (A3.15) is a complete devil's stair case. The width of the step of the devil's stair case at ℓ_{com} is

$$S_{\ell_{com}} = \psi'_+(\ell_{com}) - \psi'_-(\ell_{com}) = \frac{2a^2\lambda^2}{\sqrt{4\lambda+\lambda^2}} \sum_{n=1}^{\infty} sn\,\eta^{sn}$$

$$= \frac{2a^2\lambda^2}{\sqrt{4\lambda+\lambda^2}} \quad \frac{s\,\eta^{s-1}}{(1-\eta^s)^2} . \qquad (A3.16)$$

This formula (A3.16) shows that the width of step at a commensurability of order s decreases exponentially as s exp- γs with the Lyapounov exponent A3.8.

The energy barrier which is necessary to jump for the smallest phase shift of a commensurate configuration is obtained by calculating first the largest jump done by the atoms It is readily obtained from formula (A3.5) and (A3.4) as

$$\delta u_o = A\,\frac{1+\eta^s}{1-\eta^s} \qquad (A3.17)$$

where s is the order of commensurability of the configuration. The maximum energy of the configuration is obtained when this atom is at half distance between the two ends of the jump. The additional energy of this configuration where all the other atoms are in equilibrium is the energy barrier and is

$$\delta E = \frac{1}{2}\left(\frac{\delta u_o}{2}\right)^2 \frac{1}{1-\eta^2}\left[\lambda(1+\eta^2)+2(1-\eta^2)\right] = \frac{1}{2}\,\frac{a^2\lambda^2}{\sqrt{4\lambda+\lambda^2}}\left(\frac{1+\eta^s}{1-\eta^s}\right)^2 . \qquad (A3.18)$$

This formula clearly proves that the energy barrier does not vanish when the order of commensurability diverges. It vanishes only when λ goes to zero.

REFERENCES

[1] G. Toulouse, Comm. on Phys. 2, 115 (1977)
J. Villain, J. Phys. C, 10, 1717 (1977)
[2] M. Iizumi, J.D. Axe and G. Shirane, K. Shimaoka, Phys. Rev. B15, 4392.
A.M. Mouden, F. Denoyer and M. Lambert, Le Journal de Physique 39, 1323 (1978).
A.M. Mouden, F. Denoyer, M. Lambert, W. Fitzgerald, Solid State Comm. 32, 933 (1979).
[3] F.R.N. Nabaro, Theory of crystal dislocations, Oxford Clarendon Press (1967) and referencestherein.
[4] F.C. Frank and J.M. Van der Merwe, Proc. Roy. Soc. (London) A198, 205 (1949).
[5] S.C. Ying, Phys. Rev. B3, 4160 (1971).
[6] J. Friedel, Extended Defects in materials, preprint (1979).
[7] S. Aubry, On structural phase transitions. "Lattice locking and ergodic theory" preprint (1977).
unpublished.
[8] S. Aubry, G. André "Colloquium on group theoretical methods in physics". Kiryat Anavim Israël, Annals of the Israël Physical Society 3, 133 (1980)
[9] S. Aubry, in "Solitons and Condensed matter physics", Edited by A.R. Bishop and T. Schneider, Springer Verlag Solid State Sciences 8, 264 (1978).
S. Aubry, Ferroelectrics 24, 53 (1980).
[10] S. Aubry, "Intrinsic Stochasticity in Plasmas", page 63 (1979) Edition de Physique, Orsay, France, Edited by G. Laval and D. Gresillon.
[11] G. André, Thesis.
[12] S. Aubry, "Bifurcation Phenomenas in Mathematical Physics and Related Topics", p.163, 1980, Riedel Publishing Company. Edited by C. Bardos and D. Bessis.
[13] B. Mandelbrot ; Form, Chance and Dimension, W.H. Freeman and Company, San Francisco (1977).
[14] V.I. Arnold, Ann. Math. Soc. Trans. Serie 2, 46, 213 (1965).
M. Herman, Thesis (mathematics), Orsay (France (1976)).
[15] A. Niven, Diophantine approximations, Intersciences publishers (1963).
[16] J. Von Boehm and P. Bak, Phys. Rev. Letters 42, 122 (1978), Phys. Rev. B21, 5297 (1980)
[17] A. Bruce and R. Cowley, J. Phys. C,11, 3577 (1978)
A. Bruce, R. Cowley and A.F. Murray, J. Phys. C 11, 3591 (1978).
A. Bruce, R. Cowley, J. Phys. C 11, 3609 (1978).
[18] G. Toulouse, J. Vannimenus and J.M. Maillard, Journal de Phys. Lett. 38, L459 (1977).
[19] S. Aubry, "Stochastic Behavior in Classical and Quantum Systems", Lecture notes in Physics 93, 201 (1977), Springer Verlag, Edited G. Cassati and J. Ford.
[20] R. Bidaux and L. de Seze, preprint (1980).
W. Selke and M. Fisher, preprint (1980).
[21] V.I. Oseledec, Trans. Moscow Math. Soc. 19, 197 (1968).
D. Ruelle, Proceedings of the conference on "Bifurcation theory and its applications", New York (1977).
[22] D.F. Escande and F. Doveil, preprint (1980).
[23] Y. Pomeau and P. Manneville, Intrinsic Stochasticity in Plasmas, p.329 (1979), Edition de Physique, Orsay, France, Edited G. Laval and D. Gresillon.
[24] J. Villain, M. Gordon, J. Phys. C13, 3117 (1980).
[25] S. Aubry, in preparation.
[26] W. Rudin, Real and Complex Analysis, Mc Graw Hill (1970).
[27] A.M. Moudden, F. Denoyer, in preparation.
[28] H. Cailleau, F. Moussa, C.M.E. Zeyen and J. Bouillot, Solid State Communications 33, 407 (1980).
[29] R. Plumier, M. Sougi and M. Lecomte, Physics Letters 60A, 341 (1977).
[30] M. Fisher and W. Selke, Phys. Rev. Letters, 44, 1502 (1980).

Laboratoire Léon Brillouin, BP n°2, 91190 Gif-sur-Yvette, France
and DRP Université Pierre et Marie Curie, Paris.

THE CONVERGENCE OF PADÉ APPROXIMANTS AND THEIR GENERALIZATIONS

J. Nuttall

In this lecture we will survey the current state of the theory of the convergence of Padé approximants (near diagonal) to functions with branch points. To aid the potential user who is not concerned with mathematical rigor, we first give a summary of what is expected to be true. We go on to motivate these speculations and discuss their proof. Finally some ideas about the generalization of these results to the case of Hermite-Padé approximants are given.

1. Padé Convergence - Summary

The reader must note that the results of this section have not been proved for a class of functions as large as that for which they are expected to hold.

It is convenient to deal with Padé approximants to functions expanded about infinity. Thus if f(z) is analytic in a neighborhood of ∞ with expansion

$$f(z) = \sum_{j=0}^{\infty} d_m z^{-j},$$

the [n/n] Padé approximant is defined as

$$[n/n] = -P_1(z^{-1})/P_2(z^{-1})$$

where $P_1(t), P_2(t)$ are polynomials of degree n that satisfy

$$P_1(z^{-1}) + f(z)\, P_2(z^{-1}) = O(z^{-(2n+1)})$$

Suppose that f(z) is analytic throughout the complex plane except for branch points at $z = a_i$, $i=1,\dots,n$. Then we predict that $[n/n] \to f(z)$ as $n \to \infty$ for all z except those contained in a particular set S of minimum capacity containing the points $\{a_i\}$.

Remark. Because of the possible existence of spurious pole-zero pairs, convergence must be defined in a weaker sense than normal, but this usually is of little consequence in practice (see below).

The capacity of a set may be defined as $C = e^{-V}$, where a real function g(z) satisfying Laplace's equation outside the set and being zero on its boundary has the form $g(z) \sim \ell n|z| + V$ as $|z| \to \infty$. ([1]).

The particular set of minimum capacity to be used depends on the nature of the function $f(z)$. Suppose, for instance, that no closed path exists that surrounds a subset of the points $\{a_i\}$ on which $f(z)$ is is single-valued. Then we choose the connected set S of minimum capacity that contains all the points $\{a_i\}$. It may be shown [2] that this set, which is unique, is given by $S = \{z:\mathrm{Re}\phi(z) = 0\}$, where

$$\phi(z) = \int_{a_1}^{z} dt\ (Z(t)/X(t))^{\frac{1}{2}}$$

Here $X(t) = \prod_{i=1}^{n} (t-a_i)$ and $Z(t) = \prod_{i=1}^{n-2} (t-c_i)$. The points c_i are determined by the conditions

$$\mathrm{Re}\phi(a_i) = 0 \quad i = 2,\ldots, n-1$$

$$\mathrm{Re}\phi(c_i) = 0 \quad i = 1,\ldots, n-2$$

The zeros of $P_1(z^{-1})$, $P_2(z^{-1})$ will almost all approach S as $n \to \infty$, and the density of zeros of each on S will be proportional to $|Z(z)/X(z)|^{\frac{1}{2}}$. Except near spurious pole-zero pairs it is expected that the error $|[n/n]-f(z)|$ is of order $\exp(-2n\ \mathrm{Re}\phi(z))$.

For another function $f(z)$ with the same branch points but different properties with regard to single-valuedness a different set S of minimum capacity must be used. S is now the set of minimum capacity including all $\{a_i\}$ such that $f(z)$ is single-valued in the plane cut by S. Thus S may now have several components. The corresponding function $\phi(z)$ is given by

$$\phi(z) = \int_{a_1}^{z} dt\ Y(t)\ (W(t)/X(t))^{\frac{1}{2}}$$

where $W(t) = \prod_{i=1}^{\lambda} (t-c_i)$, $Y(t) = \prod_{i=1}^{\eta} (t-b_i)$, $2\eta+\lambda=n-2$.

In this case

$$\mathrm{Re}\phi(a_i) = 0 \quad i=2,\ldots,n-1$$

$$\mathrm{Re}\phi(c_i) = 0 \quad i=1,\ldots,\lambda$$

In every case S consists of several analytic arcs ending at $\{a_i\}$, $\{c_i\}$ and those b_i for which $\mathrm{Re}\ \phi(b_i) = 0$. In the simplest case $n=2$, S is the line segment joining a_1, a_2. For $n=3$, S consists of arcs from a_1, a_2, a_3 all meeting at c_1, a point inside the triangle $a_1\ a_2\ a_3$. For $n=4$, two cases are possible. If $\eta=0$, $\lambda=2$ and S has one component with arcs joining (say) a_1c_1, a_2c_1, $c_1\ c_2$, $a_3\ c_2$, $a_4\ c_2$. If $\eta=1, \lambda=0$ then S has two components, arcs joining (say) a_1a_2, $a_3\ a_4$. Examples of these cases are shown in [3].

For a function f(z) whose discontinuity across S has no zeros (except perhaps at the ends) but is adequately smooth, we expect that the number of spurious poles may be as high as $\frac{1}{2}(n+\lambda)-1$, but fewer than this will be present for some values of n. If the discontinuity of f has zeros, additional spurious poles are to be expected.

The existence of spurious poles means that the type of convergence we expect to prove is convergence in capacity, where we show that the capacity of the set on which the error is not small can be made as small as we like by increasing n.

It should be remarked that, for some functions f(z), not all branch points are to be included in the set that defines S. If by using some of the a_i we obtain an S which cuts the plane in such a way that the remaining points a_i are not on the sheet which contains the point at ∞ about which we are expanding, then this is the solution we require. Presumably such an S is unique (this remains to be proved). For example consider the function

$$f(z) = \int_{-1}^{1} dt\,(t^2-1)^{\frac{1}{2}} \left(\frac{t-i}{t+i}\right)^{\frac{1}{2}} (z-t)^{-1}.$$

The branch points at $z = \pm i$ are not to be used in defining S, which is the line segment joining -1, 1.

It is not expected that the addition to f(z) of rational functions (perhaps even meromorphic), or functions with essential singularities on a set of zero capacity, will change the above results significantly.

2. Padé Convergence-Proofs

So far proofs of the above speculations have been restricted to the case when S corresponds to the case $\lambda=0$, $n=2(\eta+1)$ and has $(\eta+1)$ separate components. The method of proof is a generalization of Szegö's [4] treatment of the asymptotic behavior of polynomials orthogonal on (-1,1) with respect to certain real weight functions. The idea is to relate the required polynomials to a polynomial that may be explicitly determined and which is orthogonal with respect to a weight that is close to the required weight.

We define

$$p(z) = z^n P_2(z^{-1})$$

from which it may be shown that, provided f(z) has no poles in the complex plane cut by S,

$$\int_S dt\ \omega(t)\ p(t)t^k = 0,\ k=0,\ldots,n-1, \tag{1}$$

where $\omega(z)$ is the discontinuity of $f(z)$ across S. We have considered the case where $\omega(t) = X(t)^{-\frac{1}{2}}\ \sigma(t)$, with $\sigma(t)$ smooth, non-vanishing [5].

To parallel Szegö's argument we first find the polynomial $q(t)$ satisfying (1) in the case where $\sigma(t) = \rho(t)^{-1}$, $\rho(t)$ a polynomial of degree $m < n$. This was done by Akhiezer [6]. On the two-sheeted Riemann surface $\mathfrak{R}$ corresponding to $y^2=X(t)$ we construct a meromorphic function $F(t)$ having an nth order pole at $\infty^{(1)}$ (∞ on the first sheet) as well as a zero of order $n-m-\eta$ at $\infty^{(2)}$ and first order zeros at the zeros of $\rho(t)$ on the second sheet. Such a function will have η additional zeros at α_i, $i=1,\ldots,\eta$ somewhere on $\mathfrak{R}$, whose location is determined by the solution of the Jacobi inversion problem [7].

The polynomial $q(t)$ may now be written ($t^{(1)}$, $t^{(2)}$ referring to the two sheets)

$$q(t) = F(t^{(1)}) + F(t^{(2)}) \tag{2}$$

This we see because first of all (2) is a polynomial of degree n. Moreover, (1) becomes

$$\int_S dt\ X(t)^{-\frac{1}{2}}\rho^{-1}(t)\left(F(t^{(1)}) + F(t^{(2)})\right)\ t^k =$$

$$\int_\Gamma dt\ X(t)^{-\frac{1}{2}}\ \rho^{-1}(t)\ F(t^{(2)})t^k$$

where Γ is a closed curve including S. To see this we have used the facts that $X_+(t)^{-\frac{1}{2}} = -X_-(t)^{-\frac{1}{2}}$, $F_+(t^{(2)}) = F_-(t^{(1)})$, where +, - refer to opposite sides of S. Because $F(t^{(2)})\rho^{-1}(t)$ is analytic in the cut plane, the contour Γ may be distorted into a circle of large radius and the result follows.

The asymptotic form of q for large n may be found by considering the function $\chi(t)=F(t)\exp(-n\phi(t))$. Because $\phi(t)\sim \ell nt, -\ell nt$ near $\infty^{(1)}$, $\infty^{(2)}$ respectively, we see that $\chi(t)$ has a pole of order $(m+\eta)$ at $\infty^{(2)}$ and no other poles. It has zeros at those of ρ (on the second sheet) and at α_i, $i=1,\ldots,\eta$. It is not single-valued on $\mathfrak{R}$ but $\ell n\ \chi(t)$ has periods which are pure imaginary. Consequently the dominant part of the asymptotic form of $F(t)$ is $\exp(n\phi(t))$.

Just as in Szegö [4] it may be shown [5] that $p(t)$ satisfies the following integral equation.

$$\text{const. } p(t)=q(t)+\int_S dt' X(t')^{-\frac{1}{2}}(\sigma(t)-\rho^{-1}(t))$$

$$(q(t)q'(t')-q'(t)q(t'))\quad (t'-t)^{-1}p(t') \tag{3}$$

where $q'(t)$ is the polynomial of degree $(n+1)$ orthogonal with respect to ρ^{-1}. The argument is concluded by choosing m large enough so that $(\sigma-\rho^{-1})$ is small and showing that the kernel of (3) is therefore small enough to solve (3) by iteration.

With the previous assumptions it may be shown that [3]

$$f(z)-[n/n] = (2\pi i p(z))^{-1} \int_S dt \frac{\omega(t)p(t)}{t-z}$$

Now orthogonality shows that

$$\int_S dt\ \omega(t)\ p(t)\ \left(\frac{p(t)-p(z)}{t-z}\right) = 0$$

since the expression in brackets is a polynomial in t of degree $(n-1)$. Thus we have

$$f(z)-[n/n] = (2\pi i\ p(z)^2)^{-1} \int_S dt \frac{\omega(t)\ p(t)^2}{t-z}$$

and the stated results on the convergence of the Padé approximants become plausible.

This argument applies only to the case where the dominant singularity of $f(z)$ at each branch point is of square root type, but we expect the results to be true for other types of singularity. This is certainly the case for Jacobi polynomials [8] and the lecture of Gammel and Nuttall [9] indicates that the same holds for some functions with three non-square root singularities. In fact we can make a more detailed prediction. In the example above where $\sigma=\rho^{-1}$, we see that [7], $F(t)$ being evaluated on the first sheet,

$$\sigma\ F_+(t)F_-(t) = \text{const.} \prod_{i=1}^{\eta} (t-\alpha_i). \tag{4}$$

This is a Riemann-Hilbert problem for $F(t)$, analytic and single-valued in the plane cut by S with a pole of order n at ∞.

It may be checked that, for $z \notin S$, $p \sim F$ in the case of Jacobi polynomials where σ has end-point singularities, and the same thing holds in the example of Gammel and Nuttall [9]. In addition, for $z \in S$, $p \sim F_+ + F_-$ except when z is near one of the points a_i or c_i. Thus it is natural to speculate that the solution of (4) gives the asymptotic form of p for a class of σ defined on S, with end point singularities, that are otherwise smooth and non-vanishing. We are hopeful of being able to modify the above proof to make it apply to this case.

3. Hermite-Padé Approximants

One way of generalizing the notion of Padé approximants was first discussed by Hermite [10] and Padé [11]. In the diagonal case, which we consider for simplicity, we have m functions $f_i(z)$, analytic near ∞ and m polynomials of degree n, $p_i(z)$, which satisfy

$$\sum_{i=1}^{m} p_i(z) f_i(z) = O(z^{-(m-1)n-m+1}). \tag{5}$$

The study of the asymptotic form of $p_i(z)$ as $n \to \infty$ is in its infancy. It has been possible to generalize the results of Akhiezer [6] and to construct $p_i(z)$ in the case when each $f_i(z)$ is meromorphic on a Riemann surface $\mathcal{R}$ with m sheets that corresponds to the equation $R(y,z)=0$, R being an irreducible polynomial in y,z [12]. In this case we have shown that the dominant part of the remainder function

$$Q(z) = \sum_{i=1}^{m} p_i(z) f_i(z) \tag{6}$$

is $\exp(n\phi(z))$ where $\phi(z)$ is a generalization of the function $\phi(z)$ of Sec. 1. We must now define $\phi(z)$ as the unique Abelian integral of the third kind [7] with periods which are pure imaginary that has poles at $\infty^{(i)}$, $i=1,\ldots,m$, where the residues are (m-1) at $\infty^{(1)}$ and -1 at $\infty^{(i)}$, $i=2,\ldots,m$. If k is the sheet for which $\mathrm{Re}\,\phi(z^{(j)})$ is largest, $j=1,\ldots,m$ for fixed z then $\exp(n\phi(z^{(k)}))$ is the dominant part of $p_i(z)$ for large n. Almost all the zeros of $p_i(z)$ will approach the curve on which $\mathrm{Re}\,\phi(z)$ is equally large on two sheets and smaller on all other sheets.

An example should help to make the situation clearer. Consider the Riemann surface given by

$$z\, y^3 - (z-1) = 0 \tag{7}$$

This surface has two branch points, z=0,1, at each of which all three solutions of (7) for y coincide. The surface has three sheets joined across a cut which we shall choose to run along the real axis from 0 to 1.

For a given z the value of y on sheets 1,2,3 is

$$y = \left(\frac{z-1}{z}\right)^{1/3} \quad \text{sheet 1}$$

$$y = \omega\left(\frac{z-1}{z}\right)^{1/3} \quad \text{sheet 2}$$

$$y = \omega^2\left(\frac{z-1}{z}\right)^{1/3} \quad \text{sheet 3}$$

where $\left(\frac{z-1}{z}\right)^{1/3}$ is defined in the cut plane so that it takes on the value 1 as $z \to \infty$, and $\omega=\exp(\frac{2}{3}\, i\, \pi)$.

The surface has genus zero (no period loops) so that $\exp \phi$ is in this case meromorphic. This function has first order poles at $\infty^{(2)}$, $\infty^{(3)}$, a second order zero at $\infty^{(1)}$, but no other poles or zeros on $\mathcal{R}$. It may be checked that

$$\exp(\phi(z)) = z(1-y)^3$$

which, being rational in z,y is meromorphic.

Now suppose we choose $f_1=1$, $f_2=y$, $f_3=y^{-1}$. According to [12] Q(z) will be meromorphic on $\mathcal{R}$ having poles 0,1, $(\infty^{(2)})^n$, $(\infty^{(3)})^n$ and zeros $(\infty^{(1)})^{2n+2}$. (The power indicates the order of the pole of zero). Such a function is unique up to a constant factor and it is not hard to see that

$$Q(z) = \exp(n\phi(z))\,(y-1)^2\, y^{-1}.$$

Writing (6) on each sheet in turn we find

$$Q(z^{(j)}) = \sum_{i=1}^{3} p_i(z)\, f_i(z^{(j)}) \tag{8}$$

which may be solved for $p_i(z)$ to give

$$p_1(z) = \frac{1}{3}\sum_{i=1}^{3} Q(z^{(i)})$$

$$p_2(z) = \frac{1}{3}\sum_{i=1}^{3} Q(z^{(i)})\, y^{-1}(z^{(i)}) \tag{9}$$

$$p_3(z) = \frac{1}{3}\sum_{i=1}^{3} Q(z^{(i)})\, y(z^{(i)})$$

It may be checked directly that (9) satisfies (6). A very similar case was previously studied by Shafer [13].

Asymptotically, the zeros of $p_i(z)$, i=1, 2,3, will lie on the real axis from $-\infty$ to 0 and 1 to $+\infty$, for there $\mathrm{Re}\,\phi(z^{(2)})=\mathrm{Re}\,\phi(z^{(3)})>\mathrm{Re}\,\phi(z^{(1)})$, and nowhere else does such a relation hold.

After our experience with ordinary Padé approximants we are on the watch for a universal behavior of the asymptotic form of the polynomials, depending only on the location of the singularities of $f_i(z)$. Chudnovsky [14] (Sec. 3.4) has shown that the leading term in the asymptotic behavior of $p_i(z)$ constructed from functions $f_i(z) = 1$, $f_2(z) = {}_2F_1(1,\omega_1;\gamma;z^{-1})$, $f_3(z) = {}_2F_1(1,\omega_2;\gamma;z^{-1})$, where $\omega_1-\omega_2$ is not an integer, is just as we have found above. This supports tne idea of universality, for on the first sheet these functions have branch points only at $z=0,1$. The branch point at ∞ on another sheet must not be affecting the asymptotic form in this case.

It must not be thought that the leading part of the asymptotic form of $p_i(z)$ is the same for all sets of functions $f_i(z)$ which have the same singularities on the first sheet. Consider the Riemann surface $\mathcal{R}_1$ given by

$$R \equiv y^3 - 3yz + 2z = 0 \tag{10}$$

The branch points are found by solving (10) simultaneously with $\frac{\partial R}{\partial y} = 0$, and occur at $z=0,1,\infty$. The branch points at 1, ∞ are of square root type, and the branch point at 0 is of cube root type. Let us choose sheet 1 so that the singularity at ∞ does not occur on this sheet.

Functions $f_i(z)$ that are meromorphic on $\mathcal{R}_1$ fall within the class described above [12], and the dominant part of $p_i(z)$ for large n is again $\exp(n\phi(z))$. In this case the prescription for $\phi(z)$ leads to

$$\exp(\phi(z)) = z^{-2}\, y^6 \tag{11}$$

Thus the $p_i(z)$ have a different asymptotic behavior from the previous example even though in each case the $f_i(z)$ have the same two singularities on the first sheet. Some form of universality persists, however, because the behavior of this example is the same as that found by Chudnovsky [14] (Theorems 4.6, 6.2) for

$$f_1(z) = {}_3F_2\left(z^{-1}\,\middle|\,\begin{matrix}a_1,a_2,a_3\\ b_1,b_2\end{matrix}\right) \tag{12}$$

and $f_i(z)$, $i\neq 1$, formed from $f_1(z)$ by adding integers to the a_j, b_k.

We speculate (without much foundation) that the reason for the different behavior is the lack of independence of the three functions $f_i(z)$ in the case when the asymptotic behavior is associated with $\mathcal{R}_1$. Thus for the meromorphic $f_i(z)$, it is possible to write on sheet 1

$$f_i(z) = A_i(z)\,(z-1)^{\frac{1}{2}} + B_i(z)$$

where $A_i(z)$, $B_i(z)$ are analytic near $z=1$. This means that functions $C_i(z)$ analytic near $z=1$ exist so that

$$\sum_{i=1}^{3} C_i(z)\, f_i(z) = 0.$$

The same situation arises in Chudnovsky's example (12).

A plausible working hypothesis would seem to be that the dominant part of the asymptotic form of Hermite-Padé approximants $p_i(z)$ to the solutions of Fuchsian linear differential equations studied by Chudnovsky [14] (Sec. 5) is the same as that for the approximants to meromorphic functions on an appropriate Riemann surface. We do as yet always know which surface to choose.

All the cases discussed so far correspond to situations in which each $f_i(z)$ has the same set of singularities. The only example we know of where this is not the case was given by Chudnovsky [14] (Sec. 3.5), who studied $f_i(z) = {}_2F_1(1,b\ ;\ c;a_i z^{-1})$.

In the case of Padé approximants a heuristic understanding of the asymptotic form is gained by applying the method of steepest descent to a multiple integral formula for $p_i(z)$ [3]. A generalization of this formula has been worked out [15]. For the case $m = 3$, $f_1(z) = 1$, we have

$$p_2(z)=\int dt_1 f_2(t_1)\ldots\int dt_n f_2(t_n)\int ds_o f_3(s_o)\ldots\int ds_n f_3(s_n)\left(\prod_{i=1}^{n}(z-t_i)\right)I \tag{13}$$

where

$$I =\left(\prod_{i<j}(t_i-t_j)^2\right)\left(\prod_{i=j}(s_i-s_j)^2\right)\left(\prod_{i,j}(t_i-s_j)\right), \tag{14}$$

and a similar formula for $p_3(z)$. All integrals are taken on a large circle which encloses all singularities of $f_2(z), f_3(z)$ on the first sheet.

To illustrate our ideas, suppose that $f_2(z)$ is analytic apart from branch points at a_1, a_2 and similarly for $f_3(z)$ with branch points at b_1, b_2. Then the integrals over t_j may be taken along an arc T joining a_1 and a_2 and those over s_j along an arc S between b_1, b_2. The previous argument [3] suggests that the integral (13) is evaluated approximately for large n by choosing the arcs to minimize the maximum value of I as each t_j, s_k vary on their respective arcs. The leads to the equations

$$\frac{\partial \ell n I}{\partial t_j} = \frac{\partial \ell n I}{\partial s_k} = 0, \text{ all } j,\ k,$$

or

$$\begin{aligned} &2\sum_{j\neq k}(t_k-t_j)^{-1} + \sum_j (t_k-s_j)^{-1} = 0 \quad k=1,\ldots,n \\ &2\sum_{j\neq k}(s_k-s_j)^{-1} + \sum_j (s_k-t_j)^{-1} = 0 \quad k=0,\ldots,n \end{aligned} \tag{15}$$

If we assume that the t,s arcs do not intersect and that for large n the t_j, s_k are distributed smoothly along the arcs with normalized densities σ, ρ, we may replace (15) by

$$
\begin{aligned}
&2\, P\int_T |dt'|\sigma(t')\ (t-t')^{-1} + \int_S |ds'|\rho(s')(t-s')^{-1} = 0,\ t\varepsilon T \\
&2P\int_S |ds'|\rho(s')(s-s')^{-1} + \int_T |dt'|\sigma(t')\ (s-t')^{-1} = 0,\ s\ \varepsilon\ S
\end{aligned}
\tag{16}
$$

In terms of functions $g(z)$, $h(z)$ analytic in the plane cut by T, S respectively,

$$
\begin{aligned}
g(z) &= \int_T |dt'|\sigma(t')\ (z-t')^{-1} \\
h(z) &= \int_S |ds'|\rho(s')\ (z-s')^{-1},
\end{aligned}
\tag{17}
$$

equation (16) may be written

$$
\begin{aligned}
g_+(z) + g_-(z) + h(z) &= 0 \qquad z\ \varepsilon\ T \\
h_+(z) + h_-(z) + g(z) &= 0 \qquad z\ \varepsilon\ S
\end{aligned}
\tag{18}
$$

This Riemann-Hilbert problem is solved by constructing a three-sheeted Riemann surface $\mathcal{R}_2$ with branch points at a_1, a_2, b_1, b_2. Sheet 1 is joined to sheet 2 across T and sheet 2 to sheet 3 across S. Each branch point is of square root type. The surface is of genus zero. On $\mathcal{R}_2$ there is a unique meromorphic function F with zeros at ∞ on all sheets and first order poles at the branch points that has near $z=\infty$ the behavior $F \approx z^{-1}$ on sheets 1,3 and $F \approx -2z^{-1}$ on sheet 2. It is easy to see that

$$
F(z^{(1)}) + F(z^{(2)}) + F(z^{(3)}) = 0
$$

and that (18) is solved by

$$
g(z) = F(z^{(1)}),\ h(z) = F(z^{(3)}). \tag{19}
$$

The Riemann surface $\mathcal{R}_2$ and the function $F(z)$ do not depend on the precise location of the cuts. These are determined after finding $g(z)$, $h(z)$ from (19) by the requirement that σ, ρ from (17) should be real. We see that $F(z) = \phi'(z)$, where $\phi(z)$ is the function constructed as before that is appropriate to $\mathcal{R}_2$.

The approximate form of $p_2(z)$ is proportional to $\prod_{i=1}^{n}(z-t_i)$, where the t_i satisfy (15), and so the zeros of $p_2(z)$ lie near to T, and those of $p_3(z)$ are near to S. As in the previous cases the zeros of $p_2(z)$ lie on a curve where Re $\phi(z)$ is equal on two sheets, and similarly for

$p_3(z)$, but now the pair of sheets is not the same for the two polynomials. Further analysis shows that the zeros of $p_1(z)$ lie on a similar curve for the third pair of sheets.

The above procedure fails if the solution leads to a σ or ρ that change sign. This signals that one of these functions is zero for part of the arc. In this case a different Riemann surface, of the same topology as $\mathfrak{R}_2$ but with one branch point moved, must be used. The Chudnovsky [14] (Sec. 3.5) example is a situation where this occurs.

We have performed numerical calculations which support the above predictions, and they will appear elsewhere along with further details [16].

It is clear that Hermite-Padé approximants have a rich structure that is related in more ways than one to the Riemann-Hilbert problem, but that much remains to be discovered.

REFERENCES

1. E. Hille, "Analytic Function Theory", Vol.II. Ginn and Co., Waltham, Mass. (1962) p 275.
2. J. Nuttall, "On Sets of Minimum Capacity" preprint, 1980.
3. J. Nuttall, "Sets of Minimum Capacity, Padé Approximants and the Bubble Problem", Cargese Summer School on Bifurcation Phenomena and Related Topics" (1979).
4. G. Szego, "Orthogonal Polynomials", American Mathematical Society, New York, (1959).
5. J. Nuttall and S.R. Singh, "Orthogonal Polynomials and Padé Approximants Associated with a System of Arcs", J. Approx. Theory 21, (1977), 1-42.
6. N.I. Akhiezer, "Orthogonal Polynomials on General Intervals", Soviet Math. Doklady, 1 (1960), 989-992.
7. C.L. Siegel, "Topics in Complex Function Theory", Interscience, New York, (1971).
8. J. Nuttall and C.J. Wherry, "Gaussian Integration for Complex Weight Functions," J. Inst. Maths Applics 21, (1978), 165-170.
9. J.L. Gammel and J. Nuttall, "Note on Generalized Jacobi Polynomials," lecture in this volume.
10. C. Hermite, "Sur la generalisation des fractions continues algebriques". Jour. de Math., ser. 4,10 (1894), 291-329.
11 H. Padé, "Sur la generalisation des fractions continues algebriques", Annali di Math. ser. 2, 21 (1893), 289-308.
12. J. Nuttall, "Hermite-Padé Approximants to Meromorphic Functions" preprint, 1980.
13. R.E. Shafer, "On Quadratic Approximation", SIAM J. Numer. Anal. 11, (1974), 447-460.
14. G.V. Chudnovsky, "Padé Approximation and the Riemann Monodromy Problem," Cargese Summer School on Bifurcation Phenomena and Related Topics (1979).

15. S.K. Burley, S.O. John and J. Nuttall, "Vector Orthogonal Polynomials" preprint, 1980.

16. R.T. Baumel, J.L. Gammel and J. Nuttall, "Asymptotic Form of Hermite-Padé Polynomials" in preparation.

Supported in part by Natural Sciences and Engineering Research Council Canada

Department of Physics, University of Western Ontario
London, Ontario, Canada. N6A 3K7

NOTE ON GENERALIZED JACOBI POLYNOMIALS

J.L. Gammel† and J. Nuttall

1. Introduction

Suppose we have 3 distinct points a_1 a_2 a_3 in the complex plane and three complex numbers ν_1, ν_2, ν_3 ($\Sigma\ \nu_i$=integer).
Set

$$\omega(t) = \prod_{i=1}^{3} (t-a_i)^{\nu_i}$$

(We would eventually like to increase the number three). A generalized Jacobi polynomial of degree n, $p_n(t)$, is defined by

$$\int_C dt\ \omega(t) p_n(t)\ t^k = 0,\ k=0,\ldots,n-1 \tag{1}$$

where the integral is taken on a closed curve C surrounding all a_i.

A related function f(t) is

$$f(t) = \frac{1}{2\pi i} \int_C \frac{dt'\ \omega(t')}{t-t'},\quad t \text{ outside C.} \tag{2}$$

The diagonal [n/n] Padé approximant to f(t) expanded about $t=\infty$ has denominator $t^{-n}p_n(t)$. The function f satisfies the differential equation

$$Xf' + (X'-Y)f + U = 0 \tag{3}$$

where

$$X(t) = \prod_{i=1}^{3} (t-a_i)$$

and

$$(X\omega)' = Y\omega$$

so that Y is a polynomial of degree 2. Also U is a polynomial.

Approximants to the continued fraction of functions satisfying a differential equation of the form (3) were studied by Laguerre [1] (See also Perron [2] and Bessis [3]). He showed that p_n satisfies a differential equation with polynomial coefficients. Laguerre also obtained recurrence relations for the coefficients in these polynomials and the coefficients in the three-term recurrence relation

$$p_n(t)=(t+Q_n)p_{n-1}(t) + R_n p_{n-2}(t), \tag{4}$$

†Permanent address. Department of Physics, St. Louis University, St. Louis, Missouri, U.S.A.

but it appears that he was unable to solve these recurrence relations for the case of interest here.

In this note we derive analogous recurrence relations for Q_n, R_n and show how to obtain the other interesting coefficients from these quantities. We give a heuristic argument that asymptotically Q_n, R_n are doubly periodic functions of n, and show how the periods are related to the points a_i but are independent of the quantities ν_i. The asymptotic form of $p_n(t)$ follows (non-rigorously) by applying the WKB method to some of the differential equations. The conclusion is that almost all the zeroes of $p_n(t)$ approach the set of minimum capacity containing the points a_i (see Nuttall [4]). The diagonal Padé approximants to f no doubt converge in capacity away from the set of minimum capacity.

We have some confidence that the results hypothesized here are correct because they are supported by numerical calculations.

2. Differential Equation

We assume throughout that $p_n(t)$ is actually of degree n and is unique up to a constant factor. The alternative corresponds to poles in the solution of the recurrence relations, which complicates their rigorous discussion.

Let us define

$$L_1\phi \equiv X\phi' + \tfrac{1}{2}Y\phi \tag{5}$$

Then $L_1\, p_n$ is a polynomial of degree n+2. If π is any polynomial of degree n-3 then

$$\begin{aligned}\int_C dt\ \omega\ \pi\ L_1 p_n &= - \int_C dt\,(\pi\omega X)' p_n + \tfrac{1}{2}\int_C dt\ \omega\pi Y p_n \\ &= -\tfrac{1}{2}\int_C dt\ \omega\pi Y p_n - \int_C dt\ \omega\pi' X p_n \\ &= 0 \quad \text{from (1).}\end{aligned}$$

Thus we may write, with A_n, B_n, D_n, E_n constants,

$$L_1 p_n = A_n p_{n+2} + B_n p_{n+1} + D_n p_{n-1} + E_n p_{n-2} \tag{6}$$

We have omitted a term containing p_n, because, as the above argument shows,

$$\int_C dt\ \omega\ h\ L_1 g = - \int_C dt\ \omega\ g L_1 h \tag{7}$$

for any polynomials h, g and therefore, in particular,

$$\int_C dt\ \omega p_n L_1 p_n = 0 \tag{8}$$

In the same way we obtain

$$\begin{aligned} E_n I_{n-2} &= - A_{n-2} I_n \\ D_n I_{n-1} &= - B_{n-1} I_n \end{aligned} \tag{9}$$

where

$$I_n = \int_C dt\ \omega\ p_n^2 \tag{10}$$

From (4) we obtain

$$I_{n-1} + R_n I_{n-2} = 0 \tag{11}$$

so that (9) become

$$\begin{aligned} E_n &= -R_n R_{n+1}\ A_{n-2} \\ D_n &= R_{n+1} B_{n-1} \end{aligned} \tag{12}$$

To obtain (11) we have chosen the normalization $p_n(t)=t^n + \ldots$.

Equating powers of t^{n+2} in (6) gives

$$n + \tfrac{1}{2}(\nu_1 + 1 + \nu_2 + 1 + \nu_3 + 1) = A_n$$

which we write as

$$A_n = n + \mu \tag{13}$$

3. Recurrence Relations

Differentiate (4) to obtain

$$p'_n = p_{n-1} + (t+Q_n) p'_{n-1} + R_n p'_{n-2} \tag{14}$$

which gives

$$L_1 p_n = X p_{n-1} + (t+Q_n) L_1 p_{n-1} + R_n L_1 p_{n-2} \tag{15}$$

If we use (6) we find, with $X = t^3 - t^2\Sigma + t\pi - \tau$

$$\begin{aligned} A_n p_{n+2} + B_n p_{n+1} + D_n p_{n-1} + E_n p_{n-2} &= (t^3 - t^2\Sigma + t\pi - \tau) p_{n-1} \\ &+ (t+Q_n)(A_{n-1} p_{n+1} + B_{n-1} p_n + D_{n-1} p_{n-2} + E_{n-1} p_{n-3}) \\ &+ R_n (A_{n-2} p_n + B_{n-2} p_{n-1} + D_{n-2} p_{n-3} + E_{n-2} p_{n-4}) \end{aligned} \tag{16}$$

Equation (4) tells us

$$tp_{n-1}=p_n-Q_np_{n-1}-R_np_{n-2}$$

$$t^2p_{n-1}=p_{n+1}-(Q_{n+1}+Q_n)p_n+(Q_n^2-R_{n+1}-R_n)p_{n-1}+(Q_nR_n+Q_{n-1}R_n)p_{n-2} + R_nR_{n-1}\,p_{n-3}$$

$$\begin{aligned} t^3p_{n-1}=p_{n+2}-Q_{n+2}p_{n+1}-R_{n+2}p_n &- (Q_{n+1}+Q_n)(p_{n+1}-Q_{n+1}p_n-R_{n+1}p_{n-1}) \\ &+ (Q_n^2-R_{n+1}-R_n)(p_n-Q_np_{n-1}-R_np_{n-2}) \\ &+ (Q_nR_n+Q_{n-1}R_n)(p_{n-1}-Q_{n-1}p_{n-2}-R_{n-1}p_{n-3}) \\ &+ R_nR_{n-1}(p_{n-2}-Q_{n-2}p_{n-3}-R_{n-2}p_{n-4}) \end{aligned} \tag{17}$$

We insert (17) in (16) and equate coefficients of $p_{n+1},\ldots,p_{n-1}$ to obtain

$$-Q_{n+1}-\Sigma - A_nQ_{n+2}+A_{n-2}Q_n-B_n+B_{n-1} = 0 \tag{18}$$

$$R_{n+1}-Q_n^2-Q_{n+1}^2-Q_nQ_{n+1}-\Sigma(Q_{n+1}+Q_n)-\pi-B_{n-1}(Q_{n+1}-Q_n)+A_nR_{n+2}-A_{n-3}R_n = 0 \tag{19}$$

$$-Q_n^3+Q_{n+1}R_{n+1}+2Q_nR_{n+1}+2Q_nR_n+Q_{n-1}R_n-\Sigma(Q_n^2-R_{n+1}-R_n) - \pi Q_n - \tau -2R_{n+1}B_{n-1}+2R_nB_{n-2} = 0 \tag{20}$$

The coefficients of the other p_i give no additional information.

Given the values of the five quantities $Q_n, Q_{n+1}, R_n, R_{n+1}, B_{n-1}$ we can find R_{n+2} from (19) and solve (18) and (20) (with n advanced by 1) which are linear equations for Q_{n+2}, B_n. The determinant of these equations is $(2A_n+1)R_{n+2}$. If R_{n+2} is zero, difficulties will arise.

4. Asymptotic Solution

To obtain an approximate form of the solution of these recurrence relations we assume that, for large n,

$$\begin{aligned} Q_n &= \bar{Q}_n+\gamma_n \\ R_n &= \bar{R}_n+\delta_n \\ B_{n-2} &= nb_n,\quad b_n=\bar{b}_n+\varepsilon_n \end{aligned} \tag{21}$$

where $\gamma_n, \delta_n, \varepsilon_n$ are of order n^{-1}. Inserting (21) into (18)-(20) and equating to zero the coefficient of n in each equation gives

$$-\bar{Q}_{n+2}+\bar{Q}_n-\bar{b}_{n+2}+\bar{b}_{n+1} = 0 \tag{22}$$

$$-\bar{b}_{n+1}(\bar{Q}_{n+1}-\bar{Q}_n)+\bar{R}_{n+2}-\bar{R}_n=0 \tag{23}$$

$$\bar{R}_{n+1}\bar{b}_{n+1}-\bar{R}_n\bar{b}_n=0 \tag{24}$$

With the help of (22), we may rewrite (23) as

$$\bar{b}_{n+2}\bar{Q}_{n+1}-\bar{b}_{n+1}\bar{Q}_n-\bar{Q}_{n+1}\bar{Q}_n+\bar{Q}_{n+2}\bar{Q}_{n+1}+\bar{R}_{n+2}-\bar{R}_n=0 \tag{25}$$

Integrating (22),(25) and (26) shows that

$$\bar{Q}_{n+2}+\bar{Q}_{n+1}+\bar{b}_{n+2}=G \tag{26}$$

$$\bar{b}_{n+2}\bar{Q}_{n+1}+\bar{Q}_{n+2}\bar{Q}_{n+1}+\bar{R}_{n+2}+\bar{R}_{n+1}=H \tag{27}$$

$$\bar{R}_n\bar{b}_n=J \tag{28}$$

Using (28) in (23) gives

$$J(\bar{Q}_{n+1}-\bar{Q}_n)-\bar{R}_{n+2}\bar{R}_{n+1}+\bar{R}_{n+1}\bar{R}_n=0 \tag{29}$$

which integrates to

$$\bar{R}_{n+2}\bar{R}_{n+1}+J\bar{Q}_{n+1}=K \tag{30}$$

Here, G,H,J,K are constants. Since n does not appear now except as an index, we can add an arbitrary constant to n. Thus there are five constants of integration as expected.

Later, we will show that the solution of (26),(27),(28) and (30) is an elliptic function of n (a different one for $\bar{Q}_n,\bar{R}_n,\bar{b}_n$), with periods related to the 4 constants. Now we wish to relate these constants to Σ,π,τ.

The equations (18),(20) and (19)(in two different ways) may be written as the form

$$nW_n-(n-1)W_{n-1}=F_n \tag{31}$$

where W_n and F_n are of order unity in the sense used above. If we replace in W_n the quantities Q_n etc by their first approximations $\bar{Q}_n$ etc., we obtain $\bar{W}_n$, the expressions on the left hand sides of (26), (27),(28) and (30). Thus, for example, take (18) which has the form (31) with

$$\begin{aligned} W_n &= Q_{n+2}+Q_{n+1}+b_{n+2} \\ F_n &= \mu(Q_n-Q_{n+2})-Q_n-\Sigma \end{aligned} \tag{32}$$

If we now write $W_n=\bar{W}_n+w_n$, w_n of order n^{-1}, we have, summing (31),

$$n\bar{W}_n + nw_n = \sum_{j=1}^{n} F_j + g \qquad (33)$$

with g constant. From (26) we have $\bar{W}_n = G$, so that we deduce that

$$\sum_{j=1}^{n} F_j = nG + O(1).$$

From (32) we find $G = -(Q_{av} + \Sigma)$ (34)

where

$$Q_{av} = \lim_{n\to\infty}\left(\frac{1}{n}\sum_{j=1}^{n} Q_j\right) = \lim_{n\to\infty}\left(\frac{1}{n}\sum_{j=1}^{n} \bar{Q}_j\right) \qquad (35)$$

Following a similar analysis on the other equations gives

$$H = \lim_{n\to\infty}\left(\frac{1}{n}\{-\sum_{j=1}^{n}(\bar{R}_{j-1}+\bar{R}_j-\bar{Q}_{j-1}^2 - \Sigma\bar{Q}_{j-1})\}\right) + \pi \qquad (36)$$

$$J = \lim_{n\to\infty}\Bigl(\frac{1}{n}\{\tfrac{1}{2}\sum_{j=1}^{n}(-\bar{Q}_{j+1}^3+\bar{Q}_{j+2}\bar{R}_{j+2}+2\bar{Q}_{j+1}\bar{R}_{j+2}+2\bar{Q}_{j+1}\bar{R}_{j+1} + \bar{Q}_j\bar{R}_{j+1}$$
$$- \Sigma(\bar{Q}_{j+1}^2 - \bar{R}_{j+2} - \bar{R}_{j+1}) - \pi\bar{Q}_{j+1} - \tau)\}\Bigr) \qquad (37)$$

$$K = -\lim_{n\to\infty}\Bigl(\frac{1}{n}\{\sum_{j=1}^{n}(2\bar{R}_{j-1}\bar{R}_j+\bar{R}_j^2 - \bar{Q}_{j-1}^2\bar{R}_j-\bar{Q}_j^2\bar{R}_j-\bar{Q}_{j-1}\bar{Q}_j\bar{R}_j- \Sigma(\bar{Q}_j+\bar{Q}_{j-1})\bar{R}_j$$
$$-\pi R_j-\tfrac{1}{2}\bar{Q}_j[-\bar{Q}_j^3+\bar{Q}_{j+1}\bar{R}_{j+1}+2\bar{Q}_j\bar{R}_{j+1}+2\bar{Q}_j\bar{R}_j-\bar{Q}_{j-1}\bar{R}_j \qquad (38)$$
$$-\Sigma(\bar{Q}_j^2 - \bar{R}_{j+1}-\bar{R}_j) - \pi\bar{Q}_j - \tau\,])\}\Bigr)$$

Using the relations (26)-(28),(30), we find

$$H = \tfrac{1}{2}(Q_{av}^2 - \pi) \qquad (39)$$

$$J = -\tfrac{1}{4}(Q_{av}^3 + \Sigma Q_{av}^2 + \pi Q_{av} + \tau) \qquad (40)$$

$$K = \frac{1}{16}(5Q_{av}^4+8\Sigma Q_{av}^3+ (4\Sigma^2 + 2\pi)Q_{av}^2- (4\tau-4\Sigma\pi)Q_{av} + \pi^2) \qquad (41)$$

An alternative method of obtaining (39)-(41) is given in Sect. 6.

5. Nature of Asymptotic Solution

From (26) and (30) we find

$$\bar{R}_{n+1}+\bar{R}_{n-1} - J^2\bar{R}_n^{-2} = (2K-JG)\bar{R}_n^{-1} \qquad (42)$$

The solution of (42) that is required may be shown to be

$$\bar{R}_n = \alpha(P(n+\beta) - P(1)) \tag{43}$$

where P(z) is the Weierstrass elliptic function. That (43) satisfies (42) follows because the l.h.s. of (42) is an elliptic function with no poles, provided the conditions below hold.

$$\alpha^3 = J^2 P'(1)^{-2}$$

$$\alpha^2 P'(1) P''(1) = 4K - 2GJ \tag{44}$$

In this case the left-hand side of (42) must be constant and one more condition on α and the two periods of P(z) is found.

The general solution of (42) involves another arbitrary constant, which may be evaluated by using another second order difference equation

$$R_{n+1} + R_{n-1} + 2R_n = \frac{J^2}{2R_n^2} - \frac{G^2 - 4H}{2} + \frac{R_n^2}{2J^2}(R_{n+1} - R_{n-1})^2$$

Viewed as an equation for R_{n+1} this has two roots, one of which is R_{n+1} and the other R_{n-2}. It may be verified that (43) satisfies the above equation as well as (42).

From (30), following the same sort of argument, we deduce

$$\bar{Q}_n = -\frac{\alpha^2 P'(1)}{J}(\zeta(n) - \zeta(n+1)) + \text{const.} \tag{45}$$

Thus we have

$$\bar{Q}'_n = \frac{d}{dn}\bar{Q}_n = \frac{\alpha^2 (P'(1))}{J}(P(n) - P(n+1))$$

$$= \frac{\alpha P'(1)}{J}(\bar{R}_n - \bar{R}_{n+1}) \tag{46}$$

Now from substitution in (27) we find

$$\bar{R}_{n+1} + \bar{R}_n = \bar{Q}_n^2 - G\bar{Q}_n + H \tag{47}$$

and

$$\bar{R}_n^2 - \bar{R}_n(\bar{Q}_n^2 - G\bar{Q}_n + H) - (J\bar{Q}_n - K) = 0 \tag{48}$$

Thus we deduce, with $\rho = \alpha P'(1) J^{-1}$

$$\bar{Q}_n'^2 = \rho^2((\bar{Q}_n^2 - G\bar{Q}_n + H)^2 + 4(J\bar{Q}_n - K)) \tag{49}$$

Using (34), (39)-(41) gives

$$\bar{Q}_n'^2 = \rho^2(\bar{Q}_n - G - c) X(\bar{Q}_n - G) \tag{50}$$

where

$$c = -2Q_{av} + \Sigma = -2G - \Sigma \tag{51}$$

Integrating (50) gives

$$\rho n = \int^{\bar{Q}n} dt\ Y^{-\frac{1}{2}}(t - G) \tag{52}$$

where

$$Y(t) = (t-c)X(t) \tag{53}$$

This shows that $\bar{Q}_n$ is an elliptic function related to the Riemann surface $y^2 = Y(t)$. Because adjacent poles of $\bar{Q}_n$ are distance 1 apart, we conclude that

$$\rho = 2\int_{a_1}^{\infty} dt\ Y^{-\frac{1}{2}}(t) \tag{54}$$

If we replace the sum in (35) by an integral, than (35) becomes

$$\begin{aligned}
Q_{av} &= \lim_{n\to\infty} \frac{1}{n} \int^{n} dm\ \bar{Q}_m \\
&= \lim_{n\to\infty} \frac{1}{\rho n} \int^{\bar{Q}_n} dt\ t\ Y^{-\frac{1}{2}}(t-G) \\
&= \lim_{n\to\infty} \frac{1}{\rho n} \left\{ \int^{\bar{Q}_n - G} dt'\ (t'-c)^{\frac{1}{2}}\ X^{-\frac{1}{2}}(t') + Q_{av} \int^{\bar{Q}_n - G} dt'\ Y^{-\frac{1}{2}}(t') \right\} \\
&= Q_{av} + \lim_{n\to\infty} \frac{1}{\rho n} \int^{\bar{Q}_n - G} dt'(t'-c)^{\frac{1}{2}}\ X^{-\frac{1}{2}}(t')
\end{aligned} \tag{55}$$

Thus

$$\lim_{n\to\infty} \frac{1}{\rho n} \int^{\bar{Q}_n - G} dt(t-c)^{\frac{1}{2}}\ X^{-\frac{1}{2}}(t) = 0 \tag{56}$$

The t, t' contours in (56),(55) etc must be chosen to correspond to n running along the real axis. Suppose the half-periods of the elliptic integral are defined by

$$\omega_1 = \int_{a_1}^{a_2} dt\ Y^{-\frac{1}{2}}(t), \quad \omega_2 = \int_{a_1}^{a_3} dt\ Y^{-\frac{1}{2}}(t) \tag{57}$$

and that $\frac{1}{2}\rho = p_1\omega_1 + p_2\omega_2$, p_1, p_2 real. Then the path in the t-plane in (56) should encircle a_1a_2 roughly np_1 times and encircle a_2a_3 roughly np_2 times. If we write

$$\phi_1 = \int_{a_1}^{a_2} dt(t-c)^{\frac{1}{2}}\ X^{-\frac{1}{2}}(t), \quad \phi_2 = \int_{a_1}^{a_3} dt\ (t-c)^{\frac{1}{2}}\ X^{-\frac{1}{2}}(t) \tag{58}$$

then (56) becomes

$$p_1\phi_1 + p_2\phi_2 = 0 \tag{59}$$

Equation (59) constitutes an equation for c, and thus Q_{av}, which means that $\bar{Q}_n$ is determined (up to the addition of a constant to n) completely once (59) is solved.

Apart from the degenerate case $c=a_1$ or a_2 or a_3, we believe (59) has a unique solution. It obeys (see Appendix)

$$\mathrm{Re}\ \phi_1 = \mathrm{Re}\ \phi_2 = 0 \tag{60}$$

and we call c the 'center of capacity' for the triangle $a_1a_2a_3$. The connected set of minimum capacity including $a_1a_2a_3$ is $\mathrm{Re}\ \phi(t)=0$, where (see Nuttall [4])

$$\phi(t) = \int_{a_1}^{t} dt'(t'-c)^{\frac{1}{2}} X^{-\frac{1}{2}}(t') \tag{61}$$

6. Relation to the work of Laguerre

Laguerre [1] used the relation (4) to rewrite (6) in the form

$$Xp'_n = (\Omega_n - V)p_n + \theta_n p_{n-1} \tag{62}$$

where Ω_n, θ_n are polynomials of degree 2,1 respectively. We have set $V=\frac{1}{2}(Y-X')$.

A procedure similar to that used here leads to the relations (using our notation)

$$(t+Q_{n+1})(\Omega_{n+1}-\Omega_n) + \theta_{n+1} - \frac{R_{n+1}}{R_n}\theta_{n-1} = X \tag{63}$$

$$\Omega_{n+1} - \Omega_{n-1} = -\frac{(t+Q_n)\theta_{n-1}}{R_n} + \frac{(t+Q_{n+1})\theta_n}{R_{n+1}} \tag{64}$$

Equation (64) may be integrated to give

$$\Omega_n + \Omega_{n-1} = \frac{Q_{n-1}(t+Q_n)}{R_n} \tag{65}$$

If we differentiate (62) and use (4), (62) and (65), we obtain the following second order differential equation satisfied by p_n

$$\begin{aligned} X^2p''_n = {} & Xp'_n\left[\frac{X\theta'_n}{\theta_n} - 2V - X'\right] \\ & + p_n\left[X\left(\Omega'_n - V' - (\Omega_n - V)\frac{\theta'_n}{\theta_n}\right) + \left(\frac{\theta_n\theta_{n-1}}{R_n} + \Omega_n^2 - V^2\right)\right] = 0 \end{aligned} \tag{66}$$

We conclude that

$$\Omega_n^2 - V^2 + \frac{\theta_n\theta_{n-1}}{R_n} = XS_n \tag{67}$$

where S_n is a polynomial of degree 1. Further manipulation shows that

$$S_{n+1}-S_n = \frac{\theta_n}{R_{n+1}} \tag{68}$$

Laguerre [1] writes

$$\Omega_n = (n+\mu')t^2 + \alpha_n t + \beta_n \tag{69}$$

$$\theta_n = a_n t + b_n$$

where $\mu' = \mu - \frac{3}{2}$

By carrying out the procedure outlined at the beginning of the section, we find

$$\begin{aligned} a_n &= (A_n + A_{n-2})R_{n+1} \\ b_n &= (A_n Q_{n+2} + B_n + A_{n-2}Q_n + B_{n-1})R_{n+1} \\ \alpha_n &= A_n(Q_{n+2} + Q_{n+1}) + B_n \\ \beta_n &= A_n(Q_{n+2}Q_{n+1} + R_{n+2}) + B_n Q_{n+1} - A_{n-2}R_{n+1} - \tfrac{1}{2}\pi \end{aligned} \tag{70}$$

From (67) Laguerre [1] obtains the equations (choosing $\Sigma=0$ for convenience)

$$\begin{aligned} -R_n^{-1}\, a_n a_{n-1} &= \alpha_n^2 + 2(n+\mu')\beta_n - \pi n^2 + \ldots \\ -R_n^{-1}(a_n b_{n-1} + b_n a_{n-1}) &= 2\alpha_n\beta_n - 2\pi(n+\mu')\alpha_n + \tau n^2 + \ldots \\ -R_n^{-1} b_n b_{n-1} &= \beta_n^2 + 2\tau(n+\mu')\alpha_n + \ldots \end{aligned} \tag{71}$$

where in each case ... represents a polynomial in n of first degree.

If we make the approximations of (21) in (70), and use (26) we find that the $O(n^2)$ part of (71) is equivalent to the equations (27), (28) and (30) with the expressions (39)-(41) for H,J,K. This seems to be a simpler way to obtain these relations.

7. Asymptotic form of p_n

We can learn about the form of $p_n(t)$ for large n by studying the differential equation (66). For large n, the coefficients of p_n', p_n are of order 1, whereas the coefficient of p_n is of order n^2. The dominant part of the coefficient of p_n is XS_n. If we set $S_n=\gamma_n t+\delta_n$, then we find, by equating powers of t^4, t^3 in (67), that

$$\begin{aligned} (n+\mu')^2 - \mu'^2 &= \gamma_n \\ 2(n+\mu')\alpha_n + \text{const.} &= \delta_n - \Sigma\gamma_n \end{aligned} \tag{72}$$

For large n this gives

$$\gamma_n \approx n^2$$
$$\delta_n \approx n^2(2G+\Sigma), \qquad (73)$$

and so

$$S_n \approx n^2(t-c) \qquad (74)$$

The ansatz $p_n = e^{n\chi(t)}$ in (66) gives, on equating coefficients of n^2,

$$\chi'^2 = (t-c)\ \chi(t)^{-1} \qquad (75)$$

so that

$$\chi(t) = \pm\phi(t) + \text{const.} \qquad (76)$$

We conclude that the dominant part of the asymptotic form is

$$p_n(t) = \text{const. } e^{n\phi(t)} \qquad (77)$$

We expect from this argument that most of the zeros of $p_n(t)$ will occur near to where $|e^{n\phi}|=|e^{-n\phi}|$, i.e. $\mathrm{Re}\phi=0$, which is the set of minimum capacity including $a_1a_2a_3$.

A more detailed picture of $p_n(t)$ is obtained by setting

$$p_n(t) = g(t)\ e^{n\phi(t)} \qquad (78)$$

with g(t) of order 1. Substitution into (66) gives, on equating coefficients of n,

$$2g'/g = (t-z_n)^{-1}\left(1-\frac{t^2+Gt+H-2\bar{R}_{n+1}}{X\phi'}\right)+\frac{1}{X\phi'}(2t+G+\ell)-\frac{2V+X'}{X}-\phi''/\phi' \qquad (79)$$

where

$$z_n = -\frac{b_n}{a_n} \simeq \bar{Q}_{n+1} - G \qquad (80)$$

and ℓ is a linear function of t, that part of S_n that is proportional to n.

From (48), (49) and (50) we see that, when $t=\bar{Q}_{n+1}-G$, the coefficient of $(t-z_n)^{-1}$ in (79) is 0 or 2 depending on whether, as a solution of (52), $\bar{Q}_{n+1}-G$ lies on the first or second Riemann sheet. Integrating (79) shows that p_n will have a zero near z_n in the second case, but not in the first, always assuming that z_n is not on $\mathrm{Re}\phi=0$. Such a zero will give rise to a 'spurious' pole in the Padé approximant.

We have previously speculated (Nuttall [4]) that $F(t)=g(t)\exp(n\phi(t))$ is the solution of the following problem.

i) F(t) is single-valued and analytic outside the set of minimum capacity $\mathrm{Re}\phi = 0$.

ii) On the set of minimum capacity

$$\omega F_+F_- = \text{const.}(t-z_n)X^{\frac{1}{2}}(t)\ (t-c)^{\frac{1}{2}}$$

iii) $F(t) \sim t^n$ as $t \to \infty$

iv) Outside the set of minimum capacity $(t-z_n)/F(t)$ is analytic.

The coefficient of t in ℓ is known, but our arguments do not tell us how to find the coefficient of t^o in ℓ. However, we have shown that there is a choice of this unknown coefficient which gives a solution of (79) which satisfies the above conditions. This provides support for the speculation, which would carry more weight if the results obtained in this work could be made rigorous.

References

1. E. Laguerre, Journal de Math. 1, 135 (1885).
2. O. Perron, Die Lehre von dem Kettenbrüchen.
3. D. Bessis, "A new method in the combinatonics of the topological expansion" preprint CEN Saclay (1979).
4. J. Nuttall, "Sets of minimum capacity, Padé approximants and the bubble problem", Cargese Summer School on Bifurcation Phenomena and related topics (1979).
5. J. Nuttall and S.R. Singh, J. Approx. Theory 21, 1 (1977).

Appendix

We wish to show that, if c is chosen so that $p_1\phi_1+p_2\phi_2=0$, then $\mathrm{Re}\phi_1=\mathrm{Re}\phi_2=0$. We suppose that $c\neq a_1,a_2,a_3$. Define the elliptic function $\Theta(u)$ by

$$u = \int_{a_1}^{\Theta(u)} dt\ Y^{-\frac{1}{2}}(t) \tag{A1}$$

If $t=\Theta(u)$, we have

$$\frac{du}{dt} = Y^{-\frac{1}{2}}(t) \tag{A2}$$

Also, if for d constant

$$\psi(t) = \int_{a_1}^{t} dt'(t'-d)Y^{-\frac{1}{2}}(t') \tag{A3}$$

then

$$\frac{d\psi}{dt} = (t-d)Y^{-\frac{1}{2}}(t) = (t-d)\frac{du}{dt} \tag{A4}$$

so that

$$\frac{d\psi}{du} = t-d \tag{A5}$$

If we write $b=\frac{1}{2}\rho=\int_{a_1}^{\infty} dt\ Y^{-\frac{1}{2}}(t)$, then $u=\pm b$ corresponds to $t=\infty_1, \infty_2$ and it is not difficult to show that

$$t-d = \zeta(u+b) - \zeta(u-b) + C \tag{A6}$$

for some constant C. Integrating (A5) with (A6) gives

$$\psi = \ln \sigma(u+b) - \ln \sigma(u-b) + uC \tag{A7}$$

Now suppose d(and therefore C) is chosen so that at $t=a_2, a_3, \mathrm{Re}\psi=0$. We know (Nuttall and Singh [5]) that such a d is unique. Corresponding to $t=a_2, a_3$ we have $u=\omega_1, \omega_2$, and (A7) leads to, using the 'period' relations for the Weierstrass σ-function,

$$\mathrm{Re}\psi_1 = \mathrm{Re}(C\omega_1 + 2\eta_1(p_1\omega_1 + p_2\omega_2)) = 0 \tag{A8}$$

$$\mathrm{Re}\psi_2 = \mathrm{Re}(C\omega_2 + 2\eta_2(p_1\omega_1 + p_2\omega_2)) = 0$$

where $\eta_i = \zeta(\omega_i)$. Using the relation

$$\eta_1\omega_2 - \eta_2\omega_1 = i\pi/2 \tag{A9}$$

we find

$$C = -2(p_1\eta_1 + p_2\eta_2) \tag{A10}$$

We may write

$$\phi(t) = \psi(t) + (d-c)\int_{a_1}^{t} dt\ Y^{-\frac{1}{2}}(t) \tag{A11}$$

so that $p_1\phi_1 + p_2\phi_2 = 0$ becomes

$$(d-c)(p_1\omega_1 + p_2\omega_2) = 0$$

or

$$d = c\,. \tag{A12}$$

Supported in part by Natural Sciences and Engineering Research Council

Department of Physics
University of Western Ontario
London, Ontario, Canada. N6A 3K7

Multidimensional Hermite Interpolation and Padé Approximation

by

D.V. and G.V. Chudnovsky

Abstract. The paper considers the construction of multidimensional Padé approximation and estimates of the remainder function using the global Grothendieck residue symbol. The introduction of Grothendieck residu symbol provides the possibility of constructing effectively multidimensional Padé approximations.

The multidimensional Hermite interpolation problem is discussed. Then Padé approximation and Hermite interpolation formulae are applied to computation of Feynman integrals.

The connection between the computation of Feynman integrals and Riemann problems is pointed out.

§1. The formulation of the Riemann monodromy problem.

§2. An effective solution of Riemann problem in the terms of polylogarithmic functions by Lappo-Danilevsky.

§3. Polylogarithmic functions as the basis for representation of the Feynman integrals (examples of Veltman and t'Hooft).

§4. General problem of Padé approximation.

§5. Hermite and Lagrange interpolation formulae.

§6. Multidimensional interpolation. Grothendieck residue symbol.

§7. Padé approximation to polylogarithm functions.

Classical Riemann problem and the monodromy group in the one dimensional case.

The Riemann problem for $\mathbb{C} \cup \{\infty\}$ is always formulated for Fuchsian linear differential equations (i.e. for linear differential equations with regular singularities only).

Thus we start with a matrix differential equation having the simplest singularities (so-called canonical or normal form):

$$\frac{d\bar{y}}{dx} = \Sigma_{i=1}^{n} \frac{A_i}{x - a_i} \bar{y}, \tag{1.1}$$

where $\bar{y} = {}^t(y_1,\ldots,y_m)$ and A_i are $m \times m$ constant matrices: $i = 1,\ldots,n$.

The system (1.1) is the subject of all our discussions and applications. Also from the point of view of Riemann problem (1.1) is the best example for all the construction.

For an arbitrary system of linear differential equations the Riemann problem is formulated as follows.

Let

$$\frac{d\bar{y}}{dx} = A(x)\bar{y}, \quad \bar{y} = {}^t(y_1,\ldots,y_m) \tag{1.2}$$

be a system of equations with a matrix $A(x)$ with rational coefficients. Let $Y = Y(x)$ be a fundamental solution matrix of (1.2) and let

$$\{a_1,\ldots,a_n\}$$

be the set of poles of $A(x)$. In general the fundamental matrix $Y(x)$ is a multivalued function having $a_1,\ldots,a_n$ and $a_\infty = \infty$ as its branch points and

$$Y(x) \to Y(x)\,M(\gamma) \text{ when prolonged along } \gamma.$$

Here $M(\gamma) \in GL(m,\mathbb{C})$ and $M(\gamma)$ depends only on the homotopy class of γ. Then, of course,

$$M(\gamma_1\gamma_2) = M(\gamma_1)M(\gamma_2),$$

so the set $\mathfrak{M} = \{M(\gamma)\}$ is a group. It is called the monodromy group of the equation (1.2).

This group is generated by the matrices $M_i = M(\gamma_i)$: $i = 1,\ldots,n$, where γ_i is a clockwise circuit around a_i which does not contain other singular points inside and we have

$$M_1\ldots M_n M_\infty = I. \tag{1.3}$$

§1. Classical Riemann Problem:

1.1 Given branch points $a_1,\ldots,a_n \in \mathbb{C}$ and matrices $M_1,\ldots,M_n \in GL(m,\mathbb{C})$ find a linear differential equation (1.2) or, better, (1.1) whose monodromy group M coincides with the group generated by $M_1,\ldots,M_n$.

We also naturally demand in the classical Riemann problem that the fundamental matrix Y(x) of (1.2) to be at most regularly singular at the points $a_1,\ldots,a_n,\infty$. This means that

$$(1.4)\qquad Y(x) = H_i(x)(x-a_i)^{-L_i}: i = 1,\ldots,n,\infty$$

(natural parameter near ∞ is $1/x$). In (1.4) $H_i(x)$ is an invertible holomorphic matrix at $x = a_i$ and L_i (the exponent) is a constant matrix such that

$$(1.5)\qquad e^{2\pi i L_j} = M_j: j = 1,\ldots,n.$$

Since Riemann-Hilbert this problem was solved in different ways by Plemelj, Birkhoff (reduction to the singular integral equations); by Röhrl (for an arbitrary Riemann surface using fiber bundles); by Lappo-Danielevsky (using series expansions in hyper-logarithms). From our present point of view the most important was the reduction of the linear Riemann problem to a non-linear "completely integrable" system of equations, namely Schlesinger's equations.

1.2 Schlesinger's Theorem: Let $Y = Y(x_0;x)$ be the fundamental matrix solution of (1.1); $Y(x_0;x_0) = I$. The necessary and sufficient conditions for $A_i: i = 1,\ldots n$ as the function of the parameters $a_1,\ldots,a_n,x_0$ to have the fixed monodromy group of Y(x) is the following completely integrable system of total differential equations

$$(1.6)\qquad dA_j = -\Sigma_{i\neq j}\ [A_j,A_i]d\log\frac{a_j-a_i}{x_0-a_i}: j = 1,\ldots,n.$$

This Schlesinger system can be written in a classical form [1]:

$$(1.7)\qquad \begin{aligned} \frac{\partial A_j}{\partial a_i} &= \frac{[A_i,A_j]}{a_j-a_i}: i \neq j,\quad i,j = 1,\ldots,n \\ \Sigma_{j=1}^n \frac{\partial A_j}{\partial a_i} &= 0: i = 1,\ldots,n. \end{aligned}$$

§2. Backgrounds and the analytic construction of the Schlesinger's system.

Let $a_1,\dots,a_n$ and x_0 now be distinct points on $\mathbb{PC}^1$ and let $L_1,\dots,L_n$ be $m \times m$ matrices satisfying the natural condition

$$(2.1)\qquad e^{2\pi i L_1}\dots e^{2\pi i L_n} = 1.$$

We consider the following precise version of the Riemann problem: Find a matrix $Y(x)$ with the properties

a) $Y(x)$ is a multi-valued analytic matrix on $\mathbb{PC}^1\setminus\{a_1,\dots,a_n\}$;

b) $Y(x) = H_i(x)\cdot(x-a_i)^{-L_i}$ at $x = a_i$ $(i = 1,\dots n)$, where $H_i(x)$ is an invertible holomorphic matrix at $x = a_i$;

c) $\det Y(x) \neq 0$ for $x \neq a_1,\dots,a_n$;

d) $Y(x_0) = 1$.

Such a matrix $Y(x)$ is unique (but does not always exist). Lappo-Danielevsky have proved that for sufficiently small $|L_i|: i = 1,\dots,n$ (excluding $i = \infty$) such a matrix

$$y = Y(x_0;x;{}^{a_1,\dots\, a_n}_{L_1,\dots,L_n})$$

exists and can be written as a series in $L_i: i = 1,\dots,n$.

Let us suppose now that $a_i \neq \infty: i = 1,\dots,n$. Then the function $Y(y;x;{}^{a_1\dots a_n}_{L_1\dots L_n})$ gives us simultaneously 1) the Fuchsian linear differential system with the monodromy group, generated by $\{e^{2\pi i L_1},\dots,e^{2\pi i L_n}\}$ and 2) the Schlesinger system of the equation (1.6) for the coefficients $A_1,\dots A_n$ of (1.1).

We can write this for a function

$$Y = Y(y;x;{}^{a_1\ \dots,a_n}_{L_1,\dots,L_n})$$

as a linear total differential equation

$$dY = \Omega Y$$

$$(2.2)\qquad \Omega = \Sigma_{i=1}^n A_i\ d.\log\frac{x-a_i}{y-a_i} = \Sigma_{i=1}^n A_i\left(\frac{d(x-a_i)}{x-a_i} - \frac{d(y-a_i)}{y-a_i}\right),$$

where

$$(2.3)\quad A_i = A_i(y;{}^{a_1,\dots,a_n}_{L_1,\dots,L_n}) = -H_i(a_i)L_iH_i(a_i)^{-1}: i = 1,\dots,n$$

are matrices independent of x satisfying

$$\Sigma_{i=1}^n A_i = 0.$$

The equation (2.2) means that Y as a function of x satisfies the Fuchsian system of linear ordinary differential equations

$$\frac{dY}{dx} = \Sigma_{i=1}^n \frac{A_i}{x-a_i} Y.$$

Now the coefficients A_i: $i = 1,\dots,n$ as functions of $y = x_0$ and $\bar{a} = (a_1,\dots,a_n)$ satisfy the Schlesinger's equation (1.6):

$$dA_j = -\Sigma_{j\neq i}\ [A_j,A_i]d\log\frac{a_i-a_j}{a_i-y}: j = 1,\dots,n.$$

Sato + Miwa + Jimbo [2] in a series of papers describe a quantum field theory approach of the construction of $Y(y;x;{}^{a_1,\dots,a_n}_{L_1,\dots,L_n})$ and the solution of the Riemann problem. They represent Y and A_i in the terms of classical field theory operators and then apply such representations to the explicit expressions for the n-th correlation functions in the two-dimensional Ising model. It should be noted, however, that the series expansions for $Y(y;x;{}^{a_1,\dots,a_n}_{L_1,\dots,L_n})$ proposed by Sato + Miwa + Jimbo are basically the same as in the papers of Lappo-Danielevsky [3].

§3.

There exists an effective method for solving the Riemann problem and for explicit construction of the fundamental solution in terms of the monodromy group and the singularities. This method, belonging to Poincaré and Lappo-Danielevski, uses matrix series in terms of polylogarithmic functions.

These polylogarithm functions were introduced by Poincaré and studies by Lappo-Danilevsky in the connection with the Fuchsian equation. However these functions were known already for a long time (especially dilogarithms were known to Euler and then studied e.g. by Abel and Lobachevsky).

In the notation of Lappo-Danilevsky for a fixed x_0 we define these polylogarithms functions by induction:

$$L_{x_0}(a_{j_1},|x) = \int_{x_0}^{x} \frac{dx}{x-a_{j_1}} = \log \frac{x-a_{j_1}}{x_0-a_{j_1}};$$

$$L_{x_0}(a_{j_1},\ldots,a_{j_\nu}|x) = \int_{x_0}^{x} \frac{L_{x_0}(a_{j_1},\ldots,a_{j_{\nu-1}}|x)}{x - a_{j_\nu}} dx.$$

Then we can define a fundamental matrix $Y_{x_0}(x)$ as follows:

$$(3.1)\quad Y_{x_0}(x) = \varphi_{x_0}\begin{pmatrix} A_1,\ldots,A_n \\ a_1,\ldots,a_n \end{pmatrix} x\Big)$$

$$= I + \Sigma_{\nu=1}^{\infty} \Sigma_{(j_1,\ldots,j_\nu)}^{1,\ldots,n} A_{j_1},\ldots,A_{j_\nu} L_{x_0}(a_{j_1},\ldots,a_{j_\nu}|x).$$

The series defines an entire function of A_i and is uniformly convergent with respect to x in any finite domain $\mathfrak{D}$ (of Riemann surface corresponding to cuts $(a_1,\infty);\ldots;(a_n,\infty)$) having no points a_j in $\mathfrak{D}$ or on $\partial\mathfrak{D}$-boundary of $\mathfrak{D}$.

The series (3.1) is chosen in such a way that $Y_{x_0}(x)$ satisfies a linear differential equation with regular singularities at a_i and coefficients A_i:

$$(3.2)\qquad \frac{dY_{x_0}}{dx} = \Sigma_{j=1}^{n} \frac{Y_{x_0} A_j}{x - a_j} \quad \text{and} \quad Y_{x_0}(x_0) = I.$$

It is now possible to rewrite the monodromy matrices M_i (or integral substitutions in the terminology of Lappo-Danielevsky) in terms of matrix series of polylogarithms. For this let us consider a closed contour γ_i from x_0 traveling around a_i and returning back to x_0. We put

$$P_j(a_{j_1}|x_0) = \int_{\gamma_i} \frac{dx}{x - a_{j_1}} = \begin{cases} 2\pi i: & j = j_1 \\ 0 & : j \neq j_1 \end{cases}$$

$$P_j(a_{j_1},\dots,a_{j_\nu}|x_0) = \int_{\gamma_i} \frac{L_{x_0}(a_{j_1},\dots,a_{j_{\nu-1}}|x)}{x - a_{j_\nu}} dx.$$

An "integral substitution" M_i corresponding to a_i of a regular matrix Y normed at x_0 and having the coefficients of the differential equations A_j at singularities a_j: $j = 1,\dots n$ is represented as an entire function of the coefficients $A_1,\dots,A_n$:

(3.3) $$M_j = I + \Sigma_{\nu=1}^{\infty} \Sigma_{(j_1,\dots,j_\nu)}^{1,\dots,n} A_{j_1},\dots,A_{j_\nu} P_j(a_{j_1},\dots,a_{j_\nu}|x_0).$$

Now $\{M_j\}$ gives us the monodromy group of the equation (3.2). In order to look on the behavior of the fundamental matrix $Y_{x_0}(x)$ of (3.2) defined in (3.1) it is better instead of monodromy matrices M_i to consider the corresponding exponents L_i: $i = 1,\dots,n$.

For this, following Lappo-Danilevsky, we suppose first that A_i: $i = 1,\dots,n$ are in the neighborhood of a zero matrix.

Then the corresponding monodromy matrices $M_1,\dots,M_n$ from (3.3) belong to the neighborhood of unit matrix I and we can introduce "exponents" $L_1,\dots,L_n$ as

(3.4) $$L_j = \ell n\, M_j = \frac{1}{2\pi i} \Sigma_{\nu=1}^{\infty} \frac{(-1)^{\nu-1}}{\nu} (M_j - I)^{\nu}$$

and by (3.3) they are holomorphic functions of $A_1,\dots,A_n$ in that neighborhood of a zero matrix: $j = 1,\dots,n$.

Moreover, Lappo-Danilevsky have shown how to present L_j as an entire function of $A_1,\dots,A_n$ divided by another entire function of $A_1,\dots,A_n$, i.e. a meromorphic function. Moreover, it follows that L_j as a function of A_i has a singularity iff among the eigenvalues of A_ν there are such two that their difference is a non-zero integer.

Now in terms of the "exponents" L_j: $j = 1,\dots,n$ (3.4) we can present precisely the monodromy properties of the fundamental solution $Y_{x_0}(x)$. We have for $Y_{x_0}(x)$ from (3.1):

(3.5) $$Y_{x_0}(x) = \varphi_{x_0}\left(\begin{matrix}A_1 \ldots, A_n \\ a_1, \ldots, a_n\end{matrix}\Big| x\right)$$

$$= \left(\frac{x-a_j}{x_0-a_j}\right)^{L_j} \tilde{\varphi}^{(j)}_{x_0}\left(\begin{matrix}A_1, \ldots, A_n \\ a_1, \ldots, a_n\end{matrix}\Big| x\right) : j = 1, \ldots, n,$$

here

(3.6) $$\tilde{Y}^{(j)}_{x_0}(x) = \tilde{\varphi}^{(j)}_{x_0}\left(\begin{matrix}A_1, \ldots, A_n \\ a_1, \ldots, a_n\end{matrix}\Big| x\right)$$

and $[\tilde{Y}^{(j)}_{x_0}(x)]^{-1}$ are holomorphic with respect to x at $x = a_j$: $j = 1, \ldots, n$. If now $A_1, \ldots, A_n$ are in the neighborhood of the zero matrix, then there exist <u>no</u> <u>other</u> exponents L_j, different from those given by formula (3.4) with the property that in (3.5)-(3.6) both

$$[\tilde{Y}^{(j)}_{x_0}(x)] \quad \text{and} \quad [\tilde{Y}^{(j)}_{x_0}(x)]^{-1}$$

are holomorphic at $a_j = x$: $j = 1, \ldots, n$.

In other words, the fundamental matrix $Y_{x_0}(x)$ defined in (3.1) has indeed regular singularities at $x = a_j$: $j = 1, \ldots, n$ and exponents $L_1, \ldots, L_n$ from (3.4) and is, what we called before

$$Y\left(x_0; x; \begin{matrix}a_1, \ldots, a_n \\ L_1, \ldots, L_n\end{matrix}\right).$$

We now observe that the formalism of Lappo-Danilevsky also solves the Riemann monodromy problem. Instead of constructing the fundamental solution $Y_{x_0}(x)$ and "exponents" L_j as meromorphic functions of the coefficients $A_1, \ldots, A_n$ and $a_1, \ldots, a_n$; we can invert the problem and represent the coefficients $A_1, \ldots, A_n$ of the differential equation (3.2) and the fundamental solution $Y_{x_0}(x)$ of (3.2) in terms of composition (matrix) series in exponents $L_1, \ldots, L_n$ and $a_1, \ldots, a_n$. (So-called "effective" solution of Riemann-Hilbert problem).

This problem is solved by an inversion of the previous series. We start with the exponent matrices $L_1, \ldots, L_n$ in the neighborhood of zero matrix. Then inverting the previous series we obtain the following representations for the coefficients $A_1, \ldots, A_n$ (again close to a zero matrix):

$$(3.7)\quad A_j(x_0) = \Sigma_{\nu=1}^{\infty} \Sigma_{(j_1,\ldots,j_\nu)}^{(1,\ldots,n)} L_{j_1} \ldots L_{j_\nu} R_j(a_{j_1},\ldots,a_{j_\nu}|x_0).$$

Here the coefficients $R_j(a_{j_1},\ldots,a_{j_\nu}|x_0)$ are defined by induction in terms of the values of the polylogarithm functions $P_j(\ldots|\ldots)$:

$$R_j(a_{j_1}|b) = \delta_{jj_1},$$

$$R_j(a_{j_1},\ldots,a_{j_\nu}|x_0)$$

$$= - \Sigma_{\mu=2}^{\infty} \Sigma_{(h_1,\ldots,h_\mu)}^{(1,\ldots,n)} \Sigma_{1\leq k_1<\ldots<k_{\mu-1}<\nu} \times$$

$$\times R_{h_1}(a_{j_1},\ldots,a_{j_{k_1}}|x_0) R_{h_2}(a_{j_{k_1+1}},\ldots,a_{j_{k_2}}|x_0)\ldots$$

$$\ldots\times R_{h_\mu}(a_{j_{k_{\mu-1}+1}},\ldots,a_{j_\nu}|x_0) Q_j(a_{h_1},\ldots,a_{h_\mu}|x_0),$$

for

$$Q_j(a_{j_1},\ldots,a_{j_\nu}|x_0) = \frac{1}{2\pi i} \Sigma_{\mu=1}^{\nu} \Sigma_{0<k_1<\ldots<k_{\mu-1}<\nu} \frac{(-1)^{\mu-1}}{\mu} \times$$

$$\times P_j(a_{j_1},\ldots,a_{j_{k_1}}|x_0)\ldots P_j(a_{j_{k_{\mu-1}+1}},\ldots,a_{j_\nu}|x_0).$$

The series (3.7) are holomorphic functions of the exponents $L_1,\ldots,L_n$ in the neighborhood of zero matrix. Substituting (3.7) into (3.1), one now obtains effectively the whole fundamental solution

$$Y(x_0;x;\begin{matrix}a_1,\ldots,a_n\\ L_1,\ldots,L_n\end{matrix})$$

of the Riemann-Hilbert monodromy problem (together with the corresponding equations (3.2)).

§4.

Such polylogarithmic functions, as we see, form natural basis for representation of solutions of Fuchsian equations. Moreover it so happens that in many interesting physical applications the corresponding series terminates and solutions are presented as a finite linear combination of polylogarithmic functions. Conditions for termination can be in principle expressed in terms of the monodromy group only.

However the most important problem in these cases is the effective expression of corresponding functions as linear combinations of polylogarithms and determination of the corresponding Riemann surface.

The functions, for which the people are looking for such representations are Feynman integrals. The presence of polylogarithmic functions immediately follows from the occurence (rather often) of such constants as

$$\log 2 = L_1(-1), \ \frac{\pi^2}{6} = \zeta(2) = L_2(1) \quad \zeta(3) = L_3(1).$$

See for example the paper of Levine and Wright [4] on approximate calculation of the sixth-order magnetic moment of the electron.

There is not only numerical but also analytical evidence of this, see e.g. the papers, that were among the first in this subject, of Karplus and Neuman [5], where certain Feynman integrals were expressed explicitly as dilogarithms. However the major progress was made only recently by Veltman and t'Hooft who represented rather general class of Feynman integrals as combination of different logarithm and dilogarithm functions of parameters.

Let us present the corresponding results contained in the paper of Veltman and t'Hooft "Scalar one-loop integrals", [6].

The basic integral that was treated by them is the following nasty

(4.1) $D(p_1,p_2,p_3,p_4,m_1,m_2,m_3,m_4) =$

$$= \int d_n q \frac{1}{(q^2+m_1^2)((q+p_1)^2+m_2^2)((q+p_1+p_2)^2+m_3^2)((q+p_1+p_2+p_3)^2+m_4^2)}$$

This integral is naturally rewritten in the following form using Feynman variables

$$D = i\pi^2 \int d_4 u \frac{\delta(\Sigma u-1)\theta(u_1)\theta(u_2)\theta(u_3)\theta(u_4)}{[\Sigma m_i^2 u_i + \Sigma_{i<j} p_{ij}^2 u_i u_j]^2} \tag{4.2}$$

with constants p_{ij} in natural way connected with p_i.

After the linear change of variable we obtain the following integral, which is the most pleasant from mathematical point of view

$$(4.3)\quad \frac{D}{i\pi^2} = \int_0^1 dx \int_0^x dy \int_0^y dz\,[ax^2 + by^2 + gz^2 + cxy + hxz + jyz + dx + cy + kz + f]^{-2}$$

Now we can make in the integral (4.3) some projective transformations in order to kill the coefficients of z^2, xz, yz. It is possible after this to integrate over z. The integral becomes much easier and can be written in the form

$$(4.4)\quad \frac{D}{i\pi^2} = A\int_0^1 dx \int_0^x dy\, \frac{1}{k}[\{ax^2 + by^2 + cxy + dx + ey + f\}$$

$$- \{ax^2 + by^2 + cxy + dx + (e+k)y + f\}^{-1}].$$

Now the integration over x, which can be performed due to the following

Remark: Consider the integral

$$\int_0^1 dx \int_0^x dy \frac{1}{[ax^2 + by^2 + cxy + \ldots]}.$$

By the substitution $y = y + \alpha x$ we can get rid of the term x^2, if we choose α as a root of the equation $b\alpha^2 + c\alpha + a = 0$. Then this integral is transformed into

$$\int_0^1 dx \int_{-\alpha x}^{(1-\alpha)x} dy \frac{1}{[by^2 + (c+2\alpha)xy + \ldots]}.$$

Now in order to reduce the last integral to the integral from logarithm, i.e. to the dilogarithm function, we need only to perform such change of variables

$$\int_0^1 dx \int_{-\alpha x}^{(1-\alpha)x} dy = \int_0^1 dx \int_0^{(1-\alpha)x} dy - \int_0^1 dx \int_0^{-\alpha x} dy = \int_0^{(1-\alpha)} dy \int_{\frac{y}{1-\alpha}}^1 dx - \int_0^{-\alpha} dy \int_{\frac{y}{-\alpha}}^1 dx.$$

Now the integration over x can be performed. Using this remark the integral (4.4) can be reduced to the sum of 12 integrals in the logarithms of quadratic polynomials in y.

In other words the integral (4.4) and so the initial integral (4.1) can be represented as a sum of 24 dilogarithms. This number cannot be decreased. However a lot of problems remain even with such a representation, because for complex parameters p_i, m_i different dilogarithms have different domains of definition, sometimes non-intersecting. Veltman and t'Hooft have shown that in order to have a consistent representation it is necessary to have 128 dilogarithms. The dilogarithm here is taked in the following form

$$L_2(x) = -\int_0^x \frac{\ln(1-t)}{t} dt = -\int_0^1 dt\, \frac{\ln(1-xt)}{t}.$$

Remark: The story is not finished here but only beginning. How does one compute dilogarithmic and polylogarithmic functions taking into account that the corresponding series converge very slowly? For this we use Padé approximation.

§5.

It is just a good time to give a general definition of N-point Padé approximation.

Definition 5.1: Let $z_1,\ldots,z_n$ be N distinct points in $\mathbb{C}^1$ (or $\mathbb{C}^k$) and $f_1(z),\ldots,f_n(z)$ be functions analytic in the neighborhood of $z = z_1,\ldots,z = z_n$. For n non-negative integers $m_1,\ldots,m_n$, we consider such polynomials $P_1(z),\ldots,P_n(z)$ of degrees $\leq m_1,\ldots, \leq m_n$, respectively, that the function

$$R(z) = P_1(z)f_1(z) +\ldots+ P_n(z)f_n(z)$$

has zeroes at z_i: $i = 1,\ldots,N$:

$$\Sigma_{i=1}^N \operatorname{ord}_{z_i}(R) \geq M,$$

where

$$M = \Sigma_{i=1}^n (m_i + 1) - 1$$

in the $\mathbb{C}^1$-case and

$$M = \Sigma_{i=1}^{n} \binom{m_i + k}{k} - 1$$

in the $\mathbb{C}^k$-case.

Then $(P_1(z),\dots,P_n(z))$ is called an $(m_1,\dots,m_n)$ Padé approximation to $(f_1(z),\dots,f_n(z))$ at the points $\{z_1,\dots z_N\}$. A Padé approximation is perfect, if always $\Sigma_{i=1}^{N} \text{ord}_{z_i}(R) = M$ and almost perfect if $|\Sigma_{i=1}^{N} \text{ord}_{z_i}(R) - M| \le C(\bar{f})$ for some absolute constant $C(\bar{f}) \ge 0$.

We consider the sequence of divided differences, corresponding to a given function $f(x)$. Let $\lambda_1,\lambda_2,\lambda_3,\dots$ be the sequence of distinct points; we set

$$[\lambda_1]_f = f(\lambda_1);$$

$$[\lambda_1,\lambda_2]f = \frac{f(\lambda_1) - f(\lambda_2)}{\lambda_1 - \lambda_2}$$

and inductively

$$[\lambda_1,\dots,\lambda_{k+1}]_f = \frac{[\lambda_1,\dots,\lambda_k]_f - [\lambda_2,\dots,\lambda_{k+1}]_f}{\lambda_1 - \lambda_{k+1}} \dots.$$

Of course, if $f(x)$ is smooth enough (e.g. C^∞), then this formula becomes a continuous function of the variables λ_i. In this case we can write an expression for divided differences when some of λ_i are equal. Suppose that in sequence

$$(x_1,\dots,x_n)$$

we have only m distinct numbers $\xi_1,\dots,\xi_m$, where ξ_j is repeated ν_j times: $j = 1,\dots m$;

$$\nu_1 + \dots + \nu_m = n$$

and

$$P(x) = \Pi_{i=1}^{n}(x-x_i) = \Pi_{j=1}^{m}(x-\xi_j)^{\nu_j}.$$

Then

$$[x_1,\ldots,x_n]_f = \Sigma_{j=1}^{m} \frac{1}{(\nu_j-1)!}\left(\frac{\partial}{\partial x}\right)^{\nu_j-1}\left\{\frac{f(x)(x-\xi_j)^{\nu_j}}{P(x)}\right\}\Big|x = \xi_j.$$

In this case we have the following useful results (Hermite interpolation):

Lemma 5.2: <u>For sufficiently smooth</u> $f(x)$ <u>and arbitrary</u> $\lambda_1,\ldots,\lambda_n$ <u>we have the following representation</u>

$$f(x) = \Sigma_{i=0}^{n-1}[\lambda_1,\ldots,\lambda_{i+1}]_f\ P_i(x) + R_n(x),$$

where

$$R_n(x) = [\lambda_1,\ldots,\lambda_n,x]_f\ P_n(x)$$

and

$$P_0(x) = 1, \quad P_1(x) = x - \lambda_1,\ldots,P_k(x) = \Pi_{j=1}^{k}(x-\lambda_j),\ldots.$$

In particular, $R_n(x)$ vanishes at any point λ_i: $i = 1,\ldots,n$ with the multiplicity equal to the number of occurrence of λ_i in $\lambda_1,\ldots,\lambda_n$.

In other words we obtain closed expression for polynomial Padé approximation to $f(x)$ at any set of points; this Padé approximation is, of course, unique.

Under some conditions on $f(x)$ and $\lambda_1,\ldots,\lambda_n,\ldots$ we can prove that for an infinite system of points $\lambda_1,\ldots,\lambda_n,\ldots$ the series

$$\Sigma_{i=1}^{\infty}[\lambda_1,\ldots,\lambda_{i+1}]_f\ P_i(x)$$

converges to $f(x)$. For an entire function $f(x)$ this can be uniform convergence; for meromorphic ones only convergence in measure.

In particular, if $f(z)$ is not a polynomial, then $a_n \neq 0$ for infinitely many n.

Now we realize that in the special case

$$f(z) = e^{xz}$$

the expression $a_{k-1}(x)$,

$$a_{k-1}(x) = \frac{1}{2\pi i} \int_C \frac{e^{xz}dz}{P_k(z)}$$

is exactly the function $R(x)$ of Hermite constructed for the Padé approximation to $(e^{\omega_1 x},\ldots,e^{\omega_m x})$. This corresponds to the case

$$K = \Sigma_{i=1}^{m}(n_i+1) = N + 1$$

when we take for $\{z_1,\ldots,z_{N+1}\}$ the set containing $n_i + 1$ times ω_i: $i = 1,\ldots,m$:

$$(\underbrace{\omega_1,\ldots,\omega_1}_{n_1+1};\ldots;\underbrace{\omega_m,\ldots,\omega_m}_{n_{m+1}})$$

Why should the coefficient $a_{k-1}(x)$ be the remainder function for one-point Padé approximation?

Let us present the Hermite interpolation formula in terms of the residue theorem. Let our function $f(x)$ be regular in a domain D of the complex z-plane and n points $z_1,\ldots,z_n$ are given. We want (as in all approximation problems) to determine a polynomial $H_{n-1}(z)$ of degree $\leq$ n-1 such that the fraction

$$T_n(z) = \frac{f(z) - H_{n-1}(z)}{\Pi_{i=1}^{n}(z-z_i)}$$

is regular in D. It's clear that the solution is unique. We put

$$P_k(z) = (z-z_1)\ldots(z-z_k): k = 0,1,\ldots,n.$$

Then

$$P_k(z) = (z-z_k)P_{k-1}(z)$$

and

$$-P_k(\zeta)+(z-z_k)P_{k-1}(\zeta) = (z-\zeta)P_{k-1}(\zeta),$$

so

(5.1) $$\frac{1}{z-\zeta}\left(\frac{P_{k-1}(\zeta)}{P_{k-1}(z)} - \frac{P_k(\zeta)}{P_k(z)}\right) = \frac{P_{k-1}(\zeta)}{P_k(z)}.$$

Suppose that ζ lies in D and consider a simply closed curve C in D whose interior lies in D and contains the $n + 1$ points $z_1,\ldots,z_n,\zeta$. We define then

(5.2) $$a_{k-1} = \frac{1}{2\pi i}\int_C \frac{f(z)}{P_k(z)}dz: \quad k = 1,\ldots,n$$

and

(5.3) $$R_k(\zeta) = \frac{1}{2\pi i}\int_C \frac{P_k(\zeta)}{P_k(z)} \cdot \frac{f(z)}{z-\zeta}dz.$$

Then, by the Cauchy theorem,

$$R_0(\zeta) = f(\zeta)$$

and by (5.1) - (5.3) we have

$$R_{k-1}(\zeta) - R_k(\zeta) = a_{k-1}P_{k-1}(\zeta): \; k = 1,\ldots,n.$$

Adding these expressions we obtain, at last,

(5.4) $$H_{n-1}(\zeta) = a_0P(\zeta) +\ldots+ a_{n-1}P_{n-1}(\zeta)$$

and

$$f(\zeta) = H_{n-1}(\zeta) + R_n(\zeta),$$

where

(5.5) $$T_n(\zeta) = \frac{R_n(\zeta)}{P_n(\zeta)} = \int_C \frac{f(z)}{P_n(z)}\frac{dz}{z-\zeta}.$$

Here $H_{n-1}(z)$ is an approximation polynomial and $R_n(z)$ is the remainder function. If the sequence $z_1,\ldots,z_n,\ldots$ is an infinite one and $\lim_{n\to\infty} R_n(z) = 0$ for all $z \in D_0 \subset D$, then

$$f(z) = a_0P_0(z) + a_1P_1(z) +\ldots \qquad (z \in D_0).$$

The connection between N-point Padé for one function and one-point Padé for many functions lies in the following trivial one-dimensional (in $\mathbb{C}^1$).

Corollary: Let f(z) be regular in the interior of the closed curve C; where the interior of C contains the origin z = 0 and n points $z_1,\ldots,z_n$. Then for the function

$$R(x) = \frac{1}{2\pi i}\int_C \frac{f(xz)}{P_n(z)}dz,$$

(5.6)

$$P_n(z) = \Pi_{i=1}^n (z-z_i),$$

R(x) has a zero at x = 0 of multiplicity $\geq n - 1$.

Also we have the finite expression for R(z) in terms of $f(xz_i)$ and their derivatives. Let $\{z_1,\ldots,z_n\}$, as before, consist of m different points $\omega_1,\ldots,\omega_m$; where ω_j is repeated $n_j + 1$ times: $j = 1,\ldots,m$:

$$P_n(z) = \Pi_{i=1}^n (z-z_i) = \Pi_{j=1}^m (z-\omega_j)^{n_j+1}.$$

Then we have

$$R(x) = \Sigma_{j=1}^m \frac{1}{n_j!}\left(\frac{\partial}{\partial z}\right)^{n_j}\left\{\frac{f(xz)\cdot(z-\omega_j)^{n_j+1}}{P_n(z)}\right\}\Big|_{z=\omega_j}$$

Now R(x) is a linear combination of $\frac{\partial^k}{\partial z^k} f(\omega_j x)\cdot x^k$.

§6.

Let us turn now to the multidimensional interpolation problems.

The situation in the n-dimensional space $P\mathbb{C}^n$ (or simply $\mathbb{C}^n$) is totally different from that in the one-dimensional case.

Let us start with the finite set S of points in $\mathbb{C}^n$. In order to start with the interpolation problem in $\mathbb{C}^n$, corresponding to a set S we need first of all to have some natural representation of S as a set of zeroes of some algebraic expression.

For n = 1 this is trivial: if $S = \{x_1,\ldots,x_n\}$, then $P_S(x) = \Pi_{i+1}^n (x-x_i)$ is the unique (up to multiplicative constant) polynomial of degree n having x_i as zeroes of multiplicity (≥ 1)

counted as the time of occurence of x_i at S.

This simple fact is the basis for all interpolation methods and formulae. However for $\mathbb{C}^k$: $k > 1$ the situation is changed. As codimension of S in $\mathbb{C}^k$ is k, we can try to represent S only as an intersection of k hypersurfaces

$$P_1 = 0, \ldots, P_k = 0$$

for $P_i \in \mathbb{C}[x_1, \ldots, x_k]$: $i = 1, \ldots, k$. How does one find such P_i? Do they exist at all?

Let us take $k = 2$, e.g., the two curves $f = 0$, $g = 0$ of degrees n, m. Then, by Bezout's theorem, if these curves have no common components, then the <u>number</u> of points of intersections (counted <u>with</u> multiplicities) is mn (in $P\mathbb{C}^2$).

There arises an immediate difference with the 1-dimensional situation:

a) only that set S can be represented as a set of <u>simple</u> intersections of two curves of degrees n,m, if $|S| = n.m$;

b) moreover only <u>some</u> sets S, $|S| = n.m$, can be written in the form of simple intersections of curves of degrees n and m.

In order that mn given points S should be the intersections of two curves of orders m and n, with $m > n$, the coordinates of the mn must be connected by $mn - 3n + 1$ conditions; when $m = n$, they must be connected by $n^2 - 3n + 2$ conditions.

E.g. 9 points, to be common to two cubic must have their coordinates subject to 2 conditions.

However it's possible to present <u>any</u> set S in $\mathbb{C}^n$ as an intersection of n hypersurfaces (though not only as simple intersections). We have

Theorem 6.1: <u>Let</u> S <u>be a inifinite in</u> $\mathbb{C}^n$. <u>Then these are</u> n <u>hypersurfaces of degrees</u> $\leq |S|$:

$$P_1 = 0, \ldots, P_n = 0,$$

$P_i \in \mathbb{C}[x_1, \ldots, x_n]$: $i = 1, \ldots, n$, <u>such that the set</u> S <u>is the set</u>-

<u>theoretic intersection of the hypersurfaces</u> $P_1 = 0,\dots,P_n = 0$.

The proof (rather simple and based on the dimensional considerations mainly) was proposed as an answer to our question by J.P. Serre as an extension of the work of Mme Poitou.

We know now that the degrees of P_i can be dropped further: $d(P_i) \underset{n}{\ll} \sqrt[n]{|S|}$ and we can take P_i in such a form that $P_1 = 0,\dots,P_n = 0$ have non-simple intersection only at <u>one</u> of the points of S ("exceptional" point) [7].

Now instead of polynomials $P_S(x)$ for $\mathbb{C}^1$ we use as a basis set of interpolation polynomials the following expressions

$$P_1^{i_1} \dots P_n^{i_n} \quad : \quad i_1 \geq 0,\dots,i_n \geq 0.$$

The first non-trivial attempt to do something with the interpolation in two-dimensional case belongs to Angelesco [8] where he considered the case of the intersections of two quadratic polynomials.

Now we can solve this problem completely. For this it should be noted that the polynomials $P_1(x_1,\dots,x_n),\dots,P_n(x_1,\dots,x_n)$ are polynomials in the coordinates of the points of S. Now let's take Z as a 0-cycle (divisor of codim k in $P\mathbb{C}^k$) corresponding to an element $\vec{x}_1,\dots,\vec{x}_m$ of S:

$$Z = \tilde{m}_1\vec{x}_1 + \dots + \tilde{m}_m \cdot \vec{x}_m.$$

Then Z can be represented for <u>some</u> values of $\tilde{m}_i$ corresponding to the multiplicities of the intersections of $P_1 = 0,\dots,P_n = 0$ at $\bar{x}_i$ as the intersection of hypersurfaces

$$R_1 = 0,\dots,R_n = 0,$$

where the multiplicity of the intersection of $R_1 = 0,\dots,R_n = 0$ in $\bar{x}_j$ is $\tilde{m}_j$ precisely, and $R_1,\dots,R_n$ are composed from $P_1^{i_1}\dots P_n^{i_n}$.

Now instead of Hermite interpolation formula, having $S_0 = \{\bar{x}_1^0,\dots,\bar{x}_{m-1}^0\}$ fixed and $\bar{x}_m = \bar{x}$ varying we can write

$$Q(\bar{x}) = \frac{1}{(2\pi i)^n} \oint \frac{dz_1 \wedge \ldots \wedge dz_n}{R_1 \ldots R_n} .$$

We present now some versions of multidimensional applications of the Residue Formula.

There exists a very nice residue formalism connected with the Grothendieck residue symbol.

Let U be the ball $\{x \in \mathbb{C}^n : |z| < \varepsilon\}$ and $f_1, \ldots, f_n \in o(\bar{U})$ functions holomorphic in a neighborhood of the closure $\bar{U}$ of U. We assume that

$$D_i = (f_i) = \text{divisors of } f_i : \ i = 1, \ldots, n$$

have the origin as their set-theoretic intersection,

$$f^{-1}(0) = \{\bar{0}\}$$

for

$$f = (f_1, \ldots, f_n) : U^* = U \setminus \{\bar{0}\} \to \mathbb{C}^n \setminus \{0\}.$$

We are interested in residues associated with a meromorphic n-form

$$\omega = \frac{g(z) dz_1 \wedge \ldots \wedge dz_n}{f_1(z) \ldots f_n(z)} : g \in o(\bar{U})$$

having polar divisor

$$D = D_1 + \ldots + D_n .$$

In order to define the Grothendieck residue symbol we take the cycle of integration

$$\Gamma = \{z : |f_i(z)| = \varepsilon\}$$

(with the orientation, say $d(\arg f_1) \wedge \ldots \wedge d(\arg f_n) \geq 0$). Then the <u>residue</u> of ω at $\bar{0}$ is

(6.1) $$\operatorname{Res}_{\{\bar{0}\}}\omega = \left(\frac{1}{2\pi i}\right)^r \int_\Gamma \omega.$$

First of all, $\operatorname{Res}_{\{\bar{0}\}}\omega$ possesses all "normal" local properties

Lemma 6.2: (Local properties of residues.)

1) In the generic case, when D_i are smooth and meet transversely, i.e. Jacobian of f

(6.2) $$J_f(\bar{0}) = \frac{\partial(f_1,\dots,f_n)}{\partial(z_1,\dots,z_n)}(\bar{0}) \neq 0$$

then

(6.3) $$\operatorname{Res}_{\{\bar{0}\}}\omega = g(\bar{0})/J_f(\bar{0}).$$

2) (Transformation formula.) Suppose that $f = (f_1,\dots,f_n)$ and $g = (g_1,\dots,g_n)$ give holomorphic maps $f, g\colon \bar{U} \to \mathbb{C}^n$ with $f^{-1}(0) = g^{-1}(0) = \{\bar{0}\}$. Suppose that for ideals we have

$$\{g_1,\dots,g_n\} \subset \{f_1,\dots,f_n\},$$

i.e.

$$g_i(z) = \Sigma_{j=1}^n a_{ij}(z) f_i(z)$$

for holomorphic matrix $A(z) = a_{ij}(z)$. Then for $h(z) \in o(\bar{U})$ we have:

$$\operatorname{Res}_{\{\bar{0}\}}\left(\frac{h dz_1 \wedge\dots\wedge dz_n}{f_1 \dots f_n}\right) = \operatorname{Res}_{\{\bar{0}\}}\left(\frac{h \det A\, dz_1 \wedge\dots\wedge d\, z_n}{g_1 \dots g_n}\right).$$

Residues can be also used for an analytic formula of local intersection number of $f = 0$ at $\bar{0}$. For this we define

$$(D_1,\dots,D_n)_{\{\bar{0}\}} = \operatorname{Res}_{\{\bar{0}\}}\left(\frac{df_1}{f_1} \wedge\dots\wedge \frac{df_n}{f_n}\right).$$

Then $(D_1,\dots,D_n)_{\{\bar{0}\}}$ has indeed sense as the local intersection number of $f = 0$:

a) For the local ring o_0 at the origin and $I_f \subset o_0$ the ideal, generated by the f_i, we have

$$(D_1,\ldots,D_n)_{\{\bar{0}\}} = \dim_{\mathbb{C}} \mathcal{O}_0/I_f;$$

b) $f\colon U^* \to \mathbb{C}^n \setminus \{0\}$ <u>has topological degree</u>

$$D_1,\ldots,D_n)_{\{\bar{0}\}}.$$

All these assertions together with Global Residue Formula below belong to Ph. Griffiths [9] in the form presented here.

Of course, the Global Residue Formula is just kind of expression we need in the Padé approximations.

Let M be the n-dimensional compact complex manifold and ω a meromorphic differential form on M, whose polar divisor D can be expressed as a union $D = D_1 \cup \ldots \cup D_n$ of n divisors D_i with the property that their intersection

$$Z = D_1 \cap \ldots \cap D_n$$

is a finite set. Then we have

Lemma 6.3: (Global Residue Formula.)

$$\Sigma_{\bar{p}\in Z} \operatorname{Res}_{\{\bar{p}\}}\omega = 0. \tag{6.4}$$

Then most interesting applications proposed by Griffiths deal with $M = \mathbb{P}^n$. We assume that $D_1,\ldots,D_n$ are hypersurfaces of respective degrees $d_1,\ldots,d_n$ with intersections at isolated points P_ν and this intersection we present as a zero cycle

$$D_1 \ldots D_n = \Sigma_\nu \, m_\nu P_\nu,$$

where by Bezout's theorem

$$\Sigma_\nu \, m_\nu = d_1 \ldots d_n.$$

We assume below simply that all P_ν lie in $\mathbb{C}^n \subset \mathbb{P}^n$ and that D_i is defined by

$$f_i(x_1,\ldots,x_n) = 0$$

for polynomials f_i of degree d_i.

The most general meromorphic n-form on $\mathbb{P}^n$ with polar divisor $D = D_1 +\ldots+ D_n$ has in $\mathbb{C}^n$ an expression

$$\omega = \frac{g(x)dx_1 \wedge\ldots\wedge dx_n}{f_1(x)\ldots f_n(x)}$$

for a polynomial $g(x)$. Here ω doesn't have the hyperplane at infinity as a component of its polar divisor when the degree of $g(x)$ satisfy:

(6.5) $$\deg(g) \leq (d_1 +\ldots+ d_n) - (n+1).$$

Thus the Global Residue formula gives in this case

6.4 <u>Generalized</u> <u>Jacobi-Kronecker</u> <u>Formula</u> 6.4:

(6.6) $$\Sigma_\nu \operatorname{Res}_{P_\nu} \left(\frac{g(x)dx_1 \wedge\ldots\wedge dx_n}{f_1(x)\ldots f_n(x)}\right) = 0.$$

Why the Jacobi-Kronecker formula? Because in 1834 Jacobi claimed and Kronecker [13] proved using only linear algebra the following important identity

(6.7) $$\Sigma_\nu \frac{g(P_\nu)}{J_f(P_\nu)} = 0$$

if $\deg(g) \leq \Sigma_{i=1}^n d_i - (n+1)$ and D_i meet transversely at $d_1 \ldots d_n$,

$$J_f = \frac{\partial(f_1,\ldots,f_n)}{\partial(z_1,\ldots,z_n)} \text{ is the Jacobian of } f.$$

Of course (6.7) follows from (6.6) and results above.

This formula was already applied by I. Petrovsky [10] to 16th Hilbert Problem on real plane curves $(n = 2)$.

In particular, from these formulae we obtain an interesting multidimensional generalization of Hermite interpolation formula.

Corollary 6.5: Let $f(z)$ be holomorphic in $\bar{U}$ and we define in the neighborhood of $\bar{0}$ a new function $f(\bar{x})$ by

$$F(\bar{x}) = \left(\frac{1}{2\pi i}\right)^n \int_\Gamma \frac{f(\bar{x}\bar{z})dz_1 \wedge \ldots \wedge dz_n}{f_1(z)\ldots f_n(z)} .$$

Then $F(\bar{x})$ vanishes at $\bar{x} = \bar{0}$ of order $\geq \Sigma_{i=1}^n d(f_i) - n$.

We can write an explicit expression for $F(\bar{x})$ in terms of $\partial_z^k f$ in certain cases, e.g. when $f_1,\ldots,f_n$ have only simple intersection or f_i' are powers of such f_i, etc... or in any case when the singularities of the intersections are known. In each of these cases $F(\bar{x})$ is a linear form from partial derivatives $\partial_z^{k_1,\ldots,k_n} f(\bar{x}\bar{z})$ with coefficients being rational functions in partial derivatives in $f_i(\bar{x})$ at fixed $\bar{x}_0$.

Let's explain how from the generalized Jacobi-Kronecker formula it follows, e.g. Lagrange interpolation formula.

Let $n = 1$; $x_0,\ldots,x_m$ are fixed and x is a variable. We set

$$f(z) = \Pi_{i=0}^m (x-x_i)(z-x).$$

Then the formula (4.7) can be written as

$$\Sigma_{i=0}^m g(x_i)/\psi(x_i)(x_i-x) + g(x)/\psi(x) = 0$$

for

$$\psi(x) = \Pi_{i=0}^m (x-x_i) = f'(x)|_{z=x},$$

or

$$g(x) = \Sigma_{i=0}^m \frac{g(x_i)\psi(x)}{\psi(x_i)(x-x_i)}$$

for $d(g) \leq m$.

Now (6.6) is a natural generalization of the Lagrange interpolation formula, if one of the $\bar{x}_\nu$ varies.

§7.

Definition 7.1: Let $f_1(x),\ldots,f_m(x)$ be formal power series and $\rho_1,\ldots,\rho_m$ be m positive integers. We say that $\{\mathcal{A}_1(x),\ldots,\mathcal{A}_m(x)\}$ is the system of polynomials of N II type corresponding to

$(\rho_1,\ldots,\rho_m),(f_1(x),\ldots,f_m(x))$, if $\{\mathfrak{A}_i(x): i = 1,\ldots,m\}$ is a non-trivial system;

1) $\deg \mathfrak{A}_i(x) \leq \sigma - \rho_i$; $\sigma = \Sigma_{j=1}^m \rho_j$;

2) the order in $x = 0$ of

$$\mathfrak{A}_\ell(x) f_k(x) - \mathfrak{A}_k(x) f_\ell(x)$$

is at least $\sigma + 1$: $k,\ell = 1,\ldots,m$.

The first example of the system of polynomials NII was constructed in 1873 by Hermité in connection with the transcendence of e.

The polynomials NII and usual Padé approximations $A_i(x;\ \rho_1,\ldots,\rho_m)$: $i = 1,\ldots,m$ to $f_1(x),\ldots,f_m(x)$ (so-called polynomials NI) are connected, of course.

Under the conditions of perfectness and normality (see precise statements in the papers of Mahler, Jagier, Coates, de Brujn), for matrices

$$A(x;\ \rho_1,\ldots,\rho_m) = (A_i(x;\ \rho_1 + \delta_{h1},\ldots,\rho_m + \delta_{hm}))$$

$$\mathfrak{A}(x;\ \rho_1,\ldots,\rho_m) = (\mathfrak{A}_i(x;\ \rho_1 - \delta_{h1},\ldots,\rho_m - \delta_{hm})):$$

$i,h = 1,\ldots,m$ we have

$$A(x;\ \rho_1,\ldots,\rho_m)\mathfrak{A}^T(x;\ \rho_1,\ldots,\rho_m) = \begin{pmatrix} c_1 x^\sigma & & 0 \\ & \ddots & \\ 0 & & c_m x^\sigma \end{pmatrix}$$

for some constants c_i.

In other words, the systems of polynomials NI-II determine each other.

Let us present explicit formula for rational approximations to polylogarithmic functions $L_k(x)$. Our formulae give us simultaneous rational approximations (so-called polynomials NII) [11].

Here are the formulae for <u>one</u> polylogarithmic function $L_n(x)$: $n \geq 1$. We define for $k = 0,1,2,\ldots$

$$E_k = \Sigma_{i=0\ \ i+k-m\geq 0}^{m} (-1)^i \binom{m}{i} \frac{1}{(i+k-m+1)^n} \times \binom{j+s+b}{j+b}\binom{j+s+b+1}{j+b}^n.$$

Here, as before

$$s = [\frac{m-1}{n}], \quad b \text{ is an integer (say, } b = 0,1).$$

Now the approximations are defined as follows:

$$A_m(x) = \Sigma_{j=1}^{m} \binom{m}{j}(-1)^j x^{m-j} \times \binom{j+s+b}{j+b}\binom{j+s+b+1}{j+b}^n$$

$$= (s+b)\dots(s+1)[(s+2)\dots(s+b+1)]^n \times x^m \times [(b+1)!]^{-1}$$

$$\times {}_{n+2}F_{n+1}\left(x^{-1}\middle|\begin{matrix} -m, s+b+1, \dots, s+b+1 \\ b+2, \dots, b+2 \end{matrix}\right),$$

and

$$B_m(x) = \Sigma_{i=0}^{m+b-1} E_i x^{i+1}.$$

Then we have the following

Corollary 7.2: *We have for any* $n \geq 1$ *and* $m \geq 1$,

$$A_m(x)L_n(x) - B_m(x) = \Sigma_{k=s+m+b+2}^{\infty} E_k x^k = R_m(x),$$

$$s = [\frac{m-1}{n}],$$

and $d(A_m) = m$, $d(B_m) = m + b$ and $b = 0,1,\dots$.

We can present asymptotics for $|A_m(x)|$, $|B_m(x)|$ and $|R_m(x)|$ for a fixed n, b, x with $m \to \infty$.

We have, e.g. in the case of small x, $0 < |x| < 1$ the following result: the generalization of the asymptotics of the Padé approximation to the logarithm $\log(1-x) = L_1(x)$, obtained essentially by Riemann [12] (the case $n = 1$). For $n \geq 2$ the situation is not at all a trivial one.

Let $0 < |x| < 1$, n and b be fixed. Let's denote by $t_1(x)$ and $t_2(x)$ the largest (by modulus) roots of the equations

$$nt^{n+1} - (n+1)t^n + z = 0$$

and

$$nx^{\frac{1}{n}} t^{\frac{n+1}{n}} + (1-n)x^{\frac{1}{n}} t^{\frac{1}{n}} - (n+1)t + n = 0,$$

respectively. Then

$$|A_m(z)| \sim (n\ z|t_1(z)|^{n+1})^m; \quad m \to \infty$$

$$|R_m(z)| \sim \left(\frac{(1-|t_2|)^{1/n}|t_2|}{(1-\ |t_2|z)^{1/n}}\ |z|^{n+1/n}\right)^m; \quad m \to \infty.$$

In comparison with the previous results we can quote a long series of papers and books written by Luke. He constructed effectively rational approximations to hypergeometric functions. However the speed of convergence of approximations is not significantly better than for the polynomial approximations.

Our approximation that we had presented is not the best possible, however it is effective. We can present the best possible approximation for the dilogarithm that is absolutely non-effective.

Let us introduce the corresponding notations.

For a given n and any j, $1 \leq j \leq n+1$, we put

$$\Delta_{n,j}(x_1,\ldots,x_n) = \Pi_{1\leq i<k\leq n}(x_i-x_k) \times \Pi_{i=1}^{n} \Pi_{\nu=1,\nu\neq j}^{n+1}(x_i+\nu)^{-1}.$$

Then Padé approximations of $L_2(x)$ are determined in terms of

$$\delta_{n,j} = \frac{\partial^n}{\partial_{x_1}\cdots\partial_{x_n}}\ \Delta_{n,j}(x_1,\ldots,x_n)|_{x_1=1,x_2=2,\ldots,x_n=n}.$$

The coefficients of the Padé approximations to $L_2(x)$ are now

$$\frac{\delta_{n,j}}{\delta_{n+1,n+2}}; \ j = 1,\ldots,n+1.$$

References

[1] L. Schlesinger, Einführung in die theorie der gewöhnlichen differentialgleichungen auf functionentheoretischer grundlage, 3 aufl., Berlin-Leipzig, 1922.

[2] M. Sato, T. Miwa and M. Jimbo, Studies in holonomic quantum fields, Preprints of the Research Institute for Mathematical Sciences, Kyoto University, Kyoto, Japan, 1978-1979.

[3] L.A. Lappo-Danilevsky, Mémoires sur la théorie des systêmes des equations différentielles linéaires, Chelsea Publishing Company, N.Y., 1953.

[4] M. Levine, J. Wright, Phys. Rev., D8 (1973), 3171.

[5] R. Karplus, M. Neuman, Phys. Rev., 80 (1950), 380; Phys. Rev. 83 (1951), 776.

[6] M. Veltman, G. t'Hooft, Nuclear Physics B, B153 (1979), 365.

[7] G.V. Chudnovsky, Singular points on complex hypersurfaces and multidimensional Schwarz lemma, Seminaire Delange-Pirot-Poitou, 19e Annee 1977/1978, January 1978, pp. 1-40, University Publishing House, Paris, France, 1978 (to be reprinted by Birkhauser Verlag, 1980).

[8] M.A. Angelesco, Sur des polynomes généralisant les polynomer de Legendre et d'Hermité et sur le calcul approché des intégrales multiples, Thesis, Gauthier-Villars, 1916, 150 pp.

[9] Ph. Griffith, J. Harris, Principles of algebraic geometry, Wiley, N.Y. 1978.

[10] I. G. Petrowski, Ann. of Math. 39 (1938), 189-207.

[11] G. V. Chudnovsky, Journal de Mathematique Pures et Appliques, Paris, 58 (1979), 445-476.

[12] B. Riemann, Oeuvres Mathématiques, Albert Blanchard, Paris, 1968, pp. 353-363.

[13] L. Kronecker, Werke, Über einige interpolationsformeln für ganze funktionen mehrer variablen, v. 1, 1895, 133-141.

Department of Mathematics
Columbia University
New York, NY
USA

Hermite-Padé Approximations to Exponential Functions and Elementary Estimates of the Measure of Irrationality of π.

by

G.V. Chudnovsky

Introduction: This paper is devoted to the study of the most elementary example of the number theoretical aspects of Padé approximations: the study of the irrationality and transcendence of logarithms of algebraic numbers. The system of Padé approximations to exponential functions upon which all number theoretical arguments are built, was proposed by Hermite circa 1870, was used by him in 1873 to prove the transcendence of e, and then was used by Lindemann in 1882 to prove the transcendence of π. In this paper we examine irrationality and measures of irrationality of logarithms of algebraic numbers using the same Hermite system of Padé approximations. We look at this system from the point of view of the inverse Laplace transformation of rational functions. This provides us with good upper and lower bounds for values of the remainder function and polynomial coefficients, and with an explicit determination of asymptotics of Padé approximants. Our results on the measures of irrationality are the best from the point of view of existing analytic methods. They are an improvement of the previous results of K. Mahler (1930-1966) (see [1]) and M. Mignotte [8], where asymptotics had only been estimated. This paper can be consider to be an elementary introduction into number theoretical applications of methods of Padé approximations. The use of the Laplace transformation is very promising, and is applied in further publications in more complicated situations connected with Abelian functions. In the first part of this paper, before the Hermite system of Padé approximations is introduced, we discuss continued fraction expansions and diophantine approximations to irrational numbers. For the discussion of continued fraction expansions of functions and numbers we use the books of Wall [6] and Khintchine [7].

§0. We start with the definition of transcendental numbers and their measure of transcendence.

Definition 0.1: Complex number θ is called transcendental, if for any $P(x) \in \mathbb{Z}[x]$ (with rational integer coefficients)

$$P(\theta) \neq 0.$$

The bound $|P(\theta)| > \psi(d(P),H(P))$ for some positive function $\psi(d,H)$ on degree $d(P)$ of $P(x)$ and $H(P)$ -the maximum of the modulus of the coefficients of $P(x)$ -is called a measure of transcendence of θ.

If $\psi(d(P),H(P)) = H(P)^{-\psi_0(d)}$, then the measure of the transcendence is called a normal one.

In this paper we consider only measures of irrationality corresponding to the case $d(P) = 1$, i.e. to $\psi(1,H(P))$.

Historical remarks: Euler was first to start to consider problems of numbers that are not rational or not algebraic. His work in this field was summarized in his book Introductis in analysin infinitorum, Lausanne , 1748 finished in 1744. We must remember however, that Euler called algebraic numbers "irrational" and that the first proof of the existence of "many" transcendental numbers was given by G. Cantor in 1873; while the first example of transcendental number was presented by Liouville in 1844.

In any case the natural starting point of the study of irrational numbers is their irrationality. In other words, if α is irrational, then for any integers p, q

$$\alpha - \frac{p}{q}$$

is non-zero (irrational number). If we want to compare α with rational numbers, this means that we want to study how small $|\alpha - p/q|$ can be in comparison with $\max(|p|,|q|)$.

In other words we want to study the best approximations of α by rational numbers

Definition 0.2: A rational fraction a/b, $b > 0$ is called the best approximation of α if from $c/d \neq a/b$, $0 < d \leq b$ it follows

$$|d\cdot\alpha - c| > |b\alpha - a|.$$

I.e. the best approximations a/b to α give us successful minima of the linear forms

$$\{|q\cdot\alpha - p| : p,q \in \mathbb{Z}\}.$$

If we are speaking now about the arithmetic nature of a given number α we mean, in particular, the complete knowledge of the sequence of successful best approximations to α.

How do we find this sequence of best approximations?

The algorithm to find this sequence has been known for a thousand years as continued fraction algorithm [7].

Euler was the first who applied continued fraction expansion to problems of irrationality.

Definition of the regular continued fraction expansion: Let α be a real number. We put

$$a_0 = [\alpha] \qquad \text{(integral part of } \alpha\text{)}.$$

Then

$$\alpha = a_0 + \frac{1}{\alpha_1}$$

for $\alpha_1 \geq 1$ and we can define again

$$a_1 = [\alpha_1],$$

$$\alpha_1 = a_1 + \frac{1}{\alpha_2},$$

etc. if

$$a_n = [\alpha_n]$$

and α_n is not an integer. We define α_{n+1} by a relation

$$\alpha_{n+1} = a_n + \frac{1}{\alpha_n}$$

and again

$$a_{n+1} = [\alpha_{n+1}].$$

If none of α_n is an integer we get an infinite continued fraction expansion

$$\alpha = [a_0; a_1, a_2, a_3, \ldots, a_n, \ldots]$$
$$= a_0 + \cfrac{1}{a_1 + \cfrac{1}{a_2 + \cfrac{1}{a_3 + \ldots}}} .$$

If some α_n is an integer (but none of α_k: $k < n$ is), then $a_{n+1} = 0$ and for $a_n = \alpha_n$, we have a finite continued fraction

$$\alpha = [a_0; a_1, \ldots, a_n]$$
$$= a_0 + \cfrac{1}{a_1 + \cfrac{1}{a_2 + \cfrac{1}{a_3 + \ddots + \cfrac{1}{a_n}}}} .$$

Here all a_i are positive rational integers ($i = 1,2,\ldots$), which explains the name "a regular" continued fraction expansion.

<u>Theorem</u> 0.3: <u>For any real</u> α <u>there is a unique regular continued fraction expansion</u>

$$a_0 + \cfrac{1}{a_1 + \cfrac{1}{a_2 + \cfrac{1}{a_3 + \ldots}}}$$

<u>having</u> α <u>as its value (i.e. the continued fraction is convergent to</u>

α). <u>This fraction is finite if and only if</u> α <u>is rational</u>.

<u>Proof</u>: Indeed, if the fraction is finite, then α is rational. Let α be rational. Then the process is finished after a finite number of steps.

If $\alpha_n = a/b$, then $\alpha_n - a_n = a-ba_n/b = c/b$, where $c < b$, because $\alpha_n - a_n < 1$. Thus

$$\alpha_{n+1} = \frac{b}{c}.$$

In other words, denominators of rational numbers α_m: $m = 1,2,\ldots$ are decreasing and after finitely many steps we come to an integer α_n. Then $a_n = [\alpha_n] = \alpha_n$ and $a_{n+1} = 0$, i.e. α is a finite continued fraction expansion.

<u>Corollary</u> 0.4: <u>If</u> α <u>has an infinite continued fraction expansion, then</u> α <u>is an irrational number</u>.

<u>Properties of the continued fraction expansion</u>. Let α be an irrational number and $\alpha = [a_0;a_1,a_2,\ldots,a_n,\ldots]$. We define a rational approximation to α by a finite pieces of a continued fraction expansion

$$[a_0;a_1,a_2,\ldots,a_n] = \frac{p_n}{q_n},$$

where $q_n > 0$, $n = 1,2,3,\ldots$. These rational fractions satisfy very simple recurrence formulae. The most characteristic feature that is always connected with continued fraction expansion is the existence of the three-term linear recurrence relating p_n, q_n:

<u>Lemma</u> 0.5. <u>For any</u> $k \geq 2$,

$$\begin{cases} p_k = a_k p_{k-1} + p_{k-2}; \\ q_k = a_k q_{k-1} + q_{k-2}. \end{cases}$$

This is proved by induction. As a corollary we have

Corollary 0.6. *We have for* $k \geq 0$

$$q_k p_{k-1} - p_k q_{k-1} = (-1)^k$$

(where $p_{-1} = 1$, $q_{-1} = 0$) *and so*

$$\frac{p_{k-1}}{q_{k-1}} - \frac{p_k}{q_k} = \frac{(-1)^k}{q_k q_{k-1}} .$$

As a corollary we find that p_n and q_n are relatively prime, so p_n/q_n is an irreducible fraction.

Wc have

$$\frac{p_{2n}}{q_{2n}} < \frac{p_{2n+2}}{q_{2n+2}} < \alpha < \frac{p_{2n+1}}{q_{2n+1}} < \frac{p_{2n-1}}{q_{2n-1}}$$

and

$$\frac{p_n}{q_n} \to \alpha \quad \text{as} \quad n \to \infty.$$

Moreover we can find the true approximation of α by p_n/q_n. We have:

Theorem 0.7: *For any* $k \geq 0$,

$$\frac{1}{q_k(q_k + q_{k+1})} < |\alpha - \frac{p_k}{q_k}| < \frac{1}{q_k q_{k+1}}.$$

In particular,

$$|\alpha - \frac{p_k}{q_k}| < \frac{1}{q_k \cdot a_{k+1} q_k} \leq \frac{1}{q_k^2}.$$

Another statement shows us that indeed all best approximations to α are among partial fractions p_n/q_n and vice versa:

Theorem 0.8: Any best approximation a/b to α is one of the partial fractions p_n/q_n to α.

All partial fractions p_n/q_n are best approximations to α with the possible exception of $p_0/q_0 = a_0/1$, only.

The discussion above can be completed by a

Proposition 0.9: If

$$|\alpha - \frac{p}{q}| < \frac{1}{2q^2},$$

then p/q is one of the partial fractions p_n/q_n to α.

We know that for any n, $|\alpha - p_n/q_n| < 1/q_n^2$. Moreover it's known that for any $n \geq 1$, either

$$|\alpha - \frac{p_n}{q_n}| < \frac{1}{2q_n^2} \quad \text{or} \quad |\alpha - \frac{p_{n-1}}{q_{n-1}}| < \frac{1}{2q_{n-1}^2}.$$

§1. From what we see it's absolutely clear that the knowledge of rational approximations to α is determined by a continued fraction expansion of α.

In order to get a complete information about an arithmetic nature of α we need a continued fraction expansion of α.

How do we get a continued fraction expansion of α?

Euler was the first who proposed the way to get a continued fraction expansion of a number. You first get the continued fraction expansion of the functions $f(z)$ satisfying linear differential equations and then you get continued fraction expansion of

$$\alpha = f(z) \text{ for rational } z.$$

Here is Euler's method of 1737. Let $f(z) = y'(z)/y(z)$ satisfies a Riccati equation with rational function coefficients, i.e. $y(z)$ satisfies a linear differential equation of the second order

$$y = Q_0(z)y'(z) + P_1(x)y''(z)$$

for rational $Q_0(z)$, $P_1(z)$. Then Euler proposed a continued fraction expansion (not of a regular type):

$$\frac{y(z)}{y'(z)} = Q_0 + \cfrac{P_1}{Q_1 + \cfrac{P_2}{Q_2 + \dots}}$$

where P_n, Q_n are defined by induction in such a way that

$$y^{(n)} = Q_n y^{(n+1)} + P_{n+1} y^{(n+2)},$$

i.e.

$$Q_n = \frac{Q_{n-1} + P_n'}{1 - Q_{n-1}'}; \qquad P_{n+1} = \frac{P_n}{1 - Q_{n-1}'}.$$

If we define

$$\alpha_n(x) = Q_n(x) \cdot \{\Pi_{m=1}^{n} P_m(x)\}^{-1} \in \mathbb{C}(x):$$

$n = 0,1,2,\dots$, then the function y/y' can be represented as continued fraction

$$\frac{y(z)}{y'(z)} = [\alpha_0(z); \alpha_1(z), \alpha_2(z), \dots].$$

For special $Q_0(z)$, $P_1(z) \in \mathbb{Z}[z]$ and special rational values of z we get still a regular continued fraction expansion.

Euler proved in this way irrationality of e and e^2 in "De fractionibus continuis", Comment. Acad. de Petrop. v. 6 (presented to the St. Peters Academy in March, 1737; published in 1744).

Here is Euler's development:

$$\frac{e - 1}{e + 1} = [0;2,6,10,14,\dots],$$

which was then generalized by Lambert in 1770:

(1.1) $$\frac{e^{1/y} - e^{-1/y}}{e^{1/y} + e^{-1/y}} = [0;y,3y,5y,7y,\dots].$$

Here are some corollaries of Euler-Lambert continued fraction expansion:

Corollary 1.1: We have

$$e = [2;1,2,1,1,4,1,1,6,\dots] = [2;\overline{1,2m,1}]$$

and

$$e^2 = [7;\overline{3m-1,1,3m,12m+6}: m = 1,2,\dots].$$

Lambert noticed that such continued fraction development gives results for π and π^2.

Indeed we can substitute in (1.1), $y = 1/x$. Then

$$\frac{e^x - e^{-x}}{e^x + e^{-x}} = \tanh x = \frac{1}{i}\,\mathrm{tg}(xi)$$

and so

$$\mathrm{tg}\, z = [0; \frac{1}{z}, -\frac{3}{z}, -\frac{5}{z}, -\frac{7}{z}, \dots]. \tag{1.2}$$

Using continued fraction expansion (1.1), (1.2) Lambert formulated and Legendre in 1794 proved the following results:

Theorem 1.2: If x is rational and $x \neq 0$, then e^x is irrational; also $\mathrm{tg}\, x$ is irrational. If x^2 is a rational number and $x \neq 0$, then $\mathrm{tg}\, x/x$ is irrational.

In order to get irrationality and transcendence results it's better to use not only the function $f(z)$ (for $\alpha = f(z_0)$) but also its powers $f^i(z)$: $i = 0,1,\dots,m-1$. In other words, instead of constructing the best rational function approximations to the functions

$$e^x \quad \text{or} \quad \log x$$

we consider linear forms in

$$\{1, e^x, \dots, e^{x(m-1)}\} \quad \text{or} \quad \{1, \log x, \dots, \log^{m-1} x\}$$

with polynomial coefficients having the smallest value at a given point.

In general it's very difficult to obtain such explicit formulae (of Padé approximations). Now we are able to deduce these formulae from the solution of the Riemann monodromy problem [9].

However for exponents and powers of logarithms this problem was solved as early as in 1873 by Hermite [2]. All known proofs of the transcendence or irrationality of π or e are either based on Hermite formulae or can be reduced to them [1].

§2. We want to find for a given integer $n \geq 0$ a system of polynomials $P_{0,n}(v),\ldots,P_{m-1,n}(v)$ of degrees $\leq n$ such that

$$F(v) = \Sigma_{i=0}^{m-1} \log^i(1 - v)\cdot P_{i,n}(v)$$

has a zero at $v = 0$ of an order at least $(n+1)m - 1$. The existence of such a system of nontrivial polynomial follows from the fact that there is always a nontrivial solution of a system of $(n + 1)m - 1$ linear equations

$$F_j = 0: j = 0,\ldots,(n + 1)m - 2$$

for

$$F(v) = \Sigma_{j=0}^{\infty} F_j v^j$$

in $(n + 1)m$ unknowns which are the coefficients of $P_{i,n}(v)$.

Hermité constructions gives us explicitly all $P_{i,n}(v)$ and $F(v)$.

For simplicity we substitute

$$1 - v = e^t$$

and $F(v) = R(t)$, where

$$R(t) = \Sigma_{i=0}^{m-1} t^i\cdot P_i(e^t;n).$$

The remainder function $R(t)$ is represented in the form of a contour integral.

Let

$$Q(z) = \Pi_{k=0}^{n}(z - k). \tag{2.1}$$

We define

$$R(t;n) = \frac{(n!)^m}{2\pi i}\int_C \frac{e^{zt}dz}{Q(z)^m}, \tag{2.2}$$

where C is a contour with positive orientation, enclosing all the poles $z = 0, 1, \ldots, n$ of the integrand.

First of all, $R(t;n)$ has an order of zero at $t = 0$ exactly $(n + 1)m - 1$. Indeed $e^{zt} = 1 + z\cdot t + z^2t^2/2! + \ldots$; so looking at residues at $z = \infty$ we get

$$R(t) = a_0 + a_1t + a_2t^2 + \ldots$$

for

$$a_k = \frac{(n!)^m}{k!} \operatorname{Res}_{z=\infty} \frac{z^k}{Q(z)^m}.$$

Since the degree of $Q(z)^m$ is $(n + 1)m$, we have

$$a_0 = a_1 = \ldots = a_{\sigma-2} = 0, \qquad a_{\sigma-1} \neq 0 \quad \text{for} \quad \sigma = (n + 1)m.$$

Here $a_{\sigma-1} = (n!)^m/\{(n+1)m-1\}!$

Now we can represent $R(t;n)$ as a linear combination of $\{1,t,\ldots,t^{m-1}\}$ with coefficients being polynomials in e^t of the degree n.

We have

$$\frac{(n!)^m}{Q(z)^m} = \Sigma_{k=0}^{n} \frac{c_{k,m}}{(z-k)^m} + \frac{c_{k,m-1}}{(z-k)^{m-1}} + \ldots + \frac{c_{k,1}}{(z-k)}. \tag{2.3}$$

Then, immediately, computing the residue at $z = k$

$$\text{(2.4)} \qquad \operatorname*{Res}_{z=k} \frac{(n!)^m e^{zt}}{Q(z)^m} = e^{kt} \Sigma_{j=0}^{m-1} \frac{c_{k,j+1}}{j!} t^j.$$

Now we can rewrite (2.2) by Cauchy residue theorem using the formula (2.4):

$$\text{(2.5)} \qquad R(t;n) = \Sigma_{k=0}^{n} e^{kt} \Sigma_{j=0}^{m-1} \frac{c_{k,j+1}}{j!} t^j$$

$$= \Sigma_{j=0}^{m-1} t^j \Sigma_{k=0}^{n} e^{kt} \frac{c_{k,j+1}}{j!}.$$

Thus we found our remainder function

$$\text{(2.6)} \qquad R(t;n) = \Sigma_{j=0}^{m-1} t^j \cdot P_j(e^t;n)$$

for polynomials

$$\text{(2.7)} \qquad P_j(z;n) = \Sigma_{k=0}^{n} z^k \cdot \frac{c_{k,j+1}}{j!}: \; j = 0,\dots,m-1$$

of degrees n.

Here is the information about the structure of $c_{k,j+1}$ in (2.3). It's clear that $c_{k,j+1}$ are rational numbers and we can find their denominators.

We denote

$$[1,\dots,n] = \text{l.c.m.}\ \{1,\dots,n\}.$$

<u>Lemma</u> 2.1: <u>The coefficients of the polynomials</u>

$$(m-1)! \cdot [1,\dots,n]^{m-1-j} \cdot P_j(z,n): \qquad j = 0,1,\dots,m-1$$

<u>are rational integers</u>.

<u>Proof</u>: For $Q(z) = \Pi_{k=0}^{n}(z-k)$ we have $1/Q(z) = \Sigma_{k=0}^{n} 1/(z-k) \cdot 1/k!(-1)\cdots(-n+k)$. Thus we have

(2.8) $$\frac{n!}{Q(z)} = \Sigma_{k=0}^{n} \binom{n}{k} (-1)^{n-k} \cdot \frac{1}{z - k} .$$

Now in order to get from (2.8) the formulae for $c_{k,j+1}$ we use (2.3) and the following simple rule for multiplication

(2.9) $$\frac{1}{z - k} \cdot \frac{1}{z - \ell} = \frac{1}{\ell - k} \cdot \{\frac{1}{z - k} - \frac{1}{z - \ell}\}.$$

From (2.8), (2.3) and (2.9) it follows immediately that for any $k = 0,1,\ldots,n$, the number

(2.10) $[1,\ldots,n]^{m-j} c_{k,j}$ is a rational integer: $j = m,m-1,\ldots,1$.

Now lemma 2.1 follows from (2.7) and (2.10).

§3. Now let us determine the asymptotics of Padé approximations to logarithms

In order to get the asymptotics of $R(t;n)$ for a fixed $t \neq 0$ and $n \to \infty$ we use the representation of $R(t;n)$ as an inverse Laplace transform of a rational function.

Indeed, let

$$g(p) = \int_0^\infty e^{-pt} f(t)dt$$

be a Laplace transform of a function t (this is a case of a real t). Then the formula of an inverse Laplace transform gives us

$$f(t) = \frac{1}{2\pi i}\int_{c-i\infty}^{c+i\infty} e^{zt} g(z)dz.$$

This shows immediately that

(3.1) $$R(t) = \mathcal{L}^{-1}\{\frac{(n!)^m}{Q_n(p)^m}\}$$

for $Q_n(p) = \Pi_{k=0}^{n}(p - k)$. In general for $m \geq 2$, the expression for

R(t) is rather complicated. However, for m = 1 it is indeed very simple:

(3.2) $$R(t;h;m = 1) = (1 - e^{t})^{h}.$$

For the proof we use (2.8) and

$$\mathcal{L}^{-1}\left(\frac{1}{z - k}\right) = e^{kt}.$$

We try to express R(t;n;m) in terms of (3.2) only. We have obviously

(3.3) $$Q_n(p)\cdot Q_n\left(p - \frac{1}{m}\right) \cdots Q_n\left(p - \frac{m - 1}{m}\right)$$
$$= m^{-m(n+1)}\cdot Q_{nm+m-1}(mp).$$

We now use simple rules of the inverse Laplace transform:

(3.4) $$\mathcal{L}^{-1}\{g_1(p)\cdot g_2(p)\} = \mathcal{L}^{-1}\{g_1(p)\}*\mathcal{L}^{-2}\{g_2(p)\}$$
$$\overset{\text{def}}{=}\int_0^t f_1(u)\cdot f_2(t - u)du; \quad f_i(t) = \mathcal{L}^{-1}\{g_i(p)\} \quad \text{for } i = 1,2,$$

then we have

(3.5) $$\mathcal{L}^{-1}\{g(p + \alpha)\} = e^{-\alpha t}\mathcal{L}^{-1}\{g(p)\}$$

and

(3.6) $$\mathcal{L}^{-1}\{g(mp)\} = \frac{1}{m}\,\mathcal{L}^{-1}\{g\}\left(\frac{t}{m}\right).$$

We have at last

$$R(t;n) = \mathcal{L}^{-1}\left\{\left(\frac{n!}{Q_n(p)}\right)^{m}\right\}$$
$$= \mathcal{L}^{-1}\left\{\frac{n!}{Q_n(p)}\right\}*\ldots*\mathcal{L}^{-1}\left\{\frac{n!}{Q_n(p)}\right\}.$$

If $f_n(t) = \mathcal{L}^{-1}\{\frac{n!}{Q_n(p)}\}$, then by (3.5), $f_n(t) = e^{-j/m^t}\cdot\mathcal{L}^{-1}\{\frac{n!}{Q_n(p-j/m)}\}$.

We can write

$$R(t;n;m) = \int_0^t dt_1 \int_0^{t_1} \dots \int_0^{t_{m-1}} dt_{m-1}\cdot f_n(t-t_1)\cdot f_n(t_1-t_2)\times\dots$$

$$\dots\times f_n(t_{m-2}-t_{m-1})\cdot f_n(t_{m-1}).$$

We can now rewrite this as

$$R(t;n;m) = \int_0^t dt_1 \int_0^{t_1} \dots \int_0^{t_{m-1}} dt_{m-1} \times e^{-\frac{m-2}{m}(t-t_1)} \mathcal{L}^{-1}\{\frac{n!}{Q_n(p-\frac{m-1}{m})}\}(t-t_1)$$

$$\times e^{-\frac{m-2}{m}(t_1-t_2)} \times\dots\times e^{-\frac{1}{m}(t_{m-2}-t_{m-1})} \mathcal{L}^{-1}\{\frac{n!}{Q_n(p-\frac{1}{m})}\}(t_{m-2}-t_{m-1})$$

$$\times \mathcal{L}^{-1}\{\frac{n!}{Q_n(p)}\}(t_{m-1}).$$

Since for $b \geq 0$, $e^{-b} \leq 1$, we have the following upper and lower bounds for $R(t;n;m)$: For $t > 0$ we have

$$(3.7) \qquad e^{-\frac{m-1}{m}t}\cdot\mathcal{L}^{-1}\{\frac{n!}{Q_n(p)}\} *\dots* \mathcal{L}^{-1}\{\frac{n!}{Q_n(p-\frac{m-1}{m})}\}$$

$$\leq R(t;n;m) \leq \mathcal{L}^{-1}\{\frac{n!}{Q_n(p)}\} *\dots* \mathcal{L}^{-1}\{\frac{n!}{Q_n(p-\frac{m-1}{m})}\}.$$

We apply now established relation (3.3) together with rules (3.4) and (3.6). We obtain, taking into account (3.7) the following bounds:

$$e^{-\frac{m-1}{m}t}\cdot m^{m(n+1)-1}\cdot\mathcal{L}^{-1}\{\frac{n!^m}{Q_{nm+m-1}(p)}\}(\frac{t}{m})$$

$$\leq R(t;n;m) \leq m^{m(n+1)-1}\cdot\mathcal{L}^{-1}\{\frac{n!^m}{Q_{nm+m-1}(p)}\}(\frac{t}{m}).$$

Taking into account established (3.2) for $h = nm + m - 1$, we get

$$(3.8)\quad \frac{m^{m(n+1)-1}\cdot (n!)^m}{(nm+m-1)!}\, e^{-\frac{m-1}{m}t}\cdot (1 - e^{t/m})^{nm+m-1} \le R(t;n;m) \le \frac{m^{m(n+1)-1}\cdot (n!)^m}{(nm+m-1)!}\cdot (1 - e^{t/m})^{nm+m-1}$$

for $t > 0$.

Similar bounds are true for any complex t. If we are interested only in asymptotics, we have the following general result

Theorem 3.1: *Let $t \neq 0$ be real, and t and m be fixed. Then for large n we have the following asymptotics*

$$(3.9)\qquad |R(t;n;m)| \sim |1 - e^{t/m}|^{mn}$$

(i.e.

$$\frac{1}{n} \log|R(t;n;m)| \to m \log|1 - e^{t/m}|$$

as $n \to \infty$); for $t = \theta\cdot\sqrt{-1}$, $0 < \theta < \pi$ we have

$$(3.10)\qquad |R(\theta\sqrt{-1};n;m)| \sim (2 \sin \frac{\theta}{2m})^{mn}$$

as $n \to \infty$.

For a general t, let

$$(3.11)\qquad \Omega_m(t) = \{|1 - \zeta_m^L e^{t/m}| : L = 0,1,\ldots,m - 1\}$$

for $\zeta_m = \exp(2\pi i/m)$. If all elements of $\Omega_m(t)$ are distinct and if $|1 - \zeta_m^{L_0} e^{t/m}|$ is the smallest element from $\Omega_m(t)$, then for a fixed $t \neq 0$ we have

$$(3.12)\qquad \frac{1}{n} \log|R(t;n;m)| \to m \log|1 - \zeta_m^{L_0} e^{t/m}|$$

for $m \to \infty$.

There is also a very simple upper bound for the polynomial coefficients $P_j(e^t;n;m)$: $j = 0,1,\ldots,m-1$ that has the following form:

Theorem 3.2: *Let* $t \neq 0$ *be real and* m,t *be fixed. Then for a large* n,

$$\frac{1}{n}\log|P_j(e^t;n;m)| \leq m\log|1+e^{t/m}|.$$

Analogously, for $t = \theta\sqrt{-1}$, $0 < \theta < \pi$, *we have*

$$\frac{1}{n}\log|P_j(e^t;n;m)| \leq m\log|2\cos\frac{\theta}{2m}|$$

$j = 0,1,\ldots,m-1$.

For the proof we use simply the explicit expressions for the coefficients $c_{k,j+1}/j!$ of the polynomials $P_j(z;n;m)$. Using formulae (2.3) and (2.8) we obtain very simple bounds

$$(3.13)\qquad |c_{k,j+1}| \leq O(n)\cdot\binom{n}{k}^m, \qquad j = 0,1,\ldots,m-1,$$

where $O(\ )$ depends on m. Thus for a positive w,

$$|c_{k,j+1}w^k| \leq O(n)\binom{n}{k}^m w^k = O(n)\cdot\{\binom{n}{k}w^{k/m}\}^m: \quad k = 0,1,\ldots,n.$$

Now $\binom{n}{k}w^{k/m} \leq (1+w^{1/m})^n$ and so

$$|c_{k,j+1}w^k| = O(n)\cdot(1+w^{1/m})^{mn},$$

this established Theorem 3.2.

We have more precise bounds and we have also complete knowledge of the asymptotics of the $|P_j(e^t;n;m))$ for m and t fixed, $n \to \infty$: $j = 0,1,\ldots,m$, similar to that of the Theorem 3.1.

Theorem 3.3: *Let* m *be fixed and for a given* t *we define:*

$$\Omega_m(t) = \{|1-\zeta_m^L e^{t/m}|: L = 0,1,\ldots,m-1\}$$

as in (3.11) *for* $\zeta_m = \exp(2\pi i/m)$. *If all* m *elements of* $\Omega_m(t)$ *are distinct and if* $|1 - \zeta_m^{L_1} e^{t/m}|$ *is the maximal element of* $\Omega_m(t)$, *then*

$$\frac{1}{n} \log|P_j(e^t;n;m)| \to m \log|1 - \zeta_m^{L_1} e^{t/m}|$$

for $n \to \infty$.

For the proof we use the information that $P_j(z;n;m)$ satisfy a linear recurrence of the order m. Because the system of functions

$$\{1, \log(1 - v), \ldots, \log^{m-1}(1 - v)\}$$

is perfect at $v = 0$; there exists a linear recurrence

$$\Sigma_{\lambda=0}^{m} G_\lambda^k(v) \cdot F_{k+\lambda} = 0 : k = 0,1,2,\ldots ,$$

satisfied by polynomials

$$F_k = P_j(1 - v;k) : \quad j = 0,1,\ldots,m-1.$$

Here $G_\lambda^k(v)$ are polynomials in $1 - v = e^t$. Of course, the same recurrence is satisfied also by a remainder function

$$R(v;k) = \Sigma_{j=0}^{m-1} P_j(1 - v;k)\log^j(1 - v).$$

We can use now the following Poincaré lemma:

Lemma 3.4: *If* $G_\lambda^k(v) \to a_\lambda(v)$ *as* $k \to \infty$ *and if* $\xi_1,\ldots,\xi_m$ *are roots of the polynomial of the degree* m:

(C) $$\Sigma_{\lambda=0}^{m} \xi^\lambda a_\lambda(v) = 0,$$

then for any solution F_k *of the recurrence*

(R) $$\Sigma_{\lambda=0}^{m} G_\lambda^k(v) F_{k+\lambda} = 0 : \quad k = 0,1,2,\ldots$$

we have in general,

$$\frac{1}{k} \log|F_k| \to \log |\xi_L|$$

for some $L = 1,\dots,m$. *Moreover, if* $|\xi_1(v)| < \dots < |\xi_m(v)|$; *then there are* m *linearly independent solutions* $F_k^{(j)}$ *of* (R) *satisfying*

$$\frac{1}{k} \log|F_k^{(j)}| \to \log|\xi_m(v)|: j = 1,\dots,m$$

and only one solution G_k *of* (R) *such that*

$$\frac{1}{k} \log|G_k| \to \log|\xi_1(v)|.$$

We take this recurrence (R). The coefficients of (C) are polynomials in $1 - v = e^t$. In particular, roots $\xi_L(v)$ are algebraic in e^t. For real t we have at least one (smallest) zero

$$\xi_1(v) = (1 - e^{t/m})^m.$$

Thus, changing t to $t + 2\pi iL$ we get other $m - 1$ roots

$$\xi_{L+1}(v) = (1 - \zeta_m^L e^{t/m})^m: L = 1,\dots,m-1.$$

Because all zeroes are algebraic in e^t, for *any* complex t, all roots are given by

$$\xi_{L+1}(v) = (1 - \zeta_m^L e^{t/m})^m: L = 0,1,2,\dots,m-1.$$

Then largest root corresponds to the polynomial coefficients and smallest to a remainder function.

Now we can use our knowledge of asymptotics and the following simple

Lemma 3.5: *Let* θ *be a complex number and let us suppose we have the system of linear forms*

$$R_N = \Sigma_{i=0}^{m-1} \theta^i \cdot P_{i,N}$$

with rational integer coefficients $P_{i,N}$ *satisfying the following properties*:

i) *for* $N \to \infty$ *and* $|\alpha| < 1$,

$$|R_N| \sim \alpha^N \quad \text{or}$$

$$\frac{1}{N} \log|R_N| \to \log \alpha \quad \text{as} \quad N \to \infty;$$

ii) *for any* $i = 0,1,\dots,m-1$,

$$\frac{1}{N} \log|P_{i,N}| \leq \log \beta \quad \text{as} \quad N \to \infty.$$

Then for any rational integers p,q *we have*

$$|\theta - \frac{p}{q}| > |q|^{-\varkappa-\varepsilon} \quad \text{for} \quad |q| \geq q_0(\varepsilon)$$

and any $\varepsilon > 0$. *Here*

$$\varkappa = (m-1)\{-\frac{\log \beta}{\log \alpha} + 1\}$$

Proof: Let

$$|\theta - \frac{p}{q}| < |q|^{-\varkappa-3\varepsilon} \quad ; \quad |q| \geq q_0(\varepsilon).$$

We define

$$N = -[(1+\varepsilon)(m-1)^{\log|q|}/\log \alpha].$$

Let

$$\theta = \delta + \frac{p}{q}, \quad \text{so}$$

$$R_N = \Sigma_{i=0}^{m-1} \frac{p^i}{q^i} \cdot P_{i,N} + \delta \cdot C_m \cdot \max(|P_{i,N}|) + O(\delta^2).$$

Now $q^{m-1}R_N = A_N + \delta C' \cdot q^{m-1}\max(|P_{i,N}|) + O(\delta^2 q^{m-1}\max|P_{i,N}|)$ for an integer $A_N \in \mathbb{Z}$. By a choice of N,

$$|q^{m-1}R_N| < |q|^{-\varepsilon/2}.$$

If $A_N = 0$, then $|\delta| \geq |R_N| \cdot |\beta|^{-N(1+\varepsilon)}$ or

$$|\delta| \geq |q|^{-\varkappa-2\varepsilon} \qquad \text{q.e.d.}$$

If however, $|A_N| \geq 1$, then

$$|\delta| \geq q^{-m+1}\beta^{-N(4+\varepsilon)} \geq |q|^{-\varkappa-3\varepsilon} \qquad \text{q.e.d.}$$

Let's show you how this works on an example of the approximation to $\log(A/B)$. Let A and B be rational integers, $A > B > 1$ and m be an integer ≥ 2 such that

$$A^{1/m} - B^{1/m} < e^{1-1/m},$$

indeed such m exists. We take

$$t = \log(\tfrac{A}{B}).$$

Then

$$R(t;n;m) = \Sigma_{j=0}^{m-1} \log^j(\tfrac{A}{B}) \cdot P_j(\tfrac{A}{B};n;m)$$

for polynomials of order n. Then by 2.1 numbers

$$(m-1)! \cdot [1,\ldots,n]^{m-1} \cdot B^n \cdot P_j(\tfrac{A}{B};n;m) = P_{j,n}$$

$j = 0,1,\ldots,m-1$ are rational integers. We notice now that

$$[1,\ldots,n] = e^{n+o(n)} \quad \text{for} \quad n \to \infty.$$

Indeed

$$\log[1,\ldots,n] = \Sigma_{p^m \leq n} \; m \log p,$$

and by the prime number theorem

$$\Sigma_{p\leq n} \log p = n + o(n) \quad \text{as} \quad n \to \infty.$$

Thus, for $n \to \infty$ by Theorem 3.2 we have

$$\frac{1}{n} \log|P_{j,n}| \leq \log(e^{m-1}\cdot B\cdot((\frac{A}{B})^{1/m} + 1)^m)$$

as $n \to \infty$ and $j = 0,1,\ldots,m-1$ and for

$$R_n = \Sigma_{j=0}^{m-1} \log^j(\frac{A}{B})\cdot P_{j,n}$$

we have by Theorem 3.1,

$$\frac{1}{n} \log|R_n| \to \log(e^{m-1}\cdot B\cdot((\frac{A}{B})^{1/m} - 1)^m).$$

Thus by Lemma 3.5,

$$|\log(\frac{A}{B}) - \frac{p}{q}| > |q|^{-\chi-\varepsilon}$$

for

$$\chi = (m-1)\cdot\{-\frac{\log(e^{m-1}(A^{1/m} + B^{1/m})^m)}{\log(e^{m-1}(A^{1/m} - B^{1/m})^m)} + 1\}$$

and $|q| \geq q_0(m,A,B,\varepsilon)$.

Similarly we can consider $t = \theta\sqrt{-1}$ and then consider

$$\text{Re } r(e^{\theta\sqrt{-1}};n) \quad \text{and} \quad \text{Im } R(e^{\theta\sqrt{-1}};n).$$

We obtain immediately extremely important measure of irrationality of numbers connected with π[3]:

Corollary 3.6: We have for $\varepsilon > 0$ and rational integers p, q for $|q| \geq q_1(\varepsilon)$ the following bounds

$$|q\cdot\pi\sqrt{3} - p| > |q|^{-\chi_1-\varepsilon}$$

$$\chi_1 = -\frac{\log(e(2\cos\pi/12)^2)}{\log(e(2\sin\pi/12)^2)} = 7.30998634\ldots$$

(here $m = 2$, $t = \pi/3\cdot\sqrt{-1}$);

$$|q\cdot\pi - p| > |q|^{-\chi_2-\varepsilon}$$

for

$$\chi_2 = 4 - 5\,\frac{\log(e^5(2\cos\pi/24)^6)}{\log(e^5(2\sin\pi/24)^6)} = 18.88999444\ldots$$

(here $m = 6$, $t = \pi/2\sqrt{-1}$).

Moreover for any $|q| \geq 2$ we have

$$|q\cdot\pi\sqrt{3} - p| > |q|^{-7.31}$$

and

$$|q\cdot\pi - p| \geq |q|^{-18.9}.$$

References

[1] K. Mahler, Lectures on transcendental numbers, Lecture Notes in Math., v. 546, Springer, 1976.

[2] Ch. Hermite, Oeuvres, v. 3, Gauthier-Villars, Paris, 1917.

[3] G.V. Chudnovsky, C.R. Acad. Sci. Paris, Series A, v. 288 (1979), A-965-A-967.

[4] G.V. Chudnovsky, Lecture Notes in Math., v. 751, 1979, pp. 45-69.

[5] L. Euler, Mémoires de l'Académie Imperiale des sciences de St. Petersborg, v. 6 (prepared in March 1737, published in 1744).

[6] H.S. Wall, Analytic theory of continued fractions, Chelsea, N.Y. 1973.

[7] A.Y. Khintchine, Continued fractions, Univ. of Chicago Press, 1964.

[8] M. Mignotte, Bull. Soc. Math. France, Mem. 37 (1974), 121-132.

[9] G. V. Chudnovsky, Padé approximation and the Riemann monodromy problem, Cargese Lectures, June 1979; in Bifurcation phenomena in mathematical physics and related topics, D. Reidel Publishing Company, Boston, 1980, pp. 448-510.

Department of Mathematics
Columbia University
New York, NY
USA

Criteria of Algebraic Independence of Several Numbers

by

G.V. Chudnovsky

§1.

While for a single number one has a very powerful Gelfond criterion of transcendence, for more than one number there are only a few results that can be considered satisfactory. We present a brief survey of them and formulate their improvements. Let us stress one new feature of the proposed criteria. We formulate them in such a way that they immediately imply results on the measure of transcendence or the measure of algebraic independence. Such an approach explains why the Gelfond criterian must be changed to a different kind of statement.

Let us start with the well known Gelfond criterion, that we present in the case of splitting of size and degree.

Gelfond Lemma 1.1 (Brownawell [4] and Waldschmidt [6]): *Let $\theta \in \mathbb{C}$ and $a > 1$. Let δ_N and σ_N be monotonically increasing sequences of positive numbers such that $\sigma_N \to \infty$ and*

$$\delta_{N+1} \leq a\delta_N, \qquad \sigma_{N+1} < a\sigma_N.$$

If for every $N \geq N_0$ there is a non-zero polynomial $P_N(z) \in \mathbb{Z}[z]$ such that

$$\deg P_N < \delta_N, \qquad t(P_N) = \deg P_N + \log H(P_N) < \sigma_N;$$

and

$$|P_N(\theta)| < \exp(-6a\delta_N\sigma_N),$$

then θ *is algebraic and* $P_N(\theta) = 0$: $N \geq N_1$.

Usually this criterion is applied for $\delta_N = \sigma_N = O(N^\lambda)$ (with some interesting exceptions).

When one tries to generalize this criterion for two numbers, one finds the famous Cassels' [11] counterexample of pair of numbers $(\vartheta_{1,f},\vartheta_{2,f})$ associated with any monotone functions $f(h) \to \infty$ such that inequalities

$$|x\vartheta_{1,f} + y\vartheta_{2,f} + z| < \exp(-f(h));$$

$$\max(|x|,|y|,|z|) \leq h$$

have solutions in integers x,y,z for $h \geq h_0$.

However, the construction of the numbers $(\vartheta_{1,f},\vartheta_{2,f})$ by Cassels shows that both $\vartheta_{1,f}$ and $\vartheta_{2,f}$ are "too well" approximated by rational numbers.

In order to take this feature into account, one tries to use the fact that the subsequence of a given sequence $(\theta_1,\ldots,\theta_n)$ has small "transcendence type" in the sense of Lang:

<u>Definition</u> 1.2: Let $(\theta_1,\ldots,\theta_n) \in \mathbb{C}^n$ and $\tau \geq n+1$. The set $(\theta_1,\ldots,\theta_n)$ is said to have transcendence type at most τ $(\leq \tau)$ if there exists a constant $C > 0$ such that for every non-zero $P(x_1,\ldots,x_n) \in \mathbb{Z}[x_1,\ldots,x_n]$ we have

$$\log|P(\theta_1,\ldots,\theta_n)| > C\cdot t(P)^{\tau}.$$

We can't afford not to interrupt the exposition and attract attention to two problems:

<u>Problem</u> 1.3: Let $\varepsilon > 0$ and $n \geq 2$. Is it true that almost all numbers $(\theta_1,\ldots,\theta_n) \in \mathbb{C}^n$ (or $\mathbb{R}^n$) with respect to a Lebesque measure, have transcendence type $\leq n+1+\varepsilon$?

While $n = 2$ is within the possibilities of the existing methods, $n > 2$ seems a hard question.

<u>Problem</u> 1.4: To find for $n \geq 1$ an example of a set $(\theta_1,\ldots,\theta_n) \in \mathbb{R}^n$ with the type of transcendence $n+1$.

One can suspect that for $n = 1$ this is π; for $n = 2$ this is $(\pi,\Gamma(1/3))$ or $(\pi,\Gamma(1/4))$: their type of transcendence known to be $\leq n + 1 + \varepsilon$ for any $\varepsilon > 0$.

For sequences, whose subsequences have bounded types of transcendence we do have natural generalizations of Gelfond criterion.

Proposition 1.5 (Brownawell): *Let* $(\theta_1,\ldots,\theta_n)$ *have transcendence type* $\leq \tau$. *Let* $\theta \in \mathbb{C}$, $a > 1$, δ_N *and* σ_N *be monotonically increasing sequences of positive numbers such that* $\sigma_N \to \infty$ *and*

$$\delta_{N+1} < a\delta_N, \qquad \sigma_{N+1} < a\sigma_N.$$

There is a $C_1 > 0$ *such that, if for every* $N \geq N_0$ *there is a non-zero polynomial* $P_N(x_0,x_1,\ldots,x_n) \in \mathbb{Z}[x_0,x_1,\ldots,x_n]$ *with*

$$\deg_{x_0}(P_N) < \delta_N, \qquad t(P_N) < \sigma_N$$

and

$$\log|P_N(\theta,\theta_1,\ldots,\theta_n)| < -C_1(\delta_N\sigma_N)^\tau$$

then θ *is algebraic over* $\mathbb{Q}(\theta_1,\ldots,\theta_n)$ *and* $P_N(\theta,\theta_1,\ldots,\theta_n) = 0$ *for all* $N \geq N_1$.

Certainly, the Proposition 1.5 is not very sharp and one can wonder, whether the upper bound $-C_1(\delta_N\sigma_N)^\tau$ for $\log|P_N(\theta,\theta_1,\ldots,\theta_n)|$ can be substituted by $-C_2\delta_N\sigma_N^\tau$.

There is a possibility of a more careful analysis of the situation like in Proposition 1.5, which was considered by the author under the name of "colored sequences". We present the early result in this direction following the reformulation of D. Brownawell [12]:

Proposition 1.6: *Let* $(\theta_1,\ldots,\theta_n)$ *have transcendence type* $\leq \tau_1$ *and* $(\theta_2,\ldots,\theta_n)$ *have transcendence type* $\leq \tau_2$ *and let* $\theta \in \mathbb{C}$. *There exists a constant* $C_3 > 0$ *such that, if there exists a sequence of non-zero polynomials* $P_N(x_0,x_1,\ldots,x_n) \in \mathbb{Z}[x_0,x_1,\ldots,x_n]$ *for* $N \geq N_0$ *with*

$$t(P_N) \leq N$$

<u>and</u>

$$\log|P_N(\theta,\theta_1,\ldots,\theta_n)| < -C_3 N^{\max\{4\tau_2,\tau_1+3\tau_2-1\}}$$

<u>then</u> θ <u>is algebraic over</u> $\mathbb{Q}(\theta_1,\ldots,\theta_n)$.

Usually this statement is considered for $n = 1$, when $\tau_2 = 1$ and the upper bound for $\log|P_N(\theta,\theta_1)|$ is $-C_3N^{\max(4,\tau_1+2)}$ (in any case $\tau_1 \geq 2$). It is very easy to improve considerable this last result, what will be done later.

However, one wants unconditional generalizations of the Gelfond lemma [1], without any references to the diophantine properties of subsequences. For this there are two options: i) to impose some algebraic conditions on the polynomials $P_N(\bar{x})$ in the Gelfond lemma; ii) to add more analytic restrictions on $|P_N(\bar{\theta})|$, say. The possibility i) is the most interesting one and opens a big future when it will be combined with the abstract definition of the auxiliary function as an abstract Padé approximation. Nevertheless, the possibility ii) is also interesting, though less algebraic in its formulation.

The generalization of the Gelfond criterion using ii) was proposed by the author in 1975. In its initial form it was formulated as follows [2]:

<u>Proposition</u> 1.7: <u>Let</u> $n \geq 1$, $(\theta_1,\ldots,\theta_n) \in \mathbb{C}^n$, $a > 1$ <u>and let</u> σ_N <u>be monotonically increasing function with</u> $\sigma_N \to \infty$ <u>as</u> $N \to \infty$ <u>and</u>

$$\sigma_{N+1} < a\sigma_N.$$

Let us assume that for every $N \geq N_0$ there is a non-zero polynomial $P_N(x_1,\ldots,x_n) \in \mathbb{Z}[x_1,\ldots,x_n]$ such that

$$t(P_N) \leq \sigma_N;$$

$$-C_4\sigma_N^{2^n} < \log|P_N(\theta_1,\ldots,\theta_n)| < -C_5\sigma_N^{2^n}$$

for $N \geq N_0$. Then $\theta_1,\ldots,\theta_n$ are algebraically dependent (over $\mathbb{Q}$).

The exponent 2^n is the best only for $n = 1$. Naturally, the lower and upper bounds for $\log|P_N(\theta_1,\ldots,\theta_n)|$ can be improved and the statement of this criterion of algebraic independence can be reformulated in a better way.

However, this criterion still looks slightly artificial, because in practice one can't find such a sequence $(\theta_1,\dots,\theta_n)$ that $\log|P_N(\theta_1,\dots,\theta_n)|$ is bounded below and above by a function of $t(P_N)$ of an order $O(t(P_N)^{n+1})$ or $o(t(P_N)^{n+1})$ for $P_N(x_1,\dots,x_n) \in Z[x_1,\dots,x_n]$.

Moreover I don't know even, whether such a sequence $(\theta_1,\dots,\theta_n) \in \mathbb{C}^n$ exists (and if it does) what is the measure of such a set of sequences?

Essentially, all previously formulated criteria of algebraical independence are presented. Now we will present new ones. Most of them copy the style of existing criteria but with some new features. For example, the most important development we are trying to pursue, is an attempt to get at once and for all the results not in the form of the algebraic independence, but rather in the form of the measure of the algebraic independence. We take this approach because we want to present an unified approach and measure of algebraic independence.

As one sees, the changes in the criteria are rather minor.

In order to get an idea what kind of criteria of algebraic independence we can propose, I can suggest to you the following reformulation of Brownawell criterion.

<u>Lemma</u> 1.8: <u>Let</u> $\theta = (\theta_1,\theta_2)$ <u>be algebraically independent numbers,</u> $a > 1$ <u>and</u> σ_N <u>is a monotonically increasing function</u> $\sigma_N \to \infty$ <u>such that</u>

$$\sigma_{N+1} < a\sigma_N.$$

We suppose that for every $N \geq N_0$ there exists a polynomial $P_N(x,y) \in \mathbb{Z}[x,y]$, $P_N \not\equiv 0$, such that

$$t(P_N) \leq \sigma_N$$

and

$$\log|P_N(\theta_1,\theta_2)| < -\sigma_N^{\mu}.$$

Then $\bar{\theta}$ has type of transcendence $\geq \mu/2$ and, moreover there are infinitely many algebraic numbers ξ_1, ξ_2 such that

$$[\mathbb{Q}(\xi_1,\xi_2) : \mathbb{Q}] \le L;$$

$$[\mathbb{Q}(\xi_1,\xi_2) : \mathbb{Q}] \cdot \left(\frac{t(\xi_1)}{d(\xi_1)} + \frac{t(\xi_2)}{d(\xi_2)}\right) \le L$$

and

$$|\theta_1 - \xi_1| + |\theta_2 - \xi_2| < \exp(-C \cdot L^{\mu/2}).$$

In particular, one gets Brownawell statement. However one immediately sees that the simultaneous approximation is of non-trivial type. Roughly speaking, in the "generic case" $[\mathbb{Q}(\xi_1,\xi_2) : \mathbb{Q}]$ is the product $d(\xi_1) \cdot d(\xi_2)$. If this would be true, then

$$d(\xi_1)d(\xi_2) \le L;$$

$$d(\xi_2)t(\xi_1) + d(\xi_1)t(\xi_2) \le L.$$

This implies for one of θ_1, θ_2, say θ_1, the satisfaction of infinitely many inequalities in algebraic numbers ξ_1:

$$|\theta_1 - \xi_1| < \exp(-C(d(\xi_1)t(\xi_1))^{\mu/2}),$$

almost as if θ_1 is of the transcendence type $\ge \mu$! (Not $\mu/2$.) Simple counterexample will show that that's impossible (say $\mu = 3$ e.g.), which means that $[\mathbb{Q}(\xi_1,\xi_2) : \mathbb{Q}] \ll d(\xi_1)d(\xi_2)$. The real truth is of course in the statement like

$$[\mathbb{Q}(\xi_1,\xi_2) : \mathbb{Q}] = d(\xi_1) = d(\xi_2),$$

i.e. both ξ_1,ξ_2 are generators of $\mathbb{Q}(\xi_1,\xi_2)$. One knows e.g. that in this case the Liouville theorem for $|P(\xi_1,\xi_2)|$ looks much better than in a general case.

In any case, we can speak about simultaneous approximation (and promote this direction).

§2.

We propose first of all the following form of the criterion of the measure of transcendence of a single number:

Lemma 2.1: *Let* $\theta \in \mathbb{C}$ *and let* $\mathfrak{M}$ *be a linearly ordered countable set and functions* δ_η, σ_η, F_η, G_η, K_η *be monotonically increasing as functions of* $\eta \in \mathfrak{M}$ *such that* $\sigma_\eta \to \infty$, $F_\eta \to \infty$, $G_\eta \to \infty$, $K_\eta \to \infty$ *as* $\eta \to \infty$ *together with the conditions* $\delta_\eta \leq \sigma_\eta$ *and*

$$\sigma_\mu < a.\sigma_\eta$$

for a successor μ *of* η *in* $\mathfrak{M}$. *Let for any* $\eta \geq \eta_0$ *either*

i) *there exists a polynomial* $P_\eta(x) \in \mathbb{Z}[x]$ *such that*

$$d(P_\eta) \leq \delta_\eta, \qquad t(P_\eta) \leq \sigma_\eta$$

and

$$-G_\eta < \log|P_\eta(\theta)| < -F_\eta;$$

or

ii) *there is a system of polynomials* $C_\ell(x) \in \mathbb{Z}[x]: \ell \in \mathcal{L}_\eta$ *without a common factor such that*

$$d(C_\ell) \leq \delta_\eta, \qquad t(C_\ell) \leq \sigma_\eta;$$

$$\log|C_\ell(\theta)| < -K_\eta:$$

$\ell \in \mathcal{L}_\eta$. *Let* $\lim \sigma_\eta/F_\eta = 0$ *as* $\eta \to \infty$ *in* $\mathfrak{M}$. *Then for any algebraic number* ζ *of the degree* $\leq d(\zeta)$ *and height* $\leq H(\zeta)$ *we get for any* $\eta \in \mathfrak{M}$, $\eta \geq \eta_0$

$$(2.2) \qquad \log|\theta - \zeta| > -\max\{G_\eta + C_1\sigma_\eta, F_\eta + C_1\sigma_\eta, K_\eta + C_1\sigma_\eta\},$$

provided that

$$\min\{K_\eta, F_\eta\} \geq \sigma_\eta d(\zeta) + \delta_\eta \log H(\zeta) + 1.$$

In particular, let F_η, G_η, K_η be of the same order of magnitude

$$\lim G_{\eta}/F_{\eta} = \gamma, \qquad \lim K_{\eta}/F_{\eta} = \chi$$

and $\gamma,\ \neq 0,\infty$. We choose η_1 from the condition

$$C_2 F_{\eta_1} \geq \sigma_{\eta_1} d(\zeta) + \delta_{\eta_1} \log H(\zeta) \geq C_2^{-1} F_{\eta},$$

for some $C_2 > 0$. Then we have

$$(2.3) \qquad |\theta - \zeta| > \exp(-C_3 F_{\eta_1})$$

for $C_3 > 0$ depending on γ, χ and C_2.

<u>Remark</u> 2.4: This statement covers all the cases: a) when the type of the transcendence is estimated and $\mathfrak{M} = \mathbb{N}$, $\delta_N = \sigma_N$; b) when type and the degree are estimated simultaneously, $\mathfrak{M} \subseteq \{(x,y) \in \mathbb{N}^2 : x \leq y\}$ with a lexicographic order; or c) when the degree is bounded: $\mathfrak{M} = \mathbb{N} \times \{1,\ldots,d\}$. E.g. let us consider the case a). We get

<u>Corollary</u> 2.5: <u>Let</u> $\mathfrak{M} = \mathbb{N}$, $\delta_N = \sigma_N$ <u>and</u> $f(t)$ <u>be a function inverse to</u> $F(t)/\sigma(t)$, <u>where</u> $F(N) = F_N$, $\sigma(N) = \sigma_N$, <u>etc. Then under the assumption of the lemma</u> 2.1,

$$(2.6) \qquad |\theta - \zeta| > \exp\{-C_4\ F(f(t(\zeta)))\}$$

<u>where</u>

$$t(\zeta) = d(\zeta) + \log H(\zeta) \geq t_0.$$

The lemma 2.1 contains, in particular, the sharpened form of the Gelfond's criterion. Indeed, when F_{η} grows faster than $\delta_{\eta}\sigma_{\eta}$ and K_{η} grows faster than $\delta_{\eta}\sigma_{\eta}$, then the bound (2.2) contradicts to a Dirichlet's bound.

The bound (2.2) or (2.3), (2.6) can be, naturally, reformulated for $|P(\theta)|$ with $P(x) \in \mathbb{Z}[x]$ instead of $|\theta - \zeta|$:

<u>Proposition</u> 2.7: <u>Let us assume that the conditions of</u> 2.1 (<u>or</u> 2.5) <u>are satisfied. Then for</u> $P(x) \in \mathbb{Z}[x]$, $P(x) \neq 0$, <u>the same bounds</u> (2.2),

(2.3) (or (2.6)) are satisfied for $|\theta - \zeta|$ replaced by $|P(\theta)|$, if one replaces $d(\zeta)$ by $d(P)$, $H(\zeta)$ by $H(P)$ (and $t(\zeta)$ by $t(P)$). In particular, in the situation of the Corollary 2.5 we get:

$$|P(\theta)| > \exp(-C_5 F(f(t(P))))$$

for some $C_5 > 0$.

Proof of Lemma 2.1: Let us take a sufficiently large $C > 0$ and $\mathfrak{n}_0 \in \mathfrak{M}$ such that

$$\min\{F_{\mathfrak{n}}, G_{\mathfrak{n}}, K_{\mathfrak{n}}\} > C\sigma_{\mathfrak{n}}$$

if $\mathfrak{n} \geq \mathfrak{n}_0$. Let ζ be an algebraic number and $P(\zeta) = 0$ with $d(\zeta) \leq d(P)$, $H(\zeta) \leq H(P)$ for $P(x) \in \mathbb{Z}[x]$, $P(x) \not\equiv 0$. One can take $P(x)$ as a minimal polynomial of ζ, but in view of the Proposition 2.7, there is no need to do this.

Let us take $\mathfrak{n} \geq \mathfrak{n}_0$. If for a given $\mathfrak{n}$ the alternative ii) takes place, there is such a $\ell_0 \in \mathcal{L}_{\mathfrak{n}}$ that $C_{\ell_0}(\zeta) \neq 0$, because at least one of the $C_\ell(x)$ is relatively prime with the minimal polynomial of ζ. If $P(x)$ is the power of an irreducible polynomial, then again there is a $C_{\ell_0}(x)$ relatively prime with $P(x)$. However, if $P(x)$ is a reducible polynomial, some modifications of the arguments are necessary. Namely, one finds integer coefficients n_ℓ: $\ell \in \mathcal{L}_{\mathfrak{n}}$ such that

$$\Sigma_{\ell \in \mathcal{L}_{\mathfrak{n}}} |n_\ell| \ll d(P)$$

and $\Sigma_{\ell \in \mathcal{L}_{\mathfrak{n}}} n_\ell C_\ell(x)$ is relatively prime with $P(x)$. This leads only to slight changes in the estimates- in constants.

Henceforth, in the case ii) we can bound below $|C_{\ell_0}(\zeta)|$ using the Liouville theorem. An alternative approach is to bound $\mathrm{res}(C_{\ell_0}, P)$ according to a famous formula [1]:

$$\mathrm{res}(C_{\ell_0}, P) \leq \{d(C_{\ell_0})H(P)|P(\zeta)| + d(P\_H(C_{\ell_0})|C_{\ell_0}(\zeta)|\} \times H(C_{\ell_0})^{d(P)-1} \cdot H(P)^{d(C_{\ell_0})-1}.$$

When we are bounding $|C_{\ell_0}(\zeta)|$ we get:

$$|C_{\ell_0}(\zeta)| \geq \exp(-d(\zeta)t(C_{\ell_0}) - d(C_{\ell_0})\log H(\zeta)),$$

or

$$|C_{\ell_0}(\zeta)| \geq \exp(-d(\zeta)\sigma_{\eta} - \log H(\zeta)\delta_{\eta}).$$

At the same time, if

$$K_{\eta} \geq d(\zeta)\sigma_{\eta} + \log H(\zeta)\delta_{\eta} + 1,$$

then we have

$$C_{\eta}(\theta) = C_{\eta}(\zeta) + C'_{\eta}(\theta - \zeta)$$

with

$$\log|C'_{\eta}| \leq \sigma_{\eta} + c\cdot\delta_{\eta},$$

we have

$$|\theta - \zeta| \geq |C'_{\eta}|^{-1}||C_{\eta}(\theta)| - |C_{\eta}(\zeta)||$$

$$\geq \exp\{-K_{\eta} - \sigma_{\eta} - C\delta_{\eta}\}$$

by the inequality on K_{η}.

Now, let us consider the alternative i). In this case there are two possibilities that can happen: $P_{\eta}(\zeta) = 0$ or $P_{\eta}(\zeta) \neq 0$.

If $P_{\eta}(\zeta) \neq$), then the same as above, one gets

$$|P_{\eta}(\zeta)| \geq \exp(-\sigma_{\eta}d(\zeta) - \delta_{\eta}\log H(\zeta)).$$

We have again:

$$|\theta - \zeta| \geq |P'_{\eta}|^{-1}\cdot||P_{\eta}(\theta)| - |P_{\eta}(\zeta)||,$$

where

$$\log|P'_{\eta}| \leq \sigma_{\eta} + C\cdot\delta_{\eta}.$$

If

$$F_{\eta} \geq d(\zeta)\sigma_{\eta} + \log H(\zeta)\cdot\delta_{\eta} + 1,$$

then

$$|\theta - \zeta| > \exp\{-F_{\eta} - \sigma_{\eta} - C\delta_{\eta}\}.$$

Let, at last, $P_{\eta}(\zeta) = 0$. Then

$$|P_{\eta}(\theta)| \leq \exp\{\sigma_{\eta} + C\delta_{\eta}\}\cdot|\theta - \zeta|.$$

The inequalities in i) show

$$|\theta - \zeta| > \exp\{-G_{\eta} - \delta_{\eta}\cdot C - \sigma_{\eta}\}.$$

The inequality (2.2) is proved.

§3.

Let us formulate a version of the Liouville theorem in the case of an arbitrary set $S \subset \mathbb{C}^n$ of algebraic numbers of the dimension zero, i.e. being a set of common zeros of a zero-dimensional ideal in $\mathbb{Z}[x_1,\ldots,x_n]$. Namely, we consider the following situation. We have an ideal J in $\mathbb{Z}[x_1,\ldots,x_n]$ which is a zero-dimensional in the sense, e.g. that the set

$$S(J) = \{\vec{x}_0 \in \mathbb{C}^n : P(\vec{x}_0) = 0 \quad \text{for every} \quad P \in J\}$$

is a finite set. We are working now in the affine situation since projective considerations do not add anything. Every element of $S(J)$ has a prescribed multiplicity, defined e.g. in Schafarevitch's book [16].

Let $P_1,\ldots,P_k$ be certain generators of J, whose degrees and types we know:

$$d(P_i) \leq D_i, \qquad t(P_i) \leq T_i: \qquad i = 1,\ldots,k.$$

Naturally, $k \geq n$. By an intersection theory (say, Bergout theorem) we have

$$|S(J)| \leq D_1 \ldots D_k.$$

This bound is far from optimal whenever $k > n$. In this case we can use even the following bound

$$|S(J)| \leq (\max_{i=1,\ldots,k} D_i)^n.$$

Instead of treating different cases, we assume already that

$$P_1(x),\ldots,P_n(x)$$

have only finitely many common zeros. Then we consider an ideal $I = (P_1,\ldots,P_n)$ and $S(I)$ instead of $S(J)$. Let $\vec{x}_0 \in S(I)$ and $m(\vec{x}_0)$ be a multiplicity of $\vec{x}_0$ in $S(I)$. Then we have

$$\Sigma_{\vec{x}_0 \in S(I)} m(\vec{x}_0) \leq D_1 \ldots D_n.$$

We can apply to I the theory of u-resultants in the form of Kronecker (cf. Wan-der-Waerden [17] or Hodge-Pidoe [15]). One gets the following main statement:

3.1 Lemma: <u>In the notations above, the coordinates of common zeros</u> $\vec{x}_0 \in S(J)$ <u>are bounded above in terms of</u> $D_1,\ldots,D_n$ <u>and</u> $T_1,\ldots,T_n$. <u>Namely, let</u>

$$\vec{x}_0 = (x_{10},\ldots,x_{n0}) \quad \text{for} \quad \vec{x}_0 \in S(I).$$

Then for every $i = 1,\ldots,n$ there is such a rational integer A_i, $A_i \neq 0$, such that for any distinct elements $\vec{x}^1,\ldots,\vec{x}^\ell$ of $S(I)$ and $n_i \leq m(\vec{x}^i)$: $i = 1,\ldots,\ell$, the number

$$A_i \cdot \Pi_{j=1}^{\ell} (\vec{x}^j)_i^{n_j}$$

is an algebraic integer. Moreover, for any $i = 1,\ldots,n$ one has

$$|A_i| \cdot \max_{\vec{x} \in S(I)} \{1, (\vec{x})_i\}^{m(\vec{x})}$$

$$\leq \exp(C_1 \cdot \Sigma_{r=1}^n T_r \cdot \Pi_{s \neq r, s=1}^n D_s),$$

for a constant $C_1 > 0$ depending only on n.

The analogue of the Liouville theorem applied to the elements of the set $S(I)$ has the following form:

<u>Lemma</u> 3.2: <u>Let, as before,</u> $I = (P_1,\ldots,P_n)$ <u>where</u> $d(P_i) \leq D_i$, $t(P_i) \leq T_i$: $i = 1,\ldots,n$ <u>and the set</u> $s(I)$ <u>is a finite one</u>.

Let $R(x_1,\ldots,x_n) \in \mathbb{Z}[x_1,\ldots,x_n]$, $R \neq 0$. Let us assume that for several distinct $\vec{x}^1,\ldots,\vec{x}^\ell$ from $s(I)$ of multiplicities $m_1,\ldots,m_\ell$, respectively, we have

$$R(\vec{x}^j) \neq 0: j = 1,\ldots,\ell.$$

Then we have the following lower bound:

$$\Pi_{j=1}^{\ell} |R(\vec{x}^j)|^{m_j}$$

$$\geq \exp\{-C_2\{\Sigma_{i=1}^n t(P_i) d(R) \cdot \Pi_{s \neq i} d(P_i) + t(R) d(P_1) \cdots d(P_n)\}\}, \tag{3.3}$$

where $C_2 > 0$ depends only on n.

<u>Proof of Lemma 3.2</u>: We consider the following auxiliary object, taking into account the notations of the lemma 3.1:

$$\mathfrak{M} = \Pi_{i=1}^n A_i^{d(R)} \Pi_{\substack{\vec{x} \in S(I) \\ R(\vec{x}) \neq 0}} R(\vec{x})^{m(\vec{x})}, \tag{3.4}$$

in (3.4) the product is over only those elements $\vec{x}$ of $S(I)$ for which $R(\vec{x}) \neq 0$. This is a usual "semi-norm". Then the definition of the set $S(I)$ (invariant under the algebraic conjugation) and the choice of A_i: $i = 1,\ldots,n$ in Lemma 3.1 we get: $\mathfrak{M} \in \mathbb{Z}$. From the form of $\mathfrak{M}$ it follows that $\mathfrak{M} \neq 0$, so that

$$|\mathfrak{M}| \geq 1. \tag{3.5}$$

We represent $\mathfrak{M}$ as a product of two factors: $\mathfrak{M} = \mathfrak{A}\cdot\mathfrak{B}$, where $\mathfrak{A}$ is the product in the left hand side of (3.3). The product $\mathfrak{B}$ is bounded from above by the lemma 3.1:

$$|\mathfrak{B}| \leq (\Pi_{i=1}^{n} A_i \Pi_{\substack{\vec{x}\in S(I),\\ \vec{x}\notin\{\vec{x}^j:j=1,\ldots,L\}\\ R(\vec{x})\neq 0}} \max(1, |(\vec{x})_i|)^{m(\vec{x})})^{d(R)}$$

$$\times \exp(2t(R)\Pi_{i=1}^{n} d(P_i) \leq \exp(C_4\{t(R)\Pi_{i=1}^{n} d(P_i) + d(R)\Sigma_{i=1}^{n} t(P_i)\Pi_{s\neq i} d(P_s)\}). \tag{3.6}$$

Combining (3.5) and (3.6) one gets (3.3).

In order to prove our results in a straight-forward way, we make some agreements on the notations, that will simplify our symbolic mess.

If we start, in the general case with the ideal $I = (P_1,\ldots,P_n)$ in $\mathbb{Z}[x_1,\ldots,x_n]$ of the dimension zero, then the set

$$S(I) = \{\vec{x}_0 \in \mathbb{C}^n : P_i(\vec{x}_0) = 0: \ i = 1,\ldots,n\}$$

is a set of vectors in $\mathbb{C}^n$ with algebraic coordinates. The set S(I) is naturally divided into components S_α closed under the conjugation:

$$S(I) = \cup_{\alpha\in A} S_\alpha.$$

The partition into sets S_α can be made in such a way that all elements of S_α have the same multiplicity m_α of its occurring in S(I). We call S_α an irreducible component of S(I) and we have

$$\Sigma_{\alpha\in A} m_\alpha |S_\alpha| \leq d(P_1) \cdots d(P_n).$$

We can define a type and degree of the component S_α. The degree is, naturally $|S_\alpha|$ itself and the type is defined using the sizes of the coordinates of elements of S(I).

For this we remained at lemma 3.1, where we had non-zero rational integers A_i: i = 1,...,n such that

$$A_i \cdot \Pi_{\vec{x}\in S'} (\vec{x})_i^{n(\vec{x})}$$

is an algebraic integer for any $S' \subseteq S(I)$ and $n(\vec{x}) \leq m(\vec{x})$: $\vec{x} \in S'$; and we have a bound

$$|A_i| \cdot \Pi_{\vec{x}\in S(I)} \max\{1, |(\vec{x})_i|^{m(\vec{x})}\}$$

$$\leq \exp\{C_1 \cdot \Sigma_{j=1}^{n} t(P_j) \cdot \Pi_{s\neq j} d(P_s)\}: i = 1,\ldots,n.$$

Naturally, the quantity

$$\log\{\Pi_{i=1}^{n} |A_i| \cdot \Pi_{\vec{x}\in S(I)} \max\{1, |(\vec{x})_i|\}^{m(\vec{x})}\}$$

can be called <u>the</u> <u>size</u> of S(I). We can define in a similar way the size of the component S_α as

$$\log\{\Pi_{i=1}^{n} |a_i^\alpha| \cdot \Pi_{\vec{x}\in S_\alpha} \max\{1, |(\vec{x})_i|\}\},$$

where a_i^α are smallest non-zero rational integers such that

$$a_i^\alpha \cdot \Pi_{\vec{x}\in S_1} (\vec{x})_i$$

are algebraic integers for any $S_1 \subseteq S_\alpha$: i = 1,...,n.

By the <u>type</u> of the set S_α we understand the sum of the degree $|S_\alpha|$ and its size. We denote the type of S_α by <u>$t(S_\alpha)$</u>. Similarly one can define a type of s(I) as a sum of its degree and size. The type of S(I) is also denoted by t(S(I)). We note that the degree of S(I) is not |S(I)| but rather $\Sigma_{\vec{x}\in S(I)} m(\vec{x})$, when elements of S(I) are counted with multiplicities.

By the definition of types we have

$$\Sigma_{\alpha\in A}\, t(S_\alpha)\cdot m_\alpha \le t(S(I))$$

$$\le \exp\{C_2\, \Sigma_{j=1}^{n}\, t(P_j)\Pi_{s\ne j}\, d(P_s)\}.$$

In fact, in order to derive this inequality, in the part concerning relationship between a_i^α: $\alpha \in A$ and A_i one uses the Gauss lemma and properties of u-resultants [17].

The case n = 2 is very easy to understand using resultants [3]. In this case we have two relatively prime polynomials $P(x,y)$, $Q(x,y) \in \mathbb{Z}[x,y]$ and the set S of their common zeros

$$S = \{(x,y) \in \mathbb{C}^2 : P(x,y) = Q(x,y) = 0\}.$$

Their coordinates can be determined using the resultants of $P(x,y)$, $Q(x,y)$. We make a change of the coordinates to a "normal" form and get new polynomials $P(x',y')$, $Q(x',y')$. In "normal" coordinates distinct elements of S have both their coordinates distinct.

Then resultants $R_1(x')$ and $R_2(y')$ of $P'(x',y')$, $Q'(x',y')$ (in "normal" coordinates) with respect to y' and x', respectively, can be written as

$$R_1(x') = a_1\Pi_{i=1}^{k}(x' - \zeta'_{1,i})^{m_i};$$

$$R_2(y') = a_2\Pi_{i=1}^{k}(y' - \zeta'_{2,i})^{m_i},$$

where $(\zeta'_{1,i},\zeta'_{2,i})$ is an element of S of the multiplicity m_i. In particular, one can represent $R_1(x')$, $R_2(y')$ in terms of the powers of irreducible polynomials:

$$R_1(x') = \Pi_{\alpha\in A}\, P_\alpha^1(x')^{m_\alpha};$$

$$R_2(y') = \Pi_{\alpha\in A}\, P_\alpha^2(y')^{m_\alpha},$$

where $P_\alpha^1(x') = a_1^\alpha\Pi_j(x' - \zeta_{1j}^\alpha)$ and $P_\alpha^2(y') = a_2^\alpha\, \Pi_j(y' - \zeta_{2j}^\alpha)$ and $(\zeta_{1j}^\alpha,\zeta_{2j}^\alpha)$ are elements of S. We naturally define

$$S_\alpha = \{(\zeta_{1j}^\alpha, \zeta_{2j}^\alpha)\},$$

so that $\cup_{\alpha\in\beta} S_\alpha = S$. We can now look on the type of S and S_α: $\alpha \in A$ from the point of view of resultants.

E.g. one can define a size of S as the sum sizes of R_1 and R_2. This will be equivalent to the previous definition of the size.

Consequently we can say

$$t(S) \leq t(R_1) + t(R_2),$$

while for $\alpha \in A$,

$$t(S_\alpha) \leq t(P_\alpha^1) + t(P_\alpha^2).$$

This definition of the type, expressed not in the coordinate form is, as a matter of fact, more useful in the higher-dimensional case, when we are working with mixed ideals; non-normal intersections, etc.

We want to remark that our decomposition of $S(I)$ into the union of (disjoint) irreducible components is, certainly, not unique. Indeed, a given irreducible component can split; if the corresponding polynomial is decomposed over $\mathbb{Z}$. However, it is easier to work with our definition, though using a simple geometry and ideal theory, one can define a canonical decomposition of $S(I)$ into the (maximal) union of irreducible components over $\mathbb{Z}$.

§4.

We will present now a complete proof of the criterion of the algebraic independence of two numbers, that generalizes our previous statements.

Theorem 4.1: _Let_ $(\theta_1, \theta_2) \in \mathbb{C}^2$, $a > 1$ _and_ σ_N, σ'_N _be monotonically increasing functions such that_ $\sigma'_N > \sigma_N > N$;

$$\sigma_{N+1} < a\sigma_N.$$

Let for any $N \geq N_0$ _there exist either_

i) a polynomial $P_N(x,y) \in \mathbb{Z}[x,y]$ such that

$$t(P_N) \leq N,$$

$$-\sigma'_N < \log|P_N(\theta_1,\theta_2)| < -\sigma_N;$$

or there is

ii) a system $C_\ell(x,y) \in \mathbb{Z}[x,y]: \ell \in \mathcal{L}_N$ of polynomials without a common factor such that

$$t(C_\ell) \leq N,$$

$$\log|C_\ell(\theta_1,\theta_2)| < -\sigma_N: \quad \ell \in \mathcal{L}_N.$$

If now σ_N is growing faster than N^3: $\lim_{N\to\infty} \sigma N/N^3 = \infty$ and $\lim \sigma'_N/\sigma_N < \infty$, then numbers θ_1, θ_2 are algebraically dependent (over $\mathbb{Q}$).

Proof: Let us assume that θ_1, θ_2 are algebraically independent. We can also assume that $\sigma'_N = C_3 \cdot \sigma_N$. This last assumption can be lifted.

First of all we must start with the pair of relatively prime polynomials of the type i) or ii).

For simplicity in this proof we call for any given $N \geq N_0$ the case i) "blue", and if the case ii) is satisfied, the corresponding situation (or the number N) is called "red".

First of all, we use a very simple argument to get two relatively prime polynomials:

Lemma 4.2: Let us assume that there is a polynomial $P(x_1,\ldots,x_n) \in \mathbb{Z}[x_1,\ldots,x_n]$ such that for a given $(\theta_1,\ldots,\theta_n) \in \mathbb{C}^n$ we have

$$|P(\theta_1,\ldots,\theta_n)| \leq \epsilon < 1$$

and $t(P) \leq T$. Then either there exists a polynomial $Q(x_1,\ldots,x_n) \in \mathbb{Z}[x_1,\ldots,x_n]$ which is a power of an irreducible polynomial and

(4.3) $$|Q(\theta_1,\ldots,\theta_n)| \le \varepsilon^{1/3}$$

<u>and</u> $t(Q) \le 2T$, <u>or there are two relatively prime polynomials</u> $P_1(x_1,\ldots,x_n)$, $P_2(x_1,\ldots,x_n) \in \mathbb{Z}[x_1,\ldots,x_n]$ <u>such that</u>

(4.4) $$\max\{|P_1(\theta_1,\ldots,\theta_n)|,|P_2(\theta_1,\ldots,\theta_n)|\} \le \varepsilon^{1/3}$$

<u>and</u>

$$t(P_1) \le 2T, \quad t(P_2) \le 2T.$$

The proof is straightforward, follows Gelfonds arguments and was presented already many times.

Let us call the alternative (4.3) in Lemma 42 "green" and (4.4) "orange". If given N is "red", then, independently of what ("green" or "orange") occur, we have two relatively prime polynomials:

<u>Claim</u> 4.5: <u>For a</u> "red" $N \ge N_0$ <u>there are always two polynomials</u> $P_N^r(x,y)$, $Q_N^r(x,y) \in \mathbb{Z}[x,y]$ <u>that are relatively prime and satisfy</u>:

$$\max\{|P_N^r(\theta_1,\theta_2)|,|Q_N^r(\theta_1,\theta_2)|\}$$

(4.6) $$\le \exp(-\sigma_N/3);$$

$$\max\{t(P_N^r),t(Q_N^r)\} \le 2N.$$

Indeed, we use the lemma 4.2 for $n = 2$ and any of the non-zero polynomials $C_{\ell_0}(x,y) = P(x,y)$: $\ell_0 \in \mathcal{L}_N$. In the "orange" case (4.4) we get (4.6) at once. In the "green" case (4.3) we take a polynomial $Q(x,y) \overset{\text{def}}{=} Q_N^r(x,y)$ and find a polynomial $C_{\ell_1}(x,y)$: $\ell_1 \in \mathcal{L}_N$, relatively prime with $Q_N^r(x,y)$. Such a polynomial always exists, since $Q_N^r(x,y)$ is a power of an irreducible polynomial. We put $P_N^r(x,y) = C_{\ell_1}(x,y)$, so that (4.6) is satisfied.

As a consequence of Claim 4.5 we obtain that if there is any $N \ge N_1$ (for sufficiently large $N_1 > N_0$) which is "red", then for this N we have two relatively prime polynomials with the properties (4.6).

Hence, we can consider the situation when all $N \geq N_1$, for some N_1, are "blue":

Claim 4.7: Let all $N \geq N_1$ be "blue". Then for any $N_2 > N_1$ there is $N \geq N_2$ such that one has two relatively prime polynomials $P_N^b(x,y) \in \mathbb{Z}[x,y]$, $Q_N^b(x,y) \in \mathbb{Z}[x,y]$ satisfying

$$\max\{|P_N^b(\theta_1,\theta_2)|, |Q_N^b(\theta_1,\theta_2)|\}$$

$$(4.8) \qquad \leq \exp(-\sigma_N/3);$$

$$\max\{t(P_N^b), t(Q_N^b)\} \leq 3N.$$

Indeed, if for some $N \geq N_2$ we have an "orange" case (4.4), then (4.8) is true. Let, however, all $N \geq N_2$ be "green" (4.3). Then we have a polynomial $Q_N'(x,y) \in \mathbb{Z}[x,y]$, being a power of an irreducible one, such that

$$(4.9) \qquad \log|Q_N'(\theta_1,\theta_2)| < -\sigma_N/3, \quad t(Q_N') \leq 2N;$$

moreover, $Q_N' = (P_N')^{s_N}$ for an irreducible P_N' and $s_N \geq 0$. The two of the polynomials P_N' and P_{N+1}' must be different for $N > N_2$, since otherwise for $P_N' = P_{N+1}' = P_{N+2}' = \ldots$, we get from (4.9),

$$\log|P_N'(\theta_1,\theta_2)| < -\frac{\sigma_M}{3s_M}, \qquad s_M \leq 2M$$

and $M \to \infty$ gives us $P_N'(\theta_1,\theta_2) = 0$. If, however, $P_N'(x,y)$ and $P_{N+1}'(x,y)$ are different (relatively prime), then two polynomials $Q_N'(x,y)$ and $Q_{N+1}'(x,y)$ satisfy (4.8).

As a result of Claims 4.5 and 4.7 we obtain for any $N_3 \geq N_0$ the existence of $N > N_3$ such that there are two relatively prime polynomials $\bar{P}_N(x,y)$, $\bar{Q}_N(x,y) \in \mathbb{Z}[x,y]$ satisfying

$$\max\{|\bar{P}_N(\theta_1,\theta_2)|,|\bar{Q}_N(\theta_1,\theta_2)|\}$$

(4.10) $$< \exp(-\sigma_N/3);$$

$$\max\{t(\bar{P}_N),t(\bar{Q}_N)\} \leq 3N.$$

We take a sufficiently large N such that $\sigma_n/3$ is sufficiently large with respect to $(3N)^3$: symbolically $N^3 = o(\sigma_N)$; or $\sigma_N > C\cdot 3^4 N^3$ for a large constant $C > 0$ and take two polynomials $\bar{P}_N(x,y)$, $\bar{Q}_N(x,y)$ satisfying (4.10).

Our main object becomes a set $\bar{S}_N = S(\bar{P}_N,\bar{Q}_N)$ of common zeros of $\bar{P}_N(x,y)$ and $\bar{Q}_N(x,y)$. The set $\bar{S}_N$ has a degree (elements counted with multiplicities) at most $3^2\cdot N^2$, and type (estimated through the resultants) at most $8\cdot 3^2\cdot N^2$: $t(\bar{S}_N) \leq 72\ N^2$. We are looking at irreducible components $S_{N,\alpha}$: $\alpha \in A_N$ of $\bar{S}_N$:

$$\bar{S}_N = \cup'_{\alpha\in A_N} S_{N,\alpha}$$

($\cup'$ indicates that elements of the union are disjoint sets).

We have

(4.11) $$\Sigma_{\alpha\in A_N} m_\alpha d(S_{N,\alpha}) \leq d(\bar{S}_N) \leq 9N^2,$$

where $d(S_{N,\alpha}) = |S_{N,\alpha}|$ is a degree of $S_{N,\alpha}$ and m_α is a multiplicity of the component and similarly for types:

(4.12) $$\Sigma_{\alpha\in A_N} m_\alpha t(S_{N,\alpha}) \leq t(\bar{S}_N) \leq 72N^2.$$

We are looking now on those elements of $\bar{S}_N$ and $S_{N,\alpha}$ that are close to $\bar{\theta} = (\theta_1,\theta_2)$. All evaluations are made in ℓ^1-norm in $\mathbb{C}^2$.

For a given $\bar{\zeta} \in \bar{S}_N$ we want to bound above $\|\bar{\theta} - \bar{\zeta}\|$. We use for this the following convenient notations

(4.13) $$\|\bar{\theta} - \bar{\zeta}\| \leq \exp(-E(\bar{\zeta})).$$

($E(\bar{\zeta})$ is defined since $\bar{\theta} \neq \bar{\zeta}$), where, for simplicity we always assume

(4.14) $$E(\bar{\zeta}) \leq \sigma_N/3 \quad \text{for any} \quad \bar{\zeta} \in \bar{S}_N.$$

In these notations we can express the main auxiliary result that is formulated in a rather general form:

Proposition 4.15: *Let* $\bar{\theta} = (\theta_1, \theta_2) \in \mathbb{C}^2$ *(with the* l^1*-norm) and* $P(x,y)$, $Q(x,y)$ *be two relatively prime polynomials from* $\mathbb{Z}[x,y]$.

Let $S = S(P,Q)$ be the set of the zeros of an ideal (P,Q) and

$$S = \cup'_{\alpha \in A} S_\alpha$$

its representation through irreducible components. If m_α is a multiplicity of S_α, then one has

$$\Sigma_{\alpha \in A} m_\alpha t(S_\alpha) \leq t(S) \leq 4(d(P)t(Q)$$

$$+ d(Q)t(P)) \leq 8t(P)t(Q).$$

Let us assume now that

(4.16) $$\max\{|P(\bar{\theta})|, |Q(\bar{\theta})|\} \leq \exp(-E)$$

for $E > 0$. One can define the distance of $\bar{\zeta} \in S$ to $\bar{\theta}$ as

$$\|\bar{\theta} - \bar{\zeta}\| \leq \exp(-E(\bar{\zeta})).$$

Let us put

(4.17) $$\mathcal{E}(S, \bar{\theta}) = \Sigma_{\bar{\zeta} \in S, E(\bar{\zeta}) > 0} E(\bar{\zeta})$$

and

(4.18) $$\mathcal{E}(S_\alpha, \bar{\theta}) = \Sigma_{\bar{\zeta} \in S_\alpha, E(\bar{\zeta}) > 0} E(\bar{\zeta})$$

as the definition of (minus logarithm$^+$ of)distance from $\bar{\theta}$ to S or S_α.

In order to express relations between E, $\mathcal{E}(S,\bar\theta)$ and $\mathcal{E}(S_\alpha,\bar\theta)$ we put

$$(4.19)\qquad T = \gamma_0(d(P)t(Q) + d(Q)t(P) + d(P)d(Q)\log(d(P)d(Q) + 2))$$

for an absolute constant $\gamma_0 > 0$ ($\gamma_0 \leq 4$) such that

$$T(S) \leq T.$$

Similarly for every $\alpha \in A$ there is a bound T_α of $t(S_\alpha)$ of the form

$$(4.20)\qquad t(S_\alpha) \leq T_\alpha \leq t(S_\alpha) + d(S_\alpha)\log(d(P)d(Q) + 2).$$

In terms of T_α we can formulate results on $\mathcal{E}(S,\bar\theta)$, $\mathcal{E}(S_\alpha,\bar\theta)$. If we assume $E \geq 4T$, then

$$(4.21)\qquad \mathcal{E}(S,\bar\theta) = \Sigma_{\alpha\in A}\, m_\alpha\, (S_\alpha,\bar\theta) \geq E - 2T.$$

We denote the nearest to $\bar\theta$ element of S_α by $\bar\xi_\alpha$, $E(\bar\xi_\alpha) = \min_{\zeta\in S_\alpha} E(\zeta)$. Then we have

$$(4.22)\qquad E(\bar\xi_\alpha) \geq \min\{c_1\mathcal{E}(S_\alpha,\bar\theta),\ c_1 \frac{\mathcal{E}(S_\alpha,\bar\theta)^2}{d(S_\alpha)T_\alpha}\}$$

and for the other $\bar\zeta \in S_\alpha$ close to $\bar\theta$ we have

$$(4.23)\qquad \Sigma_{\bar\xi\in S_\alpha, E(\bar\xi)\geq B_\alpha}\, E(\bar\zeta) \geq \frac{\mathcal{E}(S_\alpha,\bar\theta)}{4}$$

where

$$(4.24)\qquad B_\alpha = c_2\mathcal{E}(S_\alpha,\bar\theta)^{3/2}\cdot(d(S_\alpha)T_\alpha^{1/2})^{-1}$$

for $\alpha \in A$.

As an application of these bounds we have the following result, where $d(S_\alpha)$ is replaced by its upper bound T_α. We remark, that in the addition to (4.20), (4.25) $\Sigma_{\alpha \in A} m_\alpha T_\alpha \le T$.

In particular, there exists such $\alpha_0 \in A$ such that

$$(4.26) \qquad \frac{\mathcal{E}(S_{\alpha_0}, \bar\theta)}{T_{\alpha_0}} \ge \frac{\mathcal{E}(S,\bar\theta)}{T} \ge \frac{E}{T} - 2.$$

Under the conditions (4.26) one has as a corollary of (4.22)-(4.24):

$$E(\bar\xi_{\alpha_0}) \ge \min\{c_1 \mathcal{E}(S_{\alpha_0}, \bar\theta),\ c_1 \frac{\mathcal{E}(S_{\alpha_0},\bar\theta)^2}{T^2_{\alpha_0}}\};$$

$$(4.27) \qquad \Sigma\{E(\bar\xi) : \bar\xi \in S_{\alpha_0}, E(\bar\xi) \ge c_2(\mathcal{E}(S_{\alpha_0},\bar\theta)/T_{\alpha_0})^{3/2}\}$$

$$\ge \mathcal{E}(S_{\alpha_0},\bar\theta)/4.$$

<u>Proof</u> (see [3]): First of all we must change the system of the coordinates to a "normal" [3] one with respect to a system of polynomials $P(x,y)$, $Q(x,y)$. We use for this the lemma 3.3 [3]. According to this lemma there is a <u>nonsingular</u> transformation

$$(\pi) \qquad \begin{cases} x = x'a + y'c \\ y = x'b + y'd \end{cases}$$

for rational integers a,b,c,d such that

$$(4.28) \qquad \max(|a|,|b|,|c|,|d|) \le M \le \gamma_1 d(P) d(Q),$$

and which are normal with respect to $P(x,y)$, $Q(x,y)$. It means, from our point of view first of all that for $\bar\theta$ written in new coordinates (x',y') as $\bar\theta' = (\theta'_1,\theta'_2)$ and for <u>any</u> common zero $\bar\zeta$ of $P(x,y) = 0$, $Q(x,y) = 0$, i.e. element of S, we have in new coordinates $\bar\zeta' = (\zeta'_1,\zeta'_2)$ we have

$$(4.29) \qquad \|\bar\theta' - \bar\zeta'\|_{1_1} \le 4M^2 \min\{|\theta'_1 - \zeta'_1|, |\theta'_2 - \zeta'_2|\}.$$

The property (4.29) is, certainly, central that we use from all of "normality" properties.

In new, "normal" variables we consider the resultant $R(x')$ of $P'(x',y')$ ($\equiv P(x,y)$) and $Q'(x',y')$ ($\equiv Q(x,y)$) taken with respect to y'. The polynomial $R(x')$ is a polynomial of the degree $\leq d(P)d(Q)$, but the type of $R(x')$ might be slightly higher than that of $R(x)$. This explains, why we change the type $t(S)$ be slightly higher quantity T defined in (4.19).

We prefer, however, to work with T and T_α because the proof is straightforward, though estimates are suffering.

Now irreducible components $S'_\alpha = \pi^{-1}(S_\alpha)$ (we write now in transformed coordinates) are connected with irreducible components of $R(x')$. Namely, we have

$$R(x') = \Pi_{\alpha\in A} P_\alpha(x')^{m_\alpha}$$

where $P_\alpha(x')$ is an irreducible polynomial from $\mathbb{Z}[x']$. Here zeros of $P_\alpha(x')$ are exactly x'-projection of the set S'_α: $\alpha \in A$.

Hence we can work with $P_\alpha(x')$ rather than with $S'_\alpha = \pi^{-1}(S_\alpha)$. According to (4.29) it is enough to bound above $|\zeta'_1 - \theta'_1|$ in order to bound $\|\bar{\zeta}' - \bar{\theta}'\|$ with $\bar{\zeta}' = (\zeta'_1,\zeta'_2) \in S'$. Here and everywhere in the proof of the proposition 4.15, $\bar{\zeta}' = \pi^{-1}(\bar{\zeta})$, $\bar{\theta}' = \pi^{-1}(\bar{\theta})$.

First of all we can evaluate $\log|R(\theta'_1)|$ in terms of $E \leq -\min\{|\log|P(\bar{\theta})|,\log|Q(\bar{\theta})||\}$. For this we use the property of the resultants in the classical form [1].

Namely, we can use the formula [17]

$$|\mathrm{Res}_x(p,q)| \leq \{d(p)H^+(q)|q(x_0)| + d(q)H^+(p)|p(x_0)|\}$$
$$\times H^+(q)^{d(p)-1}\cdot H^+(p)^{d(q)-1}$$

for arbitrary polynomials $p(x),q(x) \in \mathbb{C}[x]$ and $H^+(p) = \max\{1,H(p)\}$. We put $p(x') = P'(\theta'_1,x)$, $q(x') = Q'(\theta'_1,x')$ and $x_0 = \theta'_2$.

We obtain this way the inequality:

$$\log|R'(\theta_1')| \leq -E + T.$$

In particular, writing R'(x) in the form

$$R'(x') = a\,\Pi_{\alpha\in A}\ (x' - \zeta_1'^{\alpha})^{m_\alpha}$$

$$= a\,\Pi_{\bar{\zeta}'\in S'}(x' - (\bar{\zeta}')_1)$$

for $a \in \mathbb{Z}$, and using (4.29) one immediately obtains (4.21).

In order to get other statements (4.22)-(4.24) we simply use the irreducible polynomial $P_\alpha(x')$, whose zeros are x'-projection of S' and apply to the polynomial $P_\alpha(x')$ the statement of Lemma 2.3 [3].

At last, we derive the statements (4.6)-(4.27). Indeed, if (4.26) are false for every $\alpha_0 \in A$, we get

$$\Sigma_{\alpha\in A}\ \mathcal{E}(S_\alpha,\bar{\theta})m_\alpha\cdot T < \Sigma_{\alpha\in A}\ T_\alpha\cdot m_\alpha\cdot\mathcal{E}(S,\bar{\theta}),$$

which contradicts (4.25). Results (4.27) are a consequence of (4.22)-(4.24).

<u>Remark</u>: In (4.27), the quantity $E(\bar{\xi}_{\alpha_0}) \leq -\log\|\bar{\theta} - \bar{\xi}_{\alpha_0}\|$ is bounded below by $c_1\cdot\min\{\mathcal{E}(S_{\alpha_0},\bar{\theta}),\ (\mathcal{E}(S_{\alpha_0},\bar{\theta})/T_{\alpha_0})^2\}$. From these two terms, "usually" (in practice), $(\mathcal{E}(S_{\alpha_0},\bar{\theta})/T_{\alpha_0})^2$ is a smaller one. Indeed, if $\mathcal{E}(S_{\alpha_0},\bar{\theta})$ is large, then (4.27) shows simply that "the most" of the measure $\mathcal{E}(S_{\alpha_0},\bar{\theta})$ is concentrated in a single term of it: $E(\bar{\xi}_{\alpha_0})$: $E(\bar{\xi}_{\alpha_0}) \geq c_1\mathcal{E}(S_{\alpha_0},\bar{\theta})$.

However, one must take even this possibility into consideration.

Let us continue the proof of theorem 4.1. We use Proposition 4.15 to irreducible components $S_{N,\alpha}$. We introduce the quantitites T_α, T_N:

(4.30) $$T_N = \gamma_2 N^2 \log N;$$

and similarly $T_\alpha = T_{N,\alpha}$ such that

(4.31) $$t(S_{N,\alpha}) \leq T_{N,\alpha} = \gamma_2 t(S_{N,\alpha}) \log N$$

and

(4.32) $$\Sigma_{\alpha\in A_N} m_\alpha T_{N,\alpha} \leq T_N.$$

If we apply Proposition 4.15, we obtain an $\alpha_0 \in A_N$ such that the following conditions are satisfied:

Claim 4.33: There is an $\alpha_0 \in A_N$ and $\bar{\xi}_0 \in S_{N,\alpha_0}$ with $E(\bar{\xi}_0) = \max\{E(\bar{\xi}) : \bar{\xi} \in S_{N,\alpha_0}\}$ such that for

(4.34) $$\mathcal{E}_{N,0} \overset{\text{def}}{=} \Sigma\{E(\bar{\xi}) : \bar{\xi} \in S_{N,\alpha_0}, E(\bar{\xi}) > 0\}$$

we have

(4.35) $$\mathcal{E}_{N,0}/T_{N,\alpha_0} \geq \sigma_N/3T_N - 2;$$

and

$$E(\bar{\xi}_0) \geq \min\{c_1\mathcal{E}_{N,0}, c_1(\mathcal{E}_{N,0}/T_{N,\alpha_0})^2\};$$

(4.36) $$\Sigma\{E(\bar{\xi}) : \bar{\xi} \in S_{N,\alpha_0}, E(\bar{\xi}) \geq c_2(\mathcal{E}_{N,0}/T_{N,\alpha_0})^{3/2}\}$$
$$\geq \mathcal{E}_{N,0}/4.$$

Remark 4.37: It is possible to assume that the equality in (4.35) is satisfied: $\mathcal{E}_{N,0} = T_{N,\alpha_0}(\sigma_N/3T_N - 2)$ and take this as a definition of $\mathcal{E}_{N,0}$ only in terms of T_{N,α_0} (cf. (4.31), (4.32)), σ_N and T_N.

Our main data are set down and we can proceed with the actual proof. The idea of the proof is very simple. We take $\bar{\xi} \in S_{N,\alpha_0}$ and substitute it into polynomials $P_M(\bar{x})$ in the "blue" case or $C_\ell(\bar{x})$: $\ell \in \mathcal{L}_M$ in the "red" case for $M < N$, hoping to get a contradiction using the properties of S_{N,α_0} and Liouville theorem.

This can be understood better on the language of the ideal theory.

We take an ideal I_0 corresponding to a component S_{N,α_0}; $I_0 \supset I = (P_N, Q_N)$ and we want to show that the ideal (I_0, P_M) for the blue M or $(I_0, C_\ell: \ell \in \mathcal{L}_M)$ for the red M give us $\mathbb{Z}[x,y]$. Then the contradiction may follow from the evaluation of these ideals at the point $\bar{\theta}$. One understands that this is rather general program, useful in n-dimenional situation. However, it should be handled with greatest care to bounds of different "types" and "sizes", "degrees", because none of the polynomials is annihilating at $\bar{x} = \bar{\theta}$.

First of all one very useful general remark

Claim 4.38: For polynomial $R(x,y) \in \mathbb{Z}[x,y]$, if $R(\bar{\xi}) = 0$ for some $\bar{\xi} \in S_{N,\alpha_0}$, then $R(\bar{\xi}_1) = 0$ for all $\bar{\xi}_1 \in S_{N,\alpha_0}$.

This follows from the minimal irreducibility of the component S_{N,α_0}. For the future part of the proof we write

$$S_N^0 = S_{N,\alpha_0} \quad \text{and} \quad T_N^0 = T_{N,\alpha_0}.$$

We consider M_0 defined in a way, that the following condition is satisfied: here $\gamma_3 \leq 2\max\{\log|\theta_1|, \log|\theta_2|, 1\} + 2$.

$$E(\bar{\xi}_0) > \sigma'_M + \gamma_3 M. \tag{4.39}$$

Claim 4.40: If the condition (4.39) is satisfied, M is "blue" and $P_M(x,y)$ is corresponding polynomial from i), then

$$P_M(\bar{\xi}_0) \neq 0.$$

Indeed, let us assume $P_M(\xi_0) = 0$. Then we have

$$\log|P_M(\theta)| \leq \log\|\bar{\theta} - \xi_0\| + \gamma_3 M.$$

According to the definition of σ'_M in i) and (4.39), this contradicts to the definition of $E(\bar{\xi}_0)$.

Let us now consider the consequences of $P_M(\bar{\xi}_0) \neq 0$ for M satisfying different kinds of assumptions then (4.39).

Claim 4.41: Let $M \geq N_0$ and for some $K \geq 2$ the following conditions are satisfied:

$$c_2(\mathcal{E}_{N,0}/T_N^0)^{3/2} \geq \frac{2K}{K-1}\gamma_3 M;$$

(4.42)

$$K\sigma_M/3 > E(\bar{\xi}_0).$$

If $P(x,y) \in \mathbb{Z}[x,y]$ is any polynomial such that $t(P) \leq 2M$ and

$$\log|P(\theta_1,\theta_2)| < -\sigma_M/3,$$

then from $P(\bar{\xi}_0) \neq 0$ it follows

$$c_3 M t(S_N^0) > \mathcal{E}_{N,0}/K. \tag{4.43}$$

Proof of Claim 4.41: Since $P(\bar{\xi}_0) \neq 0$, by claim 4.38, $P(\bar{\xi}) \neq 0$ for any $\bar{\xi} \in S_N^0$. We denote $c_2(\mathcal{E}_{N,0}/T_N^0)^{3/2}$ by B_0. We apply now the Liouville theorem to the quantity

$$\alpha = \Pi_{\bar{\xi}\in S_N^0, E(\bar{\xi})\geq B_0} P(\bar{\xi}).$$

According to a Liouville theorem (§3) we obtain

$$|\alpha| \geq \exp(-c_4 M t(S_{N,0})) \tag{4.44}$$

for a type $t(S_{N,0})$ of the set $S_{N,0}$. Let us obtain an upper bound for $|\alpha|$. For this we notice that

$$|P(\bar{\xi})| \leq |P(\bar{\theta})| + \|\bar{\theta} - \bar{\xi}\|\cdot\exp\{\gamma_3 2M\}. \tag{4.45}$$

We have $\|\bar{\theta} - \bar{\xi}\|\cdot\exp(-E(\bar{\xi}))$. For $E(\bar{\xi}) \geq B_0$, we have according to (4.42),

$$\|\bar{\theta} - \bar{\xi}\|\exp\{2\gamma_3 M\} \leq \exp(-E(\bar{\xi})/K).$$

Since we have $|P(\bar{\theta})| < \exp(-\sigma_M/3)$ and take into account (4.42) we obtain from (4.45):

$$|P(\bar{\xi})| \leq 2 \exp(-E(\bar{\xi})/K):$$

$\bar{\xi} \in S_N^0$, $E(\bar{\xi}) \geq B_0$. In particular, by the definition of α we obtain:

$$|\alpha| \leq 2^{d(S_N^0)} \cdot \exp\{-\Sigma_{\bar{\xi}\in S_N^0, E(\bar{\xi})\geq B_0} E(\bar{\xi})/K\}.$$

Using (4.36) we get

$$(4.46) \qquad |\alpha| \leq 2^{d(S_N^0)} \cdot \exp\{-\mathcal{E}_{N,0}/4\}.$$

Comparing (4.44) and (4.46) we deduce (4.43). Claim 4.41 is proved.

In order to simplify the statement of Claim 4.41 we make the following simple

<u>Claim</u> 4.47: <u>Let for</u> $N \geq N_4$, <u>us have</u>

$$M < \frac{1}{2c_3} \cdot \frac{\sigma_N}{KN^2}.$$

Then the condition (4.43) is false, while the first of the conditions (4.42) is satisfied.

Indeed, we use (4.35) and we get by (4.30)

$$(4.48) \qquad \mathcal{E}_{N,0}/T_N^0 \geq \sigma_N/\gamma_2 N^2 \log N-2;$$

this shows that for large N, when $N^3 = 0(\sigma_N)$, the first of the conditions (4.42) is satisfied. From (4.48) it follows also

$$\mathcal{E}_{N,0}/t(S_N^0) \geq \sigma_N/N^2 - 2\gamma_2 \log N$$

according to a definition (4.31). The choice of M immediately implies that for a large M, (4.43) is false.

<u>Corollary</u> 4.49: <u>Let</u> $M < \sigma_N/2c_3KN^2$, <u>but</u> $K\sigma_M/3 > E(\bar{\xi}_0)$ <u>for</u> $K \geq 2$. <u>Then</u>

for every polynomial $P(x,y) \in \mathbb{Z}[x,y]$ *such that* $t(P) \leq 2M$ *and*

$$\log|P(\bar{\theta})| < -\sigma_M/3,$$

we have $P(\bar{\xi}_0) = 0$.

In particular we have

Corollary 4.50: *Let for some* $K \geq 2$, $M < \sigma_N/2c_3KN^2$ *and*

$$(4.51) \qquad K\,\sigma_M/3 > E(\bar{\xi}_0) > \sigma'_M + \gamma_3 M,$$

then M *is "red", but not "blue".*

Indeed, we combine Claim 4.40 and Corollary 4.49 to get for the "blue" M, both $P_M(\bar{\xi}_0) = 0$ and $P_M(\bar{\xi}_0) \neq 0$.

We now notice that for some constant $K > 2$ there is always a sufficiently large M satisfying (4.51). Indeed, let N_6 be sufficiently large so that

$$\sigma'_M \leq \gamma_5 \sigma_M, \qquad \text{if} \qquad M \geq N_6,$$

while for a large constant C_6 ($> \gamma_5$ etc.) we have $\sigma_M > C_6 M^3$ if $M \geq N_6$. Then (4.51) can be changed to

$$(4.52) \qquad K/3\ \sigma_M > E(\bar{\xi}_0) > (\gamma_5 + \gamma_3)\sigma_M.$$

Taking into account the definition of σ_M one can always satisfy these inequalities for K depending only on γ_5, γ_3 and a. One more restriction on M reads

$$(4.53) \qquad M < \frac{\sigma_N}{2c_3KN^2}.$$

However, the definition of $E(\bar{\xi}_0)$ in (4.14) requires that $E(\bar{\xi}) \leq \sigma_N/3$. This ensures that for a sufficiently large N the condition (4.53) is satisfied for M, satisfying (4.52). Indeed, for a large N, (4.53) can be substituted by $M < C_6/2c_3 \cdot N$, where C_6 is a large constant.

Let us denote M, satisfying (4.52) (and consequently, (4.53)) by $M = M_0$. Then Corollary 4.50 asserts that M_0 is "red".

<u>Claim</u> 4.54: <u>The set</u> S_N^0 <u>has the type</u> $\leq 8\gamma_0 M_0^2$.

<u>Proof of Claim</u> 4.54: According to claim 4.5 there are two relatively prime polynomials $P_{M_0}^r(x,y)$, $Q_{M_0}^r(x,y) \in \mathbb{Z}[x,y]$ such that

$$\max\{t(P_{M_0}^r), t(Q_{M_0}^r)\} \leq 2M_0;$$

$$\max\{|P_{M_0}^r(\theta)|, |Q_{M_0}^r(\bar{\theta})|\} < \exp(-\sigma_{M_0}/3).$$

Corollary 4.49 shows that $P_{M_0}^r(\bar{\xi}_0) = 0$, $Q_{M_0}^r(\bar{\xi}_0) = 0$, since all the conditions of 4.49 are satisfied for M_0 in view of (4.52)-(4.53). Now S_N^0 is a subset of a set of common zeros of $Q_{M_0}^r(x,y)$, $P_{M_0}^r(x,y)$. Because of the bounds of types of $Q_{M_0}^r$, $P_{M_0}^r$ we obtain by a standard argument that $t(S_N^0) \leq 8\gamma_0 M_0^2$.

We are entering the final part of the proof. Here is the only place where we use the trancendence of $\bar{\theta}$! First of all one should remark that the function of N, $t(S_N^0)$ is unbounded as $N \to \infty$. Indeed from (4.36) one obtains for a large N that

$$E(\bar{\xi}_0) \geq N.$$

This means

$$\|\bar{\theta} - \bar{\xi}_0\| \leq \exp(-N)$$

and since $t(\bar{\xi}_0) \leq t(S_N^0)$ and $\bar{\theta}$ is transcendental (from $\mathbb{C}^2$), $t(\bar{\xi}_0) \to \infty$ and so is $t(S_N^0) \to \infty$.

In order to finish the proof one defines a number M_1 in such a way as:

$$(4.55) \qquad M_1 = [\sqrt{t(S_N^0)/8\gamma_0}] - 1.$$

Definition (4.55) means that

$$8\gamma_0 M_1^2 < t(S_N^0) < 8\gamma_0 (M_1 + 2)^2.$$

Claim 4.54 shows that

(4.56) $$M_1 < M_0.$$

<u>Claim</u> 4.57: <u>Let</u> $R(x,y) \in \mathbb{Z}[x,y]$, $t(R) \leq 2M_1$ <u>and</u> $\log|R(\bar{\theta})| < -\sigma_{M_1}/3$. <u>Then</u> <u>we</u> <u>have</u> $R(\bar{\xi}_0) = 0$.

<u>Proof</u>: Let us assume that $R(\bar{\xi}_0) \neq 0$. Then we can use the Liouville theorem in order to estimate $|R(\bar{\xi}_0)|$ below. We get, using (4.55) from the Liouville theorem:

(4.58) $$|R(\bar{\xi}_0)| \geq \exp(-\gamma_6 M_1 t(S_{N,0})) \geq \exp(-\gamma_7 M_1^3)$$

for some absolute constant $\gamma_7 > 0$. On the other hand we can estimate $|R(\bar{\xi}_0)|$ above using the bound for $\|\bar{\theta} - \bar{\xi}_0\|$:

$$|R(\bar{\xi}_0)| \leq |R(\bar{\theta})| + \|\bar{\theta} - \bar{\xi}_0\| \exp(\gamma_3 M_1)$$

$$\leq \exp(-\sigma_{M_1}/3) + \exp(-E(\bar{\xi}_0) + \gamma_3 M_1).$$

Now by the choice of (4.56) we have, using the definition (4.52) of M_0: $E(\bar{\xi}_0) > (\gamma_5 + \gamma_3)\sigma_{M_0} \geq \gamma_7 \sigma_{M_1} + \gamma_3 M_1$, because M_0 is sufficiently large. This implies

(4.59) $$|R(\bar{\xi}_0)| < \exp(-\gamma_8 \sigma_{M_1}).$$

Now we note that, by the remark of the unboundness of $t(S_N^0)$ as $N \to \infty$ and the definition (4.55) we have $M_1 \geq N_6$ provided $N \geq N_7$. Then $\sigma_{M_1} > C_6 M_1^3$ and for $C_6 > \gamma_7 \cdot \gamma_8^{-1}$ the inequalities (4.58) and (4.59) contradicts each other. Claim 4.57 is proved.

Now it is enough to prove that M_1 is not colored: neither "blue" nor "red".

<u>Claim</u> 4.60: <u>The</u> <u>number</u> M_1 <u>is</u> <u>not</u> <u>"blue"</u>.

Proof of Claim 4.60: Let us assume that M_1 is blue and let $P_{M_1}(x,y)$ be corresponding polynomial from i). Then, by Claim 4.57, $P_{M_1}(\bar{\xi}_0) = 0$. However, by (4.56), the choice of M_0 in (4.52) the condition (4.39):

$$E(\bar{\xi}_0) > \sigma'_{M_0} + \gamma_3 M_0 \geq \sigma'_{M_1} + \gamma_3 M_1$$

is satisfied. Then by Claim 4.40, we have $P_{M_1}(\bar{\xi}_0) \neq 0$. Hence, M_1 is not "blue".

Claim 4.61: The number M_1 from (4.55) is not "red".

Proof of Claim 4.61: Let us assume that M_1 is "red". We again remained that $t(S_N^0) \to \infty$ as $N \to \infty$; so for $N \geq N_7$, $M_1 \geq N_0$. We apply the claim 4.5 and get two relatively prime polynomials $P^r_{M_1}(x,y)$ 2 $\mathbb{Z}[x,y]$, $Q^r_{M_1}(x,y) \in \mathbb{Z}[x,y]$ such that

$$\max\{t(P^r_{M_1}), t(Q^r_{M_1})\} \leq 2M_1;$$

$$\max\{|P^r_{M_1}(\bar{\theta})|, |Q^r_{M_1}(\bar{\theta})|\} < \exp(-\sigma_{M_1}/3).$$

According to claim 4.37 we have $P^r_{M_1}(\bar{\xi}_0) = 0$, $Q^r_{M_1}(\bar{\xi}_0) = 0$. Hence, by claim 4.38 the whole set S_N^0 contains in the set of common zeros of $P^r_{M_1}(x,y)$, $Q^r_{M_1}(x,y)$. In other words we can bound the type $t(S_N^0)$ in terms of $P^r_{M_1}, Q^r_{M_1}$. This way we get a bound

$$t(S_N^0) \leq 8\gamma_0 M_1^2.$$

However this contradicts to the choice (4.55) of M_1. Claim 4.61 is proved.

The number M_1 is uncolored. However $t(S_N^0) \to \infty$ as $N \to \infty$, so that $M_1 \geq N_0$ if $N \geq N_7$. This means, according to the statement of Theorem 4.1 that M_1 is, indeed, colored. Since it is not, θ_1 and θ_2 are not algebraically independent. Theorem 4.1 is proved.

Remark 4.62: There is no need to demand, in general, $\lim \sigma'_N/\sigma_N < \infty$

if $\lim \sigma_N/N^3 = \infty$. One of the possible improvements in this direction is the following:

$$\text{if } \lim_{N\to\infty} \sigma_N/N^3 = \infty,$$

then we demand only

$$\sigma'_N \ll \sqrt[\sigma]{\sigma_N/N}$$

We notice that the only place in the proof (ad absurdum) where we used any additional information about $\bar{\theta} \in \mathbb{C}^2$ was the moment, where we demand

$$\|\bar{\theta} - \bar{\xi}\| > 0$$

for any $\bar{\xi} \in \bar{\mathbb{Q}}^2$. Hence one can change the end of the statement of Theorem 4.1:

Theorem 4.1′: Under the conditions of theorem 4.1 both numbers θ_1 and θ_2 are algebraic.

Remark: In principle one can deduce this statement from theorem 4.1 directly, without looking into the proof, if they use in addition the Gelfond criterion.

§5.

Let us formulate now one of the possible generalizations of the criterion of the algebraic dependence (we should say, rather than algebraic independence) for more then two numbers. In this case the formulation is not that strong, though:

Proposition 5.1: Let $\bar{\theta} = (\theta_1,\ldots,\theta_n) \in \mathbb{C}^n$, $a > 1$ and σ_N, σ'_N be monotonically increasing functions such that $\sigma'_N > \sigma_N$ and

$$\sigma_{N+1} < a\sigma_N.$$

Let for any $N \geq N_0$ there be a polynomial $P_N(x_1,\ldots,x_n) \in \mathbb{Z}[x_1,\ldots,x_n]$

such that

$$t(P_N) \leq N;$$

$$-\sigma_N' < \log|P_N(\theta_1,\ldots,\theta_n)| < -\sigma_N.$$

We assume that as $N \to \infty$, $\lim \sigma_N'/\sigma_N < \infty$ and $\lim_{N\to\infty} \sigma_N/N^{n+1} = \infty$. Then the numbers $\theta_1,\ldots,\theta_n$ are algebraically dependent. Moreover, each of $\theta_i: i = 1,\ldots,n$ is algebraic.

Remark 5.2: The statement of Proposition 5.1 can be improved in the case $\sigma_N' \gg \sigma_N \gg N^{n+1}$. We need in general statements like this:

$$\lim_{N\to\infty} \sigma_N/N^{n+1} = \infty$$

and for $\sigma_M' \gg\ll \sigma_N$ we have

$$\frac{\sigma_N}{N^n} \gg \frac{\sigma_M'}{\sigma_M}\cdot M,$$

i.e. if $\sigma_M/M \gg\ll N^n$, then we can take σ_M' as $\sigma_M' \ll \sigma_N$ or we can take σ_M' as

$$\sigma_M' \ll \sqrt[\sigma_n]{\sigma_M/M}$$

this is the statement close to a best possible.

It is most desirable to formulate in the n-dimensional situations the statements like Theorem 5.1.

The most natural assumption can be as follows. We take a > 1 as before and formulate:

Assumption 5.3: Let $\bar{\theta} = (\theta_1,\ldots,\theta_n) \in \mathbb{C}^n$ and σ_N, σ_N' be monotonically increasing functions such that $\sigma_N' > \sigma_N < N$, $\sigma_N \to \infty$ as $N \to \infty$ and

$$\sigma_{N+1} < a\sigma_N.$$

We assume that for every $N \geq N_0$, either

i) <u>there exists a polynomial</u> $P_N(x_1,\dots,x_n) \in \mathbb{Z}[x_1,\dots,x_n]$ <u>such that</u>

$$t(P_N) \le N,$$

$$-\sigma'_N < \log|P_N(\theta_1,\dots,\theta_n)| < -\sigma_N;$$

<u>or</u>

ii) <u>there exists a system</u> $C_\ell(x_1,\dots,x_n) \in \mathbb{Z}[x_1,\dots,x_n]: \ell \in \mathcal{L}_N$ <u>of polynomials without common factor such that</u>

$$t(C_\ell) \le N,$$

$$\log|C_\ell(\theta_1,\dots,\theta_n)| < -\sigma_N$$

<u>for</u> $\ell \in \mathcal{L}_N$.

Let us assume that $\lim_{N\to\infty} \sigma_N/N^{n+1} = \infty$ and, say, $\lim_{N\to\infty} \sigma'_N/\sigma_N < \infty$.

We can assume either that $\lim \sigma_N/N^{n+1} = \infty$ and $\sigma'_N \ll {}^{\sigma}\sqrt[n]{\sigma_N/N}$

We can conjecture that under assumption 5.3 all numbers $\theta_1,\dots,\theta_n$ are algebraic (or, say, only algebraically independent).

However we can prove this only under <u>one</u> of the following two additional assumptions:

a) for $n \ge 3$, $\lim_{N\to\infty} \sigma_N/N^{n+3/2} = \infty$;

b) for $n \ge 3$ the ideal $I_N = (C_\ell: \ell \in \mathcal{L}_N)$ has the dimension at most $n/2$.

The condition b) can be improved, however.

§6.

Looking for the proper form of the criterion of the algebraic independence, one naturally looks for the criterion applicable to the usual problems of the Transcendence Number Theory. We mean first of all the criterion useful for the Gelfond-Schneider or Thue-Siegel methods. One finds at once, that there is something peculiar in the Gelfond-Schneider method that is reasonable to adopt in the criterion.

The classical Gelfond-Schneider method can be presented briefly in the following way, see [12]. One has a system of functions

$f_{\vec{\lambda}}(\vec{z})$: $\vec{\lambda} \in \mathscr{L}$ (for $\vec{z} \in \mathbb{C}^d$) such that for a given set $S \subset \mathbb{C}^d$ one has

$$\frac{\partial^{|\vec{k}|}}{\partial z_1^{k_1} \ldots \partial z_d^{k_d}} f_{\vec{\lambda}}(\vec{z}) \Big|_{\vec{z}=\vec{w}} \in \mathbb{K}$$

for all $\vec{w} \in S$ and $\vec{k} = (k_1,\ldots,k_d) \in \mathbb{N}^d$, $|\vec{k}| = k_1 + \ldots + k_d \leq N_{\vec{w}} - 1$. Here $\mathbb{K}$ is a fixed field of the finite transcendence degree n over $\mathbb{Q}$:

$$\mathbb{K} = \mathbb{Q}(\theta_1,\ldots,\theta_n,\mathfrak{D})$$

and $\mathfrak{D}$ is algebraic over $\mathbb{Q}(\theta_1,\ldots,\theta_n)$.

Usually one takes $f_{\vec{\lambda}}(\vec{z})$ in the form

$$f_{\vec{\lambda}}(\vec{z}) = f_1(\vec{z})^{\lambda_1} \ldots f_m(\vec{z})^{\lambda_m}$$

and $\vec{\lambda} = (\lambda_1,\ldots,\lambda_m) \in \mathbb{N}^m$ with $0 \leq \lambda_i \leq L_i - 1$: $i = 1,\ldots,m$ for a fixed set of functions $f_1(\vec{z}),\ldots,f_m(\vec{z})$. These functions satisfy differential (partial differential) equations as $N_{\vec{w}} > 1$: $\vec{w} \in S$ or the law of addition (when $N_{\vec{w}} = 1$). We must note that for a clever form of Gelfond-Schneider method associated with Abelian or degenerate Abelian (group) varieties, the form of $f_{\vec{\lambda}}(\vec{z})$ as powers of fixed functions is not necessary. The same is true in the Thue-Siegel method, when $f_{\vec{\lambda}}(\vec{z})$ are of the form

$$f_{\vec{\lambda}}(\vec{z}) = z_1^{L_1} \ldots z_d^{L_d} f_j(\vec{z})$$

for $0 \leq L_i \leq M_i - 1$: $i = 1,\ldots,m$ and functions $f_j(\vec{z})$ satisfy linear differential equations.

The main object, auxiliary function $F(\vec{z})$ is defined as

$$F(\vec{z}) = \Sigma_{\vec{\lambda}\in\mathscr{L}} C_{\vec{\lambda}} \cdot f_{\vec{\lambda}}(\vec{z})$$

with undetermined coefficients $C_{\vec{\lambda}}$ from the ring $\mathbb{Z}[\theta_1,\ldots,\theta_n]$ (one may take $C_{\vec{\lambda}}$ from the ring of integers of $\mathbb{K}$, but this usually creates only problems, cf. A. Gelfond [1], without additing anything new).

These undetermined coefficients $C_{\vec{\lambda}}$, written as polynomials $C_{\vec{\lambda}}(\theta_1,\ldots,\theta_n) = \Sigma_{\vec{\mu}}\, C_{\vec{\lambda},\vec{\mu}}\, \theta_1^{\mu_1} \ldots \theta_n^{\mu_n}$ with undetermined integer coefficients $C_{\vec{\lambda},\vec{\mu}}$ are defined from the system of linear equations:

$$\frac{\partial^{\vec{k}}}{\partial z_1^{k_1} \ldots \partial z_d^{k_d}} F(\vec{z})\,|_{\vec{z}=\vec{w}} = 0$$

for all $\vec{k} \in \mathbb{N}^d$, $|\vec{k}| \leq N_{\vec{w}} - 1$ and $\vec{w} \in S$.

This system of linear equations can be written explicitly. If

$$\partial_{\vec{z}}^{\vec{k}}\, f_{\vec{\lambda}}(\vec{z})\,|_{\vec{z}=\vec{w}} = \frac{R_{\vec{k},\vec{w},\vec{\lambda}}(\theta_1,\ldots,\theta_n,\mathfrak{D})}{P_{\vec{k},\vec{w}}(\theta_1,\ldots,\theta_n,\mathfrak{D})}$$

for polynomials $R_{\vec{k},\vec{w},\vec{\lambda}}(\theta_1,\ldots,\theta_n,\mathfrak{D})$, $P_{\vec{k},\vec{w}}(\theta_1,\ldots,\theta_n,\mathfrak{D})$ with integer coefficients; then the system of equations on $F(\vec{z})$ can be represented in the explicit form:

$$\Sigma_{\vec{\lambda}\in\mathcal{L}}\, C_{\vec{\lambda}}(\theta_1,\ldots,\theta_n) R_{\vec{k},\vec{w},\vec{\lambda}}(\theta_1,\ldots,\theta_n,\mathfrak{D}) = 0$$

for $\vec{k} \in \mathbb{N}^d$, $|\vec{k}| \leq N_{\vec{w}} - 1$, $\vec{w} \in S$. For the solution of this system of equations one uses the Thue-Siegel lemma. If one knows the bound for the degree and height of polynomials $R_{\vec{k},\vec{w},\vec{\lambda}}(\theta_1,\ldots,\theta_n,\mathfrak{D})$, then we can bound both the degree and the height of polynomials $C_{\vec{\lambda}}(\theta_1,\ldots,\theta_n,\mathfrak{D})$.

We decided to include the corresponding computations, since it will be more convenient in future to refer to a single general scheme, than to start everything each time.

We assume that $\mathfrak{D}$ has degree ν over $\mathbb{Q}(\theta_1,\ldots,\theta_n)$ and that $R_{\vec{k},\vec{w},\vec{\lambda}}(\theta_1,\ldots,\theta_n,\mathfrak{D})$ can be written in the form:

$$R_{\vec{k},\vec{w},\vec{\lambda}}(\theta_1,\ldots,\theta_n,\mathfrak{D})$$

$$= \Sigma_{j_1=0}^{D_1} \ldots \Sigma_{j_n=0}^{D_n} \Sigma_{j_{n+1}=0}^{\nu-1} R_{\vec{k},\vec{w},\vec{\lambda},\vec{j}} \cdot \theta_1^{j_1} \ldots \theta_n^{j_n}, \mathfrak{D}^{j_{n+1}},$$

where $R_{\vec{k},\vec{w},\vec{\lambda},\vec{j}}$ are rational integers (or elements of the ring of integers $\mathcal{O}$ of a fixed algebraic number field L, $[L : \mathbb{Q}] = \nu_1$). If we write polynomials $C_{\vec{\lambda}}(\theta_1,\ldots,\theta_n)$ through their coefficients,

$$C_{\vec{\lambda}}(\theta_1,\ldots,\theta_n) = \Sigma_{\mu_1=0}^{X_1-1}\cdots\Sigma_{\mu_n=0}^{X_n-1} C_{\vec{\lambda},\vec{\mu}}\,\theta_1^{\mu_1}\cdots\theta_n^{\mu_n},$$

then our system of equations takes the form of the system of linear equations on $C_{\vec{\lambda},\vec{\mu}}$:

$$\Sigma_{\vec{\lambda}\in\mathcal{L}}\Sigma_{j_1+\mu_1=i_1}\cdots\Sigma_{j_n+\mu_n=i_n} C_{\vec{\lambda},\vec{\mu}}\cdot R_{\vec{k},\vec{w},\vec{\lambda},\vec{j}} = 0$$

for all $\vec{k}\in\mathbb{N}^d$, $|\vec{k}| \le N_{\vec{w}} - 1$: $\vec{w}\in S$; $j_{n+1} = 0,1,\ldots,\nu-1$ and $i_1 = 0,1,\ldots,D_1 + X_1 - 1;\ldots;\ i_n = 0,1,\ldots,D_n + X_n - 1$.

The number of unknowns $C_{\vec{\lambda},\vec{\mu}}$ is

$$\mathcal{N}_u = |\mathcal{L}|\cdot X_1\cdots X_n,$$

while the number of equations these unknowns are supposed to satisfy is

$$\mathcal{N}_e = \Sigma_{\vec{w}\in S}\binom{N_{\vec{w}} + d-1}{d}\nu(D_1 + X_1)\cdots(D_n + X_n).$$

Then, the solution $C_{\vec{\lambda},\vec{\mu}}$ of these equations exists provided $\mathcal{N}_u > \mathcal{N}_e$. If all $R_{\vec{k},\vec{w},\vec{\lambda},\vec{j}}$ are rational integers, then the nontrivial solution (not all $C_{\vec{\lambda},\vec{\mu}}$ are zeros) exists in rational integers if $\mathcal{N}_u > \mathcal{N}_e$. If all $R_{\vec{k},\vec{w},\vec{\lambda},\vec{j}}$ are integers from the field L, $[L : \mathbb{Q}] = \nu_1$, then the solution $C_{\vec{\lambda},\vec{\mu}}$ in rational integers (not all of which are zero) exists for $\mathcal{N}_u > \nu_1\mathcal{N}_e$.

If one knows the bound for the heights of integers $R_{\vec{k},\vec{w},\vec{\lambda},\vec{j}}$, then, under stronger assumptions on $\mathcal{N}_u/\mathcal{N}_e$ one gets a reasonable bound for sizes of solution $C_{\vec{\lambda},\vec{\mu}}$.

Siegel's lemma we use has the form

<u>Lemma</u> 6.1 (Thue-Siegel): <u>Let</u> $a_{i,j}$: $i = 0,\ldots,\mathcal{N}_u-1$; $j = 0,\ldots,\mathcal{N}_e-1$ <u>be elements of an algebraic number field</u> L, $[L : \mathbb{Q}] = \nu_1$, <u>that are integers over</u> $\mathbb{Z}$ <u>and</u>

$$\max_{j=0,\dots,\eta_e-1} \Sigma_{i=0}^{\eta_u-1} \overline{|a_{i,j}|} \le A.$$

If $\eta_u > \nu_1\eta_e$, then the system of equations

$$\Sigma_{i=0}^{\eta_u-1} a_{i,j}\cdot x_i = 0: \quad j = 0,\dots,\eta_e - 1$$

has a non-zero solution $(x_0,\dots,x_{\eta_u-1})$ in rational integers such that

$$\max_{i=0,\dots,\eta_u-1} |x_i| \le (\sqrt{2}A)^{\nu_1\eta_e/(\eta_u-\nu_1\eta_e)}.$$

We apply the Siegel lemma to our system of equations. Let

$$\max_{|\vec{k}|\le N_{\vec{w}}-1,\vec{w}\in S} \Sigma_{\vec{\lambda}\in\mathscr{S}}\Sigma_{\vec{j},0\le j_1\le D_1,\dots,0\le j_n\le D_n,0\le j_{n+1}\le\nu} \overline{|R_{\vec{k},\vec{w},\vec{\lambda},\vec{j}}|} \le A_0,$$

where all $R_{\vec{k},\vec{w},\vec{\lambda},\vec{j}}$ are elements of the ring $\mathcal{O}$ of integers of L, $[L : \mathbb{Q}] = \nu_1$.

We put, without the loss of generality (but with a certain loss in values of constants)

$$X_i = D_i: \quad i = 1,\dots,n.$$

Let $\tilde{\eta}_u = |\mathscr{S}|$ and $\tilde{\eta}_e = \Sigma_{\vec{w}\in S}\binom{N_{\vec{w}}+d-1}{d}$. If $\tilde{\eta}_u > \nu\nu_1 2^n\tilde{\eta}_e$, there exists a nontrivial solution of our system of equations, satisfying the following bound on its sizes

$$\max_{\lambda\in\mathscr{S},0\le\mu_1\le D_1-1,\dots,0\le\mu_n\le D_n-1} \log|C_{\vec{\lambda},\vec{\mu}}|$$

$$\le \frac{\nu\nu_1 2^n\tilde{\eta}_e}{\tilde{\eta}_u - \nu\nu_1 2^n\tilde{\eta}_e}(\log A_0 + \frac{1}{2}\log 2).$$

If the function $F(\vec{z})$ is defined, one first performs a simple, but useful operation. The coefficients $C_{\vec{\lambda}}(\theta_1,\dots,\theta_n) = \Sigma_{\vec{\mu}} C_{\vec{\lambda},\vec{\mu}}\,\theta_1^{\mu_1}\cdots\theta_n^{\mu_n}$

are polynomials in $\theta_1,\dots,\theta_n$, and, as polynomials in n variables (note that $(\theta_1,\dots,\theta_n)$ are algebraically independent), they may have a common factor:

$$C_{\vec{\lambda}}(\theta_1,\dots,\theta_n) = C^{(0)}(\theta_1,\dots,\theta_n)\cdot C'_{\vec{\lambda}}(\theta_1,\dots,\theta_n),$$

where $C^{(0)}(\theta_1,\dots,\theta_n) \in \mathbf{Z}[\theta_1,\dots,\theta_n]$ and polynomials $C'_{\vec{\lambda}}(\theta_1,\dots,\theta_n) \in \mathbf{Z}[\theta_1,\dots,\theta_n]$ are already without a common factor. One notices immediately that a new function

$$F'(\vec{z}) = (C^{(0)}(\theta_1,\dots,\theta_n))^{-1}F(\vec{z}),$$

which differs only by a constant from $F(z)$, satisfies the same equations as $F(\vec{z})$. It has a form

$$F'(\vec{z}) = \Sigma_{\vec{\lambda}\in\mathcal{L}}\, C'_{\vec{\lambda}}(\theta_1,\dots,\theta_n)f_{\vec{\lambda}}(\vec{z}),$$

with polynomial coefficients $C'_{\vec{\lambda}}(\theta_1,\dots,\theta_n)$: $\vec{\lambda} \in \mathcal{L}$, having no common factor. The bound for degree and height of $C'_{\vec{\lambda}}(\theta_1,\dots,\theta_n)$ are similar to that of $C_{\vec{\lambda}}(\theta_1,\dots,\theta_n)$:

$$\deg_{\theta_i} C'_{\vec{\lambda}} \le \deg_{\theta_i} C_{\vec{\lambda}} \le D_i - 1:$$

$i = 1,\dots,n$ and

$$\log H(C'_{\vec{\lambda}}) \le \log H(C_{\vec{\lambda}}) + n \deg C_{\vec{\lambda}}:$$

$\vec{\lambda} \in \mathcal{L}$.

In other words, one can always choose the coefficients $C_{\vec{\lambda}}(\theta_1,\dots,\theta_n)$ of the auxiliary function $F(\vec{z})$ to be polynomials from $\mathbf{Z}[\theta_1,\dots,\theta_n]$ without a common factor.

Having constructed $F(\vec{z})$ one uses the typical machinery of the Gelfond-Schneider method. First of all, we extrapolate $F(\vec{z})$, i.e. we use the knowledge of $\Sigma_{\vec{w}\in S} N_{\vec{w}}$ zeros of $F(\vec{z})$ to get an upper bound for

$$\left|\frac{\partial^{\vec{k}'}}{\partial z_1^{k_1'}\cdots\partial z_d^{k_d'}} F(\vec{z})|_{\vec{z} = \vec{w}_1}\right|$$

for $\vec{k}' \in \mathbb{N}^d$, $|\vec{k}'| \leq N_{\vec{w}_1} - 1$: $\vec{w}_1 \in S_1$ and $S \subseteq S_1$ where

$$\Sigma_{\vec{w}_1 \in S_1} N_{\vec{w}_1} > \Sigma_{\vec{w} \in S} N_{\vec{w}}$$

(though $N_{\vec{w}_1}$ is not necessarily $\geq N_{\vec{w}_1}$ for $\vec{w}_1 \in S$, cf. Baker method [9], where $N_{\vec{w}_1} < N_{\vec{w}_1}$ but $S \subset S_1$). Such extrapolation is usually achieved using the appropriate version of the Schwarz lemma.

After extrapolation comes the end of the proof in the form of zero and/or small Value Lemma. While during the extrapolation we showed the bound of the form, say

$$\max_{\substack{|\vec{k}'| \leq N_{\vec{w}_1} - 1, \\ \vec{w}_1 \in S_1}} \log\left|\frac{\partial^{\vec{k}'}}{\partial z_1^{k_1'}\cdots\partial z_d^{k_d'}} F(\vec{z})|_{\vec{z} = \vec{w}_1}\right| < -\Psi_1,$$

we show now that, either

i) $$-\Psi_2 < \max_{\substack{|\vec{k}'| \leq N_{\vec{w}_1} - 1 \\ \vec{w}_1 \in S_1}} \log\left|\frac{\partial^{\vec{k}'}}{\partial z_1^{k_1'}\cdots\partial z_d^{k_d'}} F(\vec{z})|_{\vec{z} = \vec{w}_1}\right|,$$

or

ii) $$\max_{\vec{\lambda} \in \mathcal{L}} \log|C_{\vec{\lambda}}(\theta_1,\ldots,\theta_n)| < -\Psi_3.$$

Here, usually, Ψ_1, Ψ_2, Ψ_3 are of the same order of the magnitude. Since, in the cases we are considering,

$$\frac{\partial^{\vec{k}'}}{\partial z_1^{k_1'}\cdots\partial z_d^{k_d'}} F(\vec{z})|_{\vec{z}=\vec{w}_1} \in \mathbb{K} \quad \text{for} \quad \vec{w}_1 \in S,$$

the alternative i) looks exactly as in Proposition 1.7. This actually explains why the Criterion 1.7 arose. However, there is an alternative

ii) which is more like in Gelfond lemma [1], but with one exception: the system of polynomials $C_{\vec{\lambda}}(\theta_1,\ldots,\theta_n)$, being all small, do not have a common factor. This is the first example of an algebraic restriction that can be added to any Criterion.

We will formulate now several criterion with this duality: either there is a polynomial bounded below and above at a given point (the case i), or there is a system of polynomials without a common factor, bounded only above (the case ii). This alternative is typical for all results obtained using the Gelfond-Schneider method and, hence, it will be rather a universal Criterion of Transcendence, the measure of transcendence, algebraic independence and the measure of algebraic independence.

There is one technical point that should be clarified. In the case i), we are dealing with a polynomial from $\mathbb{Z}[\theta_1,\ldots,\theta_n,\mathfrak{H}]$ not from $\mathbb{Z}[\theta_1,\ldots,\theta_n]$.

There is a very simple trick, which shows how to avoid this obstacle and always works only with polynomials in algebraically independent numbers instead of elements of a field of a finite transcendence degree. This is a lemma from our early paper (1974), see [2], [10b].

Lemma 6.2: Let $\mathfrak{H}$ be algebraic over $\mathbb{Q}(\theta_1,\ldots,\theta_n)$ and be of the degree ν. Let $R(\theta_1,\ldots,\theta_n,\mathfrak{H})$ be an element of the field $\mathbb{K} = \mathbb{Q}(\theta_1,\ldots,\theta_n,\mathfrak{H})$ of the type $\leq T$ (with respect to a basis $(\theta_1,\ldots,\theta_n,\mathfrak{H})$ of $\mathbb{K}$). Let

$$-E_1 \leq \log|R(\theta_1,\ldots,\theta_n,\mathfrak{H})| \leq -E_2$$

and the degree of $R(\theta_1,\ldots,\theta_n,\mathfrak{H})$ is at most D. Then for some $C > 0$ there is a polynomial $P(\theta_1,\ldots,\theta_n \in \mathbb{Z}[\theta_1,\ldots,\theta_n]$ such that

$$t(P) \leq T\cdot C;$$

$$d(P) \leq D\cdot C$$

and

$$T\cdot C - \frac{E_1}{n} \leq \log|P(\theta_1,\ldots,\theta_n)| \leq T\cdot C - E_2.$$

References

[1] A. O. Gelfond, Transcendental and algebraic numbers, Moscow, 1952.

[2] G.V. Chudnovsky, Analytical methods in dophantine approximations, Preprint IM-74-9, Inst. of Math. Kiev, 1979. Translation to appear in Math. Survey, Amer. Math. Soc., Providence, R.I., 1981.

[3] G.V. Chudnovsky, Algebraic grounds for the proof of algebraic independence, I. Comm. Pure Appl. Math. 1981, no. 1.

[4] W.D. Brownewell, Sequences of diophantine approximations, 6(1974), 11-21.

[5] S. Lang, Introduction to transcendental numbers, Addison-Wesley, 1966.

[6] M. Waldschmidt, Nombres transcendants, Lect. notes Math., v. 437, Springer, 1975.

[7] M. Waldschmidt, Les travaux de G.V. Chudnovsky sur les nombres transcendants, Sem. Bourbaki, 1975-1976, No. 488, 1976.

[8] T. Schneider, Einfuhrung in die transzendenten zahlen, Berlin, Springer, 1957.

[9] A. Baker, Transcendental number theory, Cambridge University Press, 1975.

[10] a) G. V. Chudnovsky, Algebraic independence of the values of elliptic function at algebraic points. Elliptic analogue of the Lindemann-Weierstrass theorem, Invent. Math. 61 (1980), 267-290.

b) Algebraic independence of some values of the exponential functions Mathem. Not. Sov. Academy, 15 (1974), n 4, 391-398

[11] J.W.S. Cassels, An introduction to diophantine approximation, Cambridge University Press, 1957.

[12] Proceedings of the Number Theory Conference, Carbondale 1979, Lecture Notes in Mathematics, v. 751, Springer, 1979.

[13] G.V. Chudnovsky, Independence algebrique dans la methode de Gelfond-Schneider. C.R. Acad. Sci. Paris, Ser. A, 291 (1980) A-419-A-222.

[14] P. Cijsoun, Transcendence measures, Amsterdam, 1972.

[15] W. Hodge, D. Pedoe, Methods of algebraic geometry, Cambridge University Press, 1952.

[16] I.R. Shafarevich, Basic Algebraic geometry, Springer, N.Y. 1976.

[17] Transcendence theory, Proceedings of the Cambridge conference 1976, Academic Press, N.Y., 1977.

[18] B.L. Van der Waerden, Modern algebra, 2 volumes, Ungar, N.Y., 1951.

Department of Mathematics
Columbia University
New York, NY
USA

RATIONAL APPROXIMATION FOR NON-LINEAR ORDINARY DIFFERENTIAL EQUATIONS

Kevin H. Prendergast

The purpose of this note is to exhibit a novel way of approximating the solutions of certain non-linear ordinary differential equations. We illustrate the method using as an example the homogeneous Duffing equation

$$\frac{d^2x}{dt^2} + x + \varepsilon x^3 = 0 \tag{1}$$

for the function $x(\varepsilon,t)$, with the initial conditions $x(\varepsilon,0) = 1$, $\frac{dx}{dt} = 0$ at $t = 0$.

The usual techniques for finding approximate solutions of this problem (Nayfeh 1973, Bender and Orszag, 1978) is to expand x in a power series in ε. This procedure yields a sequence of approximating functions of the form

$$x_M(\varepsilon,t) = \sum_{m=1}^{M} a_{M,m}(\varepsilon) \cos [(2m+1)\omega_M(\varepsilon)t], \tag{2}$$

where $\omega_M(\varepsilon)$ and the coefficients $a_{Mm}(\varepsilon)$ are polynomials of degree M in ε. Each $x_M(\varepsilon,t)$ is clearly an entire function of t. However, the exact solution of the homogeneous Duffing equation is an elliptic function, and therefore has poles distributed on a regular lattice in the complex plane. Since none of the $x_M(\varepsilon,t)$ have poles we must expect that the sequence $x_M(\varepsilon,t)$ is at best slowly convergent, particularly if $|\varepsilon|$ is large. One might attempt to improve the convergence by taking Padé approximations to the $x_M(\varepsilon,t)$, assuming that a certain number of them were known. This would generate rational functions of ε and cosines, which would have poles at the zeros of the denominators.

An alternative procedure is to put

$$x = \frac{N}{D}, \tag{3}$$

where N and D are Fourier series, and attempt to determine the coefficients in N and D directly from the differential equation, thus bypassing the perturbation series. Let $z = e^{i\omega t}$, in terms of which the Duffing equation becomes

$$-\omega^2 z \frac{d}{dz} \left(z \frac{d}{dz} \left(\frac{N}{D}\right)\right) + \frac{N}{D} + \varepsilon \frac{N^3}{D^3} = 0 , \tag{4}$$

where

$$N = \sum_m N_m z^m , \quad D = \sum_{\ell \neq 0} D z^\ell \quad \text{and } D_0 \equiv 1 . \tag{5}$$

The summations are extended over all odd integers for N, and all even integers for D. This choice is suggested by the form of the perturbation series, but could as well have been based on the observation that only odd multiples of ω occur in x if we iterate the guess that x is proportional to cos ωt. Putting the series for N and D into the Duffing equation leads to an infinite set of non-linear algebraic equations for the coefficient. Even if this system could be solved the answer would not be unique, since we have only one differential equation, and it cannot determine both N and D. However, the ambiguity can be resolved by the same device that is used to construct Padé approximations to power series: namely, we arbitrarily decide which coefficients in N and D to keep and throw away the rest. The lowest order non-trivial approximation of the form $x = N/D$ is

$$N = b(z + z^{-1}), \quad D = 1+\eta(z^2 + z^{-2}) , \tag{6}$$

where b and η are constants to be determined. We insert these expressions in equation (4), multiply by D^3 to remove the denominator, and equate to zero the coefficients of the lowest powers of z (i.e. z and z^3) which occur.

The result is

$$1+2\eta+2\eta^2 + (-1-2\eta+22\eta^2)\omega^2 + 3\varepsilon\, b^2 = 0 \tag{7}$$

and

$$\eta(2+\eta) + (6\eta-9\eta^2)\omega^2 + \varepsilon\, b^2 = 0 . \tag{8}$$

The initial condition $x = 1$ at $t = 0$ implies

$$b = \frac{1}{2} + \eta , \tag{9}$$

and the condition $\frac{dx}{dt} = 0$ at $t = 0$ is automatically satisfied. Equations (7), (8) and (9) should be solved for η, b and ω for given ε, but it is far more convenient to solve for b, ω and ε as functions of η. We readily find that

$$\omega = \left(\frac{1-4\eta-\eta^2}{1+20\eta-49\eta^2}\right)^{1/2} , \tag{10}$$

and

$$\varepsilon = \frac{32\eta(5\eta^3+6\eta^2-\eta-1)}{(2\eta+1)^2(1+20\eta-49\eta^2)} . \tag{11}$$

The solution for $\eta(\varepsilon)$ can be found from this last equation by solving a quartic. We remark that equations (10) and (11) demonstrate that our procedure is not equivalent to finding Padé approximations from the series for $x(\varepsilon,t)$ and $\omega(\varepsilon,t)$, as these would give η and ω as rational functions of ε.

Comparison with Exact Solution. The case $\varepsilon > 0$:

The exact solution of the Duffing equation for $\varepsilon>0$ is $x = \mathrm{sn}(u,k)$ where $u = t(1+\varepsilon)^{1/2}$, $k = [\varepsilon/(2+2\varepsilon)]^{1/2}$. In our approximation the entire range of solutions for positive ε is covered by letting η range over the interval 0 to $[10 - \sqrt{149}\,]/49 \approx -0.04503\,17473$, this being the root of the denominator in equation (11). In table I we give ε, the approximate value of ω and the exact value $\omega = \pi\sqrt{1+\varepsilon}/4K$ as functions of η. It will be seen that our approximation is good to at least three significant figures. There is some deterioration of accuracy for large ε, which may lead to the suspicion that the approximation breaks down for $\varepsilon \to \infty$, but this is not true; we have studied the limiting case of eqation (1) for $\varepsilon \to \infty$, and find that our approximation is still good to three figures. The approximation to $x(t,\varepsilon)$ is about as good as the approximation to $\omega(\varepsilon)$, as is shown in table II for the case $\varepsilon = 11.8237$.

The Case $\varepsilon < 0$.

Periodic solutions of equation (1) are restricted to the range of $-1 < \varepsilon \leq \infty$, the period tending to ∞ as $\varepsilon \to -1$. The exact solution of eqn. (1) for $-1 < \varepsilon \leq 0$ is $x = sn(u,k)$, where $k = (-\varepsilon/(2+\varepsilon))^{1/2}$ and $u = K + t(1+\varepsilon/2)^{1/2}$. In table I we give ε, the approximate value of ω and the exact $\omega = \frac{2\pi}{4K}(1+\varepsilon/2)^{1/2}$ as functions of η for $0 < \eta \leq 0.19$.

Our approximation does not quite cover the entire range $-1 < \varepsilon \leq 0$. As η increases from 0 to the first positive root of the numerator of equation (10), which is $\eta_{max} = -2 + \sqrt{5} \approx 0.236068$, ω decreases monotonically from 1 to 0. However as η transverses this interval, ε decreases from 0 to a minimum value $\varepsilon_{min} \approx -0.98917\,67653$, achieved at $\eta_c \sim 0.187783$, and thereafter increases, reaching the value -0.9742854861 at $\eta = \eta_{max}$.

To put it differently, the original problem was to find an approximate $x(\varepsilon,t)$ for given $\varepsilon > -1$, and equation (11) should be regarded as a quartic, one root of which is $\eta(\varepsilon)$. For ε near -1 there are two real roots which coalesce to η_c at $\varepsilon = \varepsilon_{min}$, and thereafter (for $\varepsilon_c > \varepsilon > -1$) the solution for $\eta(\varepsilon)$ is complex, as is $\omega(\varepsilon)$. It is interesting that something of the same sort occurs in the numerical evaluation of $sn(u,k)$ as the quotient of two θ functions. Spenceley and Spenceley (1947) in remarks appended to the Smithsonian Elliptic Function tables recommend the use of Jacobi's imaginary transformation to convert the Fourier series for the θ functions to more rapidly convergent expansions in hyperbolic functions in the neighborhood of $k = 1$, which corresponds to our $\varepsilon = -1$.

REFERENCES

Bender, C.M. and Orszag, S.A., 1978, Advanced Mathematical Methods for Scientists and Engineers, McGraw-Hill Book Co., New York.

Nayfeh, Ali Hasan, 1973, Perturbation Methods, John Wiley and Sons, New York.

Spenceley, G.W. and R.M., 1947, Smithsonian Elliptic Functions Tables, Smithsonian Institution, Washington, D.C.

Department of Astronomy
Columbia University, N.Y. 10027

TABLE I

η	ε	Approximate ω	Exact ω
0	0	1.0	1.0
-0.005	0.18070	1.06529	1.06529
-0.010	0.41462	1.14363	1.14362
-0.015	1.72835	1.24024	1.24024
-0.020	1.16974	1.36385	1.36386
-0.025	1.83438	1.53043	1.53043
-0.030	2.94507	1.77325	1.77327
-0.035	5.16880	2.17839	2.17846
-0.037	6.83559	2.43795	2.43803
-0.039	9.60929	2.81712	2.81725
-0.040	11.82370	3.08647	3.08664
-0.041	15.13719	3.45039	3.45060
-0.042	20.63733	3.98163	3.98187
-0.043	31.55283	4.86701	4.86998
-0.044	63.62990	6.83436	6.83489
-0.045	2116.55924	38.98676	38.99042
+0.01	-0.25978	0.89621	0.89621
0.02	-0.43618	0.81620	0.81620
0.03	-0.56257	0.75167	0.75168
0.04	-0.65660	0.69785	0.69789
0.05	-0.72850	0.65174	0.65185
0.07	-0.82903	0.57540	0.57581
0.10	-0.91633	0.48483	0.48666
0.12	-0.95002	0.43318	0.43742
0.14	-0.97130	0.38477	0.39423
0.16	-0.98354	0.33694	0.35865
0.18	-0.98876	0.28669	0.33746
0.19	-0.98914	0.25936	0.33588

TABLE II

$\varepsilon = +11.8237$

t	Approximate x	Exact x
0.0	1.0	1.0
0.0488	0.9852	0.9848
0.0984	0.9394	0.9397
0.1493	0.8661	0.8660
0.2023	0.7664	0.7660
0.2579	0.6429	0.6428
0.3171	0.4983	0.5000
0.3784	0.3417	0.3420
0.4430	0.1735	0.1737
0.5091	0.0	0.0

$\varepsilon = -0.9163346$

t	Approximate x	Exact x
0.0	1.0	1.0
0.5887	0.9855	0.9848
1.1049	0.9418	0.9397
1.5848	0.8685	0.8660
1.8961	0.7683	0.7660
2.2089	0.6447	0.6428
2.4884	0.5027	0.5000
2.7461	0.3458	0.3420
2.9905	0.1794	0.1737
3.2286	0.0088	0.0

Vol. 759: R. L. Epstein, Degrees of Unsolvability: Structure and Theory. XIV, 216 pages. 1979.

Vol. 760: H.-O. Georgii, Canonical Gibbs Measures. VIII, 190 pages. 1979.

Vol. 761: K. Johannson, Homotopy Equivalences of 3-Manifolds with Boundaries. 2, 303 pages. 1979.

Vol. 762: D. H. Sattinger, Group Theoretic Methods in Bifurcation Theory. V, 241 pages. 1979.

Vol. 763: Algebraic Topology, Aarhus 1978. Proceedings, 1978. Edited by J. L. Dupont and H. Madsen. VI, 695 pages. 1979.

Vol. 764: B. Srinivasan, Representations of Finite Chevalley Groups. XI, 177 pages. 1979.

Vol. 765: Padé Approximation and its Applications. Proceedings, 1979. Edited by L. Wuytack. VI, 392 pages. 1979.

Vol. 766: T. tom Dieck, Transformation Groups and Representation Theory. VIII, 309 pages. 1979.

Vol. 767: M. Namba, Families of Meromorphic Functions on Compact Riemann Surfaces. XII, 284 pages. 1979.

Vol. 768: R. S. Doran and J. Wichmann, Approximate Identities and Factorization in Banach Modules. X, 305 pages. 1979.

Vol. 769: J. Flum, M. Ziegler, Topological Model Theory. X, 151 pages. 1980.

Vol. 770: Séminaire Bourbaki vol. 1978/79 Exposés 525–542. IV, 341 pages. 1980.

Vol. 771: Approximation Methods for Navier-Stokes Problems. Proceedings, 1979. Edited by R. Rautmann. XVI, 581 pages. 1980.

Vol. 772: J. P. Levine, Algebraic Structure of Knot Modules. XI, 104 pages. 1980.

Vol. 773: Numerical Analysis. Proceedings, 1979. Edited by G. A. Watson. X, 184 pages. 1980.

Vol. 774: R. Azencott, Y. Guivarc'h, R. F. Gundy, Ecole d'Eté de Probabilités de Saint-Flour VIII-1978. Edited by P. L. Hennequin. XIII, 334 pages. 1980.

Vol. 775: Geometric Methods in Mathematical Physics. Proceedings, 1979. Edited by G. Kaiser and J. E. Marsden. VII, 257 pages. 1980.

Vol. 776: B. Gross, Arithmetic on Elliptic Curves with Complex Multiplication. V, 95 pages. 1980.

Vol. 777: Séminaire sur les Singularités des Surfaces. Proceedings, 1976-1977. Edited by M. Demazure, H. Pinkham and B. Teissier. IX, 339 pages. 1980.

Vol. 778: SK1 von Schiefkörpern. Proceedings, 1976. Edited by P. Draxl and M. Kneser. II, 124 pages. 1980.

Vol. 779: Euclidean Harmonic Analysis. Proceedings, 1979. Edited by J. J. Benedetto. III, 177 pages. 1980.

Vol. 780: L. Schwartz, Semi-Martingales sur des Variétés, et Martingales Conformes sur des Variétés Analytiques Complexes. XV, 132 pages. 1980.

Vol. 781: Harmonic Analysis Iraklion 1978. Proceedings 1978. Edited by N. Petridis, S. K. Pichorides and N. Varopoulos. V, 213 pages. 1980.

Vol. 782: Bifurcation and Nonlinear Eigenvalue Problems. Proceedings, 1978. Edited by C. Bardos, J. M. Lasry and M. Schatzman. VIII, 296 pages. 1980.

Vol. 783: A. Dinghas, Wertverteilung meromorpher Funktionen in ein- und mehrfach zusammenhängenden Gebieten. Edited by R. Nevanlinna and C. Andreian Cazacu. XIII, 145 pages. 1980.

Vol. 784: Séminaire de Probabilités XIV. Proceedings, 1978/79. Edited by J. Azéma and M. Yor. VIII, 546 pages. 1980.

Vol. 785: W. M. Schmidt, Diophantine Approximation. X, 299 pages. 1980.

Vol. 786: I. J. Maddox, Infinite Matrices of Operators. V, 122 pages. 1980.

Vol. 787: Potential Theory, Copenhagen 1979. Proceedings, 1979. Edited by C. Berg, G. Forst and B. Fuglede. VIII, 319 pages. 1980.

Vol. 788: Topology Symposium, Siegen 1979. Proceedings, 1979. Edited by U. Koschorke and W. D. Neumann. VIII, 495 pages. 1980.

Vol. 789: J. E. Humphreys, Arithmetic Groups. VII, 158 pages. 1980.

Vol. 790: W. Dicks, Groups, Trees and Projective Modules. IX, 127 pages. 1980.

Vol. 791: K. W. Bauer and S. Ruscheweyh, Differential Operators for Partial Differential Equations and Function Theoretic Applications. V, 258 pages. 1980.

Vol. 792: Geometry and Differential Geometry. Proceedings, 1979. Edited by R. Artzy and I. Vaisman. VI, 443 pages. 1980.

Vol. 793: J. Renault, A Groupoid Approach to C*-Algebras. III, 160 pages. 1980.

Vol. 794: Measure Theory, Oberwolfach 1979. Proceedings 1979. Edited by D. Kölzow. XV, 573 pages. 1980.

Vol. 795: Séminaire d'Algèbre Paul Dubreil et Marie-Paule Malliavin. Proceedings 1979. Edited by M. P. Malliavin. V, 433 pages. 1980.

Vol. 796: C. Constantinescu, Duality in Measure Theory. IV, 197 pages. 1980.

Vol. 797: S. Mäki, The Determination of Units in Real Cyclic Sextic Fields. III, 198 pages. 1980.

Vol. 798: Analytic Functions, Kozubnik 1979. Proceedings. Edited by J. Ławrynowicz. X, 476 pages. 1980.

Vol. 799: Functional Differential Equations and Bifurcation. Proceedings 1979. Edited by A. F. Izé. XXII, 409 pages. 1980.

Vol. 800: M.-F. Vignéras, Arithmétique des Algèbres de Quaternions. VII, 169 pages. 1980.

Vol. 801: K. Floret, Weakly Compact Sets. VII, 123 pages. 1980.

Vol. 802: J. Bair, R. Fourneau, Etude Géometrique des Espaces Vectoriels II. VII, 283 pages. 1980.

Vol. 803: F.-Y. Maeda, Dirichlet Integrals on Harmonic Spaces. X, 180 pages. 1980.

Vol. 804: M. Matsuda, First Order Algebraic Differential Equations. VII, 111 pages. 1980.

Vol. 805: O. Kowalski, Generalized Symmetric Spaces. XII, 187 pages. 1980.

Vol. 806: Burnside Groups. Proceedings, 1977. Edited by J. L. Mennicke. V, 274 pages. 1980.

Vol. 807: Fonctions de Plusieurs Variables Complexes IV. Proceedings, 1979. Edited by F. Norguet. IX, 198 pages. 1980.

Vol. 808: G. Maury et J. Raynaud, Ordres Maximaux au Sens de K. Asano. VIII, 192 pages. 1980.

Vol. 809: I. Gumowski and Ch. Mira, Recurences and Discrete Dynamic Systems. VI, 272 pages. 1980.

Vol. 810: Geometrical Approaches to Differential Equations. Proceedings 1979. Edited by R. Martini. VII, 339 pages. 1980.

Vol. 811: D. Normann, Recursion on the Countable Functionals. VIII, 191 pages. 1980.

Vol. 812: Y. Namikawa, Toroidal Compactification of Siegel Spaces. VIII, 162 pages. 1980.

Vol. 813: A. Campillo, Algebroid Curves in Positive Characteristic. V, 168 pages. 1980.

Vol. 814: Séminaire de Théorie du Potentiel, Paris, No. 5. Proceedings. Edited by F. Hirsch et G. Mokobodzki. IV, 239 pages. 1980.

Vol. 815: P. J. Slodowy, Simple Singularities and Simple Algebraic Groups. XI, 175 pages. 1980.

Vol. 816: L. Stoica, Local Operators and Markov Processes. VIII, 104 pages. 1980.

Vol. 817: L. Gerritzen, M. van der Put, Schottky Groups and Mumford Curves. VIII, 317 pages. 1980.

Vol. 818: S. Montgomery, Fixed Rings of Finite Automorphism Groups of Associative Rings. VII, 126 pages. 1980.

Vol. 819: Global Theory of Dynamical Systems. Proceedings, 1979. Edited by Z. Nitecki and C. Robinson. IX, 499 pages. 1980.

Vol. 820: W. Abikoff, The Real Analytic Theory of Teichmüller Space. VII, 144 pages. 1980.

Vol. 821: Statistique non Paramétrique Asymptotique. Proceedings, 1979. Edited by J.-P. Raoult. VII, 175 pages. 1980.

Vol. 822: Séminaire Pierre Lelong–Henri Skoda, (Analyse) Années 1978/79. Proceedings. Edited by P. Lelong et H. Skoda. VIII, 356 pages, 1980.

Vol. 823: J. Král, Integral Operators in Potential Theory. III, 171 pages. 1980.

Vol. 824: D. Frank Hsu, Cyclic Neofields and Combinatorial Designs. VI, 230 pages. 1980.

Vol. 825: Ring Theory, Antwerp 1980. Proceedings. Edited by F. van Oystaeyen. VII, 209 pages. 1980.

Vol. 826: Ph. G. Ciarlet et P. Rabier, Les Equations de von Kármán. VI, 181 pages. 1980.

Vol. 827: Ordinary and Partial Differential Equations. Proceedings, 1978. Edited by W. N. Everitt. XVI, 271 pages. 1980.

Vol. 828: Probability Theory on Vector Spaces II. Proceedings, 1979. Edited by A. Weron. XIII, 324 pages. 1980.

Vol. 829: Combinatorial Mathematics VII. Proceedings, 1979. Edited by R. W. Robinson et al.. X, 256 pages. 1980.

Vol. 830: J. A. Green, Polynomial Representations of GL_n. VI, 118 pages. 1980.

Vol. 831: Representation Theory I. Proceedings, 1979. Edited by V. Dlab and P. Gabriel. XIV, 373 pages. 1980.

Vol. 832: Representation Theory II. Proceedings, 1979. Edited by V. Dlab and P. Gabriel. XIV, 673 pages. 1980.

Vol. 833: Th. Jeulin, Semi-Martingales et Grossissement d'une Filtration. IX, 142 Seiten. 1980.

Vol. 834: Model Theory of Algebra and Arithmetic. Proceedings, 1979. Edited by L. Pacholski, J. Wierzejewski, and A. J. Wilkie. VI, 410 pages. 1980.

Vol. 835: H. Zieschang, E. Vogt and H.-D. Coldewey, Surfaces and Planar Discontinuous Groups. X, 334 pages. 1980.

Vol. 836: Differential Geometrical Methods in Mathematical Physics. Proceedings, 1979. Edited by P. L. García, A. Pérez-Rendón, and J. M. Souriau. XII, 538 pages. 1980.

Vol. 837: J. Meixner, F. W. Schäfke and G. Wolf, Mathieu Functions and Spheroidal Functions and their Mathematical Foundations Further Studies. VII, 126 pages. 1980.

Vol. 838: Global Differential Geometry and Global Analysis. Proceedings 1979. Edited by D. Ferus et al. XI, 299 pages. 1981.

Vol. 839: Cabal Seminar 77 – 79. Proceedings. Edited by A. S. Kechris, D. A. Martin and Y. N. Moschovakis. V, 274 pages. 1981.

Vol. 840: D. Henry, Geometric Theory of Semilinear Parabolic Equations. IV, 348 pages. 1981.

Vol. 841: A. Haraux, Nonlinear Evolution Equations- Global Behaviour of Solutions. XII, 313 pages. 1981.

Vol. 842: Séminaire Bourbaki vol. 1979/80. Exposés 543–560. IV, 317 pages. 1981.

Vol. 843: Functional Analysis, Holomorphy, and Approximation Theory. Proceedings. Edited by S. Machado. VI, 636 pages. 1981.

Vol. 844: Groupe de Brauer. Proceedings. Edited by M. Kervaire and M. Ojanguren. VII, 274 pages. 1981.

Vol. 845: A. Tannenbaum, Invariance and System Theory: Alg and Geometric Aspects. X, 161 pages. 1981.

Vol. 846: Ordinary and Partial Differential Equations, Procee Edited by W. N. Everitt and B. D. Sleeman. XIV, 384 pages

Vol. 847: U. Koschorke, Vector Fields and Other Vector B Morphisms – A Singularity Approach. IV, 304 pages. 1981.

Vol. 848: Algebra, Carbondale 1980. Proceedings. Ed. by Amayo. VI, 298 pages. 1981.

Vol. 849: P. Major, Multiple Wiener-Itô Integrals. VII, 127 pages

Vol. 850: Séminaire de Probabilités XV. 1979/80. Avec table gé des exposés de 1966/67 à 1978/79. Edited by J. Azéma and M IV, 704 pages. 1981.

Vol. 851: Stochastic Integrals. Proceedings, 1980. Edited Williams. IX, 540 pages. 1981.

Vol. 852: L. Schwartz, Geometry and Probability in Banach Sp X, 101 pages. 1981.

Vol. 853: N. Boboc, G. Bucur, A. Cornea, Order and Conve Potential Theory: H-Cones. IV, 286 pages. 1981.

Vol. 854: Algebraic K-Theory. Evanston 1980. Proceedings. E by E. M. Friedlander and M. R. Stein. V, 517 pages. 1981.

Vol. 855: Semigroups. Proceedings 1978. Edited by H. Jürge M. Petrich and H. J. Weinert. V, 221 pages. 1981.

Vol. 856: R. Lascar, Propagation des Singularités des Solu d'Equations Pseudo-Différentielles à Caractéristiques de M cités Variables. VIII, 237 pages. 1981.

Vol. 857: M. Miyanishi. Non-complete Algebraic Surfaces. 244 pages. 1981.

Vol. 858: E. A. Coddington, H. S. V. de Snoo: Regular Boundary Problems Associated with Pairs of Ordinary Differential Express V, 225 pages. 1981.

Vol. 859: Logic Year 1979–80. Proceedings. Edited by M. Le J. Schmerl and R. Soare. VIII, 326 pages. 1981.

Vol. 860: Probability in Banach Spaces III. Proceedings, 1980. E by A. Beck. VI, 329 pages. 1981.

Vol. 861: Analytical Methods in Probability Theory. Proceedings Edited by D. Dugué, E. Lukacs, V. K. Rohatgi. X, 183 pages.

Vol. 862: Algebraic Geometry. Proceedings 1980. Edited by A gober and P. Wagreich. V, 281 pages. 1981.

Vol. 863: Processus Aléatoires à Deux Indices. Proceedings, Edited by H. Korezlioglu, G. Mazziotto and J. Szpirglas. V, 274 p 1981.

Vol. 864: Complex Analysis and Spectral Theory. Proceed 1979/80. Edited by V. P. Havin and N. K. Nikol'skii, VI, 480 pa 1981.

Vol. 865: R. W. Bruggeman, Fourier Coefficients of Automor Forms. III, 201 pages. 1981.

Vol. 866: J.-M. Bismut, Mécanique Aléatoire. XVI, 563 pages.

Vol. 867: Séminaire d'Algèbre Paul Dubreil et Marie-Paule Malli Proceedings, 1980. Edited by M.-P. Malliavin. V, 476 pages.

Vol. 868: Surfaces Algébriques. Proceedings 1976–78. Edite J. Giraud, L. Illusie et M. Raynaud. V, 314 pages. 1981.

Vol. 869: A. V. Zelevinsky, Representations of Finite Classical Gr IV, 184 pages. 1981.

Vol. 870: Shape Theory and Geometric Topology. Proceedings, Edited by S. Mardešić and J. Segal. V, 265 pages. 1981.

Vol. 871: Continuous Lattices. Proceedings, 1979. Edited by B. B schewski and R.-E. Hoffmann. X, 413 pages. 1981.

Vol. 872: Set Theory and Model Theory. Proceedings, 1979. Ed by R. B. Jensen and A. Prestel. V, 174 pages. 1981.